Advances in

ECOLOGICAL RESEARCH

VOLUME 20

Advances in

ECOLOGICAL RESEARCH

Edited by

M. BEGON

Department of Zoology, University of Liverpool, Liverpool, L69 3BX, UK

A. H. FITTER

Department of Biology, University of York, York, YO1 5DD, UK

A. MACFADYEN

23 Mountsandel Road, Coleraine, Northern Ireland

VOLUME 20

1990

ACADEMIC PRESS

Harcourt Brace Jovanovich, Publishers

London

San Diego New York Boston

Sydney Tokyo Toronto

ACADEMIC PRESS LTD.
24/28 Oval Road
London NW1

United States Edition published by
ACADEMIC PRESS INC.
San Diego, CA 92101

British Library Cataloguing in Publication Data
Advances in ecological research.
Vol. 20
1. Ecology
I. Begon, Michael
574.5

ISBN 0–12–013920–0

This book is printed on acid-free paper

Typeset by Latimer Trend & Company Ltd, Plymouth
Printed in Great Britain by St Edmundsbury Press Limited,
Bury St Edmunds, Suffolk.

Contributors to Volume 20

V. K. BROWN, *Imperial College Field Station, Silwood Park, Ascot, Berkshire, UK.*

S. P. COURTNEY, *Department of Biology, University of Oregon, Eugene, Oregon 97403-1310, USA.*

A. D. FRIEND, *Department of Environmental Science, Clark Hall, University of Virginia, Charlottesville, Virginia, VA22903, USA.*

A. C. GANGE, *Imperial College Field Station, Silwood Park, Ascot, Berkshire, UK.*

T. T. KIBOTA, *Department of Biology, University of Oregon, Eugene, Oregon 97403-1310, USA.*

T. A. SINGLETON, *Department of Biology, University of Oregon, Eugene, Oregon 97403-1310, USA.*

J. D. THOMAS, *School of Biological Sciences, University of Sussex, Brighton BN1 9QG, UK.*

N. E. WEST, *Department of Range Science, College of Natural Resources, Utah State University, Logan, Utah 84322-5230, USA*

F. I. WOODWARD, *Department of Botany, University of Cambridge, Cambridge CB2 3EA, UK.*

Preface

The papers in this volume can hardly fail to stimulate the imagination and broaden the horizons of the ecological reader. They all, with one exception, and in very different ways, describe ecological systems which have been studied mainly for practical reasons by well established methods, but have proved resistant to such approaches. In each case the authors are able to predict greater success through the application of broader and more "ecological" treatments.

Brown and Gange's paper on "Insect herbivory below ground" draws comparisons with leaf- and stem-feeding herbivores and emphasizes the importance of stimulatory as well as negative influences on the plant. By broadening the picture to include the reproductive capacity and competitive status of the plant in its community, the importance of the root feeders in the ecosystem as well as their potential for weed control are demonstrated.

The paper by Friend and Woodward on "Evolutionary and ecophysiological responses of mountain plants to the growing season environment" is concerned with the whole spectrum of mountain microclimate features including, for instance CO_2 concentration and the $^{13}C:^{12}C$ ratio, as well as more usual parameters. These in turn influence plant structure, leaf area index etc. which react back on the microclimate. The paper is concerned with the use of simulation models to disentangle these influences on plant performance and convincingly demonstrates the value of such an approach.

Thomas' paper on "Mutualistic interactions in freshwater modular systems with molluscan components" derives from medically important research on schistosomiasis and snail vectors. It effectively de-bunks the simplistic approach of so much earlier work, which concentrated solely on the snail and the parasite, and also more recent oversimplifications which are confined to classical food-web and energetics studies. The paper argues for studies of whole "modules" of the ecosystem, including the bacterial decomposers which reveal many mutualistic interactions which were unsuspected by previous workers. The very broad-brush treatment of the paper leads to a comparison between the linked evolutionary histories of aquatic macrophytes and molluscs which are compared with those of land insects and terrestrial plants.

West's paper on "Structure and function of microphytic soil crusts in wildland ecosystems of arid to semi-arid regions" relates to practical problems of land managers who are concerned with the influence of such crusts on water run-off and retention, stability and penetrability by high plants and other important practical effects. Once more, the need for a more "ecological" understanding of the nature and properties of the organisms concerned is emphasized and practical measures for overcoming the intractable problems of microbial taxonomy are proposed.

The one paper which does not come into quite the same category of being based primarily on a practical problem, is that of Courtney and his colleagues. Progress in ecology is often made through the application of ideas derived from work with one system or taxon to another, even when these are superficially very different. Thus Yeates' comparison (1987, Volume 17 in this series) between phytophagous nematodes and plant feeding insects proved particularly stimulating and has lead to enthusiastic comment from readers. The new paper on Mushroom-feeding Drosophilidae compares systems involving fruit-flies and fungi with those based on higher plants and illuminates analogies and differences between these "guilds." Many properties of fungi, not least the ephemeral nature of their fruiting bodies, accentuate certain kinds of adaptations in the insects but the general lesson to be derived from this paper is the outstanding suitability of the system for many types of study in population interactions, feeding behaviour habitat selection and evolution.

None of these papers fits neatly into well established areas of ecological study and all, in their different ways, demonstrate the value of a more multidisciplinary approach both to the solution of applied problems and to the deepening of ecological understanding.

Contents

Mutualistic Interactions in Freshwater Modular Systems with Molluscan Components

J. D. THOMAS

Structure and Function of Microphytic Soil Crusts in Wildland Ecosystems of Arid to Semi-arid Regions

NEIL E. WEST

Ecology of Mushroom-feeding Drosophilidae

STEVEN P. COURTNEY, TRAVIS T. KIBOTA and THOMAS A. SINGLETON

Insect Herbivory Below Ground

VALERIE K. BROWN AND ALAN C. GANGE

I. SUMMARY

Plant/herbivore interactions have been the subject of active ecological research in recent years, with studies of insect herbivory featuring prominently. However, herbivory by soil-dwelling insects has been virtually unexplored, although it has received considerable attention in agriculture, as many of the species are pests of economic importance. Relatively few insect species, restricted to only six major orders, feed on roots for part or all of

ADVANCES IN ECOLOGICAL RESEARCH VOL. 20
ISBN 0-12-013920-0

their life-cycle. This, together with their relative inaccessibility for study, has resulted in a tendency for herbivory below ground to be ignored by ecologists. However, recent evidence suggests that this omission should be rectified, because the effects of these subterranean herbivores on plants can be just as great as their foliar-feeding counterparts.

Root-feeding insects have to cope with a food source which, although abundant, may be of exceptionally low quality. As a result, they often have long life-cycles and therefore tend to be denizens of established plant communities. In these communities, they invariably show very clumped distributions, often corresponding to the negative binomial. These distributions result from the facts that the soil can be a very heterogeneous environment and that the insects are highly dependent on soil texture, moisture and temperature. Such dependency leads to aggregation in favourable situations. This clumping means that they are often difficult to detect or quantify by conventional sampling methods.

In the soil, insects principally use CO_2 emissions from roots to locate a food source. In addition, a variety of chemicals are present in the roots which serve as attractants, deterrents or phagostimulants and often result in these insects displaying marked host-plant preferences. Because the species are greatly influenced by abiotic factors in the soil, these play a large part in their mortality. However, many biotic agencies such as vertebrates, predatory insects, nematodes, fungi, bacteria and viruses are also important.

Laboratory trials have indicated that root feeding can seriously reduce the vegetative growth of plants, often in the region of a 70% loss in biomass. Such trials can overestimate damage, but frequently translate to field situations where 50% loss in crop yield may occur. Little work has been attempted on the manner in which these insects affect plant reproduction and, through seedling recruitment, the population dynamics of plants. Indeed, the effects of root-feeding insects in natural plant communities have rarely been investigated, but evidence suggests that their impact is considerable, although different to that of folivorous insects.

There are a number of ways in which plants may compensate for the loss of parts of a root system to insect feeding. Having access to an increased water or mineral supply, for example, can mitigate the effects of attack. Transport of photosynthate from the foliar parts to the roots may occur and this may aid in the proliferation of lateral roots. Indeed, at modest levels of insect attack, proliferation may over-compensate with the result that stability or water and mineral acquisition are enhanced. In this way, root-feeding insects at certain densities may even benefit the plant.

In general, however, the effects of root feeding are such that they significantly reduce plant growth. Indeed, root-feeding insects can be so destructive that they have been introduced, with some success, as biological control agents of weeds. This is an area of much current interest and one in which these herbivores are likely to have an important role in the future. In natural plant communities, manipulative field experiments have demonstrated that soil-dwelling herbivores reduce plant species richness and modify the competitive balance between plant species and life-history groupings. In so doing, they can alter the direction and rate of plant succession and may therefore play a major role in the structuring of plant communities.

II. INTRODUCTION

"Out of sight out of mind" would appear to be the attitude of many ecologists to ecological processes occurring below ground. Students of plant/herbivore interactions are certainly no exception because, despite a vast and rapidly expanding literature on the subject, root herbivory is largely ignored. Even in syntheses of herbivory (e.g. Ahmad, 1983; Crawley, 1983; Denno and McClure, 1983), below-ground consumption is largely ignored (but see Andersen, 1987; Stanton, 1988). From this dearth of attention, one is tempted to conclude that either below-ground herbivory is relatively unimportant in plant community dynamics or is sufficiently different to warrant separate treatment. One of the aims of this review is to demonstrate that neither of these are valid conclusions.

Below-ground herbivory is mainly found in the rodents, nematodes, molluscs and insects. Of these groups, herbivory by nematodes is perhaps most widely documented, and there are several reviews of nematode/plant interactions (e.g. Yeates, 1987; Dropkin, 1989). This is hardly surprising, as in some grassland communities, such as the prairies of North America, nematodes may be the dominant below-ground grazers (Ingham and Detling, 1986). Reference to the nematological literature shows that nematodes exhibit very similar adaptations to insects for life in the soil and that there are parallels between them in terms of their effects on plant physiology and growth. In the insects, the root-feeding habit has only evolved in a few orders. Despite this, the agricultural literature contains many examples of damage by subterranean insects resulting in loss in quality, yield and income (e.g. Dutcher and All, 1979; Harper, 1963; Tolman *et al.*, 1986). Reference to this diverse and extensive literature will clearly demonstrate that losses of 50% or more are not uncommon. Indeed, root-feeding insects can be so destructive to herbaceous plants that they have been successfully exploited as agents in the biological control of weeds (Harris and Myers, 1981; Harris *et al.*, 1981; Powell and Myers, 1988). Woody plants are not exempt from the effects of root-feeding insects, and are particularly vulnerable at the establishment phase (Fowler and Wilson, 1971). In mature forests, their effects are more difficult to quantify, with the result that most accounts of herbivory in forested ecosystems consider only folivorous or wood-feeding insects (e.g.

Schowalter *et al.*, 1986). However, Karban (1980) showed that cicada nymphs can reduce wood growth by as much as 30%.

The paucity of ecological studies of below-ground herbivory most likely stems from difficulties with sampling and taxonomy of the causal agents and in assessing the extent of damage. However, more experimental approaches and manipulative experiments make herbivory below ground a stimulating area for future work. Such work is essential if we are to understand herbivory as a structuring force in plant community dynamics. In many ecosystems, it is an underestimate that 50% of net primary production is allocated to below-ground plant structures and in some cases 90% may be a more realistic figure (Coleman, 1976; Fogel, 1985). It is not therefore surprising that certain insects have evolved feeding habits to utilize this resource, or that their activities may have substantial effects, at least comparable to those of their foliar-feeding counterparts.

The below-ground structures of plants include roots, stems and modified stems such as rhizomes, bulbs, corms and tubers. In addition, many plant species use the soil as a "resting site" for their diaspores. The plant root system has three major functions: water absorption, nutrient uptake and anchorage (Harper *et al.*, in press). Besides these, roots may serve as the site for the synthesis of products involved in the growth and development of the shoot, the storage of metabolites and photosynthates, the protection of dormant meristematic tissue and as the site for the initiation of vegetative reproduction (Andersen, 1987). Thus, the damage to, or removal of, any part of the underground system may well disrupt any number of physiological processes and be reflected in a variety of responses by the plant. This chapter aims to demonstrate the nature of these responses and their importance in plant community dynamics. We shall restrict ourselves to terrestrial systems and to insect herbivory on higher plants. Thus, herbivory on aquatic plants and the soil fungal community is woefully ignored (but see Addison, 1977; Parkinson *et al.*, 1979; Forno and Semple, 1987; Parkinson, 1988).

III. THE HERBIVORES

Considering the abundance and diversity of life-styles found in the insects (Southwood, 1978a), it is somewhat surprising that relatively few species have exploited the underground parts of plants as food resources. Those which have often occur at exceptionally high numbers, e.g. the lettuce root aphid, *Pemphigus bursarius* (L.), can occur at densities up to 64 500 individuals m^{-2} (Dunn, 1959a), while Dybas and Davis (1962) considered the density of cicada nymphs of 607 000 ha^{-1} in an elm–hackberry (*Ulmus–Celtis*) forest in eastern North America to represent the highest recorded biomass for any terrestrial animal. Only 6 of the 26 orders of insects are well represented as below-ground herbivores (Table 1). Among the exopterygotes, the Hemiptera, Isoptera and Orthoptera feed as immatures and adults on a wide range of plant structures. In the endopterygote orders, the larvae of Coleoptera, Diptera and Lepidoptera are subterranean herbivores (but see Kamm and Buttery, 1984). However, even in these large orders, below-ground herbivory is only well developed in restricted families or subfamilies, as Ritcher (1958) has shown in the Scarabaeidae. Undoubtedly, the habit is most widely reported in the beetles, the group which also has the most examples of economically important species (Hill, 1983, 1987). Other insect groups, not normally associated with root feeding at all, include the springtails (Collembola), lacewings (Neuroptera) and thrips (Thysanoptera). Although the indirect effects of Collembola on plants through fungal feeding are well documented (Warnock *et al.*, 1982; Coleman *et al.*, 1983; McGonigle and Fitter, 1988), some species are now known to feed on roots (Greenslade and Ireson, 1986), and particularly those of seedlings (Edwards, 1962). One

Table 1
Insect orders and principal families which display below-ground herbivory

Division	Order	Family	Common name
Apterygota	Collembola	Onychiuridae	Springtails
Exopterygota	Orthoptera	Gryllidae	True crickets
		Gryllotalpidae	Mole crickets
	Isoptera	Hodotermitidae	Termites
	Hemiptera:	Aphididae	Aphids
	Homoptera	Cercopidae	Spittle bugs
		Cicadidae	Cicadas
		Pseudococcidae	Mealy bugs
	Heteroptera	Cydnidae	Burrowing bugs
	Thysanoptera	Thripidae	Thrips
Endopterygota	Neuroptera	Ithonidae	Lacewings
	Coleoptera	Chrysomelidae	Leaf beetles
		Curculionidae	Weevils
		Elateridae	Wireworms
		Scarabaeidae	Chafers
	Diptera	Anthomyiidae	
		Bibionidae	March flies
		Otitidae	
		Psilidae	
		Syrphidae	Hover flies
		Tipulidae	Leatherjackets
	Lepidoptera	Hepialidae	Swift moths
		Noctuidae	Cutworms
		Sesiidae	Clearwings
		Tortricidae	Tortrix moths

of the most unusual examples of root feeding is seen in the neuropteran family, Ithonidae: the larvae of the Australian *Ithone fusca* Newman, which were formerly thought to be carnivorous like all other Neuroptera and to feed on the larvae of scarabaeid beetles, in fact feed on roots (Kevan, 1962). The suite of insects found by Simberloff *et al.* (1978) feeding on the aerial and submerged roots of the red mangrove, *Rhizophora mangle* L., in Florida includes a thrips, *Neothrips* n. sp. A couple of thrips species have been recorded feeding on the aerial roots of *Ficus*, whereas the only other subterranean thrips are those associated with bulbs within the Liliaceae, e.g. *Thrips tabaci* Lind. (Lewis, 1975).

Root-feeding insects display a range of host specificity similar to their foliar counterparts. Some species are highly polyphagous. The larvae of Tipulidae (leatherjackets) and Scarabaeidae (chafers) are among the most ubiquitous below-ground herbivores and will feed on most grassland species (Ueckert, 1979; Clements, 1984), including woody species occurring in the sward (Kard and Hain, 1988). Even so, they normally exhibit a preference for a particular plant family (Ueckert, 1979; White and French, 1968). At the other extreme are specialist species, restricted to a single plant species, genus or tribe, such as some of the insects associated with knapweed, *Centaurea* spp. (e.g. Müller *et al.*, 1989), or the sessiid moth larva, *Vitacea postiformis* (Harris), which bores into vine roots (Dutcher and All, 1979).

The relationship of insects with the underground parts of plants is variable. Some species are freely mobile in the soil and feed externally on the plant root system. Certain noctuid moth larvae are commonly called "cutworms" because of their habit of actively moving from plant to plant and cutting off roots (or stems) (Jones and Jones, 1984). Although it is difficult to determine the amount of movement undergone by soil-dwelling insects, intuitively it seems unlikely that they undergo more than "trivial movements" (*sensu* Southwood, 1962) and then only to locate new food resources. In a pot trial using an annual forb (*Stellaria media* (L.) Vill.) and a perennial grass (*Holcus lanatus* L.), the larvae of the garden chafer, *Phyllopertha horticola* (L.), moved only short distances and the amount of movement was negatively related to the density of the plants (R. Ayres, unpublished). Other external feeders, such as early instar cicada nymphs, make feeding tunnels in the soil associated with rootlets, whereas later instars construct permanent earthen cells close to larger roots (Dybas and Davis, 1962; White and Strehl, 1978).

Generally, the larger herbivores burrow through the soil, whereas the smaller species or early instars make use of existing pore spaces, cavities or channels in the soil. Many specialist herbivores mine, burrow or tunnel into the root cortex or vascular system. The mining habit is exploited by the unusual heteropteran, the so-called burrowing bug, *Pangaeus bilineatus* (Say), which defaces developing peanuts (Smith and Pitts, 1971; Highland

and Lummus, 1986). The larvae of many species burrow or tunnel into the tissue of the root itself, such as the larvae of corn rootworms (*Diabrotica* spp.: Riedell, 1989) and the large narcissus fly, *Merodon equestris* (F.) (Jones and Jones, 1984). The feeding of some species causes hypertrophy of the plant tissue in the form of galls, similar to those occurring on the aerial structures of plants (e.g. McCrea and Abrahamson, 1985; Wapshere and Helm, 1987; Müller, 1989b). These are often distinctive in shape, such as the marble-like swellings on the roots of most cultivated crucifers caused by the turnip gall weevil, *Ceutorhynchus pleurostigma* (Marsh). Some species change their feeding behaviour during larval life; the young larvae of the ghost swift moth, *Hepialis humuli* (L.), feed on fine rootlets, while older ones tunnel into larger roots (Anon., 1969). The restricted movement generally associated with the feeding biology of soil-dwelling taxa, coupled with the generally poor adult dispersal (e.g. Raw, 1951; Pritchard, 1983), are the most likely factors underlying the restricted occurrence of this feeding habit in insects, especially as below-ground species are less commonly attacked by parasitoids and predators than their above-ground relatives (Hawkins and Lawton, 1987; Müller, 1989a; White *et al.*, 1979).

Soil-dwelling herbivores can exploit underground plant structures as a food resource in a variety of ways. With the exception of the Hemiptera and Thysanoptera, species have chewing, mandibulate mouthparts and feed in much the same way as defoliating species. Depending on their size, they either consume whole roots (e.g. Fowler and Wilson, 1971) or merely graze the surface of roots (e.g. Greenslade and Ireson, 1986). Those living internally, although surrounded by a supply of food, experience problems similar to leaf-mining species of feeding-site location, frass removal and adult emergence (Hering, 1951). The Hemiptera, represented mainly by the homopteran groups aphids, scale insects and phylloxerids, feed on the phloem sap which they tap by means of specialized mouthparts. The cicada nymphs, on the other hand, feed on the much diluter xylem sap (White and Strehl, 1978).

The concept of phytophagous guilds and species packing, so well established for insects associated with the aerial parts of plants (e.g. Kareiva, 1986; Lawton, 1984), has only received limited and recent attention for root-feeding insects. During an extensive search in Europe for potential biological control agents for diffuse knapweed, *Centaurea diffusa* Lam., Müller *et al.* (1989) found 22 phytophagous insect species associated with three closely related *Centaurea* spp., *C. diffusa*, *C. maculosa* Lam. and *C. vallesiaca* Jordan. These insect species belonged to four orders and 12 families and were all found associated with a particular root structure. In his analysis, Müller (1989a) recognized five feeding niches which were occupied by specific herbivores. However, he included two species associated with the central meristem of the rosettes which were not found in the actual tissue of the root.

For this reason, these are omitted in Fig. 1. The larvae of four small weevil species and one syrphid feed externally on the root collar. The central vascular system of the root is attacked by three gall-forming beetles and a moth, while the root cortex is exploited by the larvae of two species of moth which feed beneath a silken web, although they mine inside smaller roots and near the tip of the larger tap root. Finally, an aphid and a collembolan feed externally on the root. This diverse entomofauna of the roots of *Centaurea* spp. allowed Müller to explore the species interactions between root herbivores in a novel way. He attributed the generally low infestation rates (less than one-third of roots attacked, with less than 17% containing more than two larvae and less than 8% containing more than one insect species) to the guilds not representing equilibrium assemblages. However, he was able to demonstrate a high predictability in guild structure. This resulted from food niche competition (involving the modification of the resource, e.g. gall formation by one of the competitors), resource segregation (through larval associations to specific food niches) and species-specific responses to plant characteristics (e.g. rosette size for oviposition). There is clearly much scope for similar detailed studies on other plant species.

Root structures utilized by phytophagous insect species	*Centaurea* mac.	diff.	vall.	Feeding niches
1. Root collar				
Apion penetrans Germer	+	+	-	
Apion onopordi Kirby	+	+	+	
Apion orientale Gerst.	+	-	-	
Apion alliariae Herbst	+	-	-	
Cheilosia sp.	+	+	-	
2. Central vascular tissue				
Cyphocleonus achates Faber	+	+	-	
Cleonus piger Scop.	+	+	-	
Sphenoptera jugoslavica Ob.	+	+	-	
Pterolonche inspersa Stg.	+	+	-	
3. Root cortex				
Agapeta zoegana L.	+	-	-	
Pelochrista medullana Stgr.	+	+	-	
4. External on root				
Trama centaureae CB	?	+	?	
Sminthurodes betae Westw.	?	+	?	

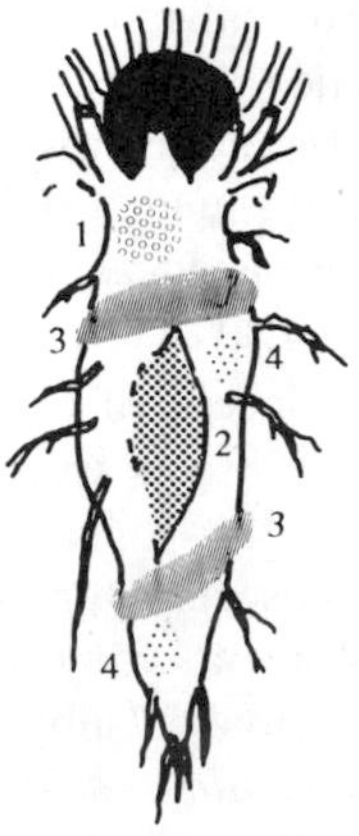

Fig. 1. Common insect species associated with the roots of *Centaurea maculosa*, *C. diffusa* and *C. vallesiaca* in Europe (redrawn from Müller, 1989a).

IV. CHARACTERISTICS OF THE HERBIVORES

A. Life Histories

A generalized life-cycle for a holometabolous root-feeding insect consists of oviposition in soil, larval feeding in or on plant roots, pupation in the soil and adult flight and mating above the ground. Such life-cycles can vary in length. In temperate climates, species tend to be univoltine or have one generation in 2–3 years. In tropical areas, there is a tendency to be multivoltine. Among Coleoptera, many species of Scarabaeidae (chafer beetles) or Elateridae (wireworms) have long life-cycles, often up to 3–4 years in temperate zones (Jones and Jones, 1984; Hill, 1987). Smaller species of scarabaeid in these zones tend to have 1-year life-cycles (Raw, 1951; Cherry, 1985), as do many tropical species (Hill, 1983). Overwintering is invariably as a larva and it is the slow growth of this stage which prolongs the life-cycle in cooler regions. Curculionoidea (weevils) and Chrysomeloidea (leaf beetles) tend to be smaller than the previous two groups. In temperate zones, these species are invariably univoltine (Hill, 1987), but multivoltine in the tropics (Hill, 1983; Sutherland, 1986). In temperate climates, overwintering may occur as adults (e.g. *Sitona* spp.), larvae (e.g. *Ceutorhynchus* spp.) or eggs (e.g. *Diabrotica* spp.) (Hill, 1987).

Parallels to these life-histories may be found among the Lepidoptera and Diptera. Among the Lepidoptera, the larvae of the Hepialidae and Sesiidae are often large and may take up to 3 years to complete their life-cycle in temperate areas (Skinner, 1984; Hill, 1987). The larvae of Noctuidae (cutworms) are univoltine in temperate regions and multivoltine in the tropics (Hill, 1983). The tropical potato tuber moth, *Phthorimaea operculella* (Zeller), a small species whose larvae mine potato tubers, may have up to 12 generations a year (Hill, 1983; Rothschild, 1986). Crane-flies (Diptera: Tipulidae) are generally temperate-zone inhabitants, have one generation a year and overwinter as larvae. The adults are very short-lived and do not feed (Pritchard, 1983). Flies of the family Anthomyiidae (e.g. Cabbage root fly, *Delia radicum* L.) may be either univoltine or multivoltine (Hill, 1983, 1987).

The hemimetabolous insects that attack roots often have very complex life-cycles, compared to those species described above. Root-feeding aphids may remain associated with the same plant roots all year (Langley, 1986) or they may show host alternation. In the latter case, generations are produced in autumn which leave the roots and migrate to woody hosts. Here, sexual reproduction occurs and the eggs overwinter (Dunn, 1959a; Foster, 1975; Moran and Whitham, 1988). Even in these species, however, some aphids may remain on roots over winter (Dunn, 1959b; Foster, 1975).

The grape phylloxera, *Daktulosphaira vitifolii* (Fitch), has possibly the most complex life-cycle of any root-feeding insect. It has two parthenogenetic phases, one in early summer on the roots of vines and one in late summer on the aerial parts of the plant. Winter is spent in the egg stage above ground. The cycle is described by Wapshere and Helm (1987).

B. Geographical Distribution

Insects feeding below ground are found on all continents, except Antarctica (Andersen, 1987). While the zoogeography of the soil-dwelling insect fauna is still far from complete, generally most insect orders which have root-feeding representatives are cosmopolitan in their distribution, while families are more restricted and genera and species often show a high degree of endemism in isolated habitats or land masses (Anderson, 1977; Hill, 1983, 1987). This is even true for islands, such as New Zealand and Fiji, which are considered to have been geographically isolated for a very long period of time (Williams, 1973) and to contain faunas which may only be explained in terms of plate tectonics (Watt, 1986). These islands have many genera and species which are endemic, good examples being the scarabaeid beetle, *Costelytra zealandica* White, and the curculionid weevil, *Irenimus compressus* (Brown), in New Zealand (Ferro, 1976) and several genera of elaterid beetles in Fiji (Watt, 1986).

Exceptions exist to confound these general rules, and are usually the result of accidental introductions of insects by man. Many root-feeding insects have wide distributions because they have been able to spread into new areas by this means. *Popillia japonica* Newman (the "Japanese beetle") is a scarabaeid, native to China and Japan (Hill, 1987), which was introduced into the eastern USA in 1916, where it spread over 100 000 square miles, achieving pest status which it did not have in Japan (Miller *et al.*, 1984). Other introductions have been even more spectacular. The curculionid weevil, *Sitona discoideus* Gyllenhall, was introduced from the Mediterranean and now occurs over most of Australia and New Zealand (Goldson and French, 1983; Aeschlimann and Vitou, 1988) and the sweet potato weevil, *Cylas formicarius* (Fab.), has a pantropical distribution, ranging over more than 50 countries (Sutherland, 1986). The grape phylloxera, *Daktulosphaira vitifolii*, which feeds on vine roots for part of its life-cycle, was native to the eastern USA but has been introduced (with disastrous consequences) into Europe, South Africa, Australia, New Zealand and Japan (Hill, 1987).

There has been little work to compare the biology of root-feeding species in different geographical areas. Some species exhibit differences in life-history, depending on geographical location. The woolly apple aphid, *Eriosoma lanigerum* (Hausmann), colonizes tree roots in South Africa, North America and Australia, but does not appear to do so in Europe (Hill, 1987). The lettuce root aphid, *Pemphigus bursarius*, generally leaves lettuce in autumn and overwinters on poplar trees (Dunn, 1959a,b). In Canada, however, the winter is frequently spent on lettuce, where apterae are able to withstand temperatures of 0°C for up to 40 weeks (Alleyne and Morrison, 1978a). In Britain, Finch *et al.* (1986) have found that there are distinct biotypes based upon the time of emergence of *Delia radicum*, the cabbage root fly, in different parts of the UK. Differences in host-plant preference may also arise because an introduced insect may suddenly be confronted with an array of acceptable hosts not encountered in its native habitat (e.g. *P. japonica*, Fleming, 1972). However, in contrast, *S. discoideus* shows the same host-plant preference in its native Mediterranean and in Australia, into which it was introduced (Aeschlimann and Vitou, 1988).

C. Spatial Distribution in the Soil

The spatial distribution of an insect is an intrinsic property of a species (Taylor, 1961) and results from interactions among individuals and between the individuals and their habitat. Analyses of these patterns may thus reveal behavioural characteristics of the species (Allsopp and Bull, 1989). An understanding of the spatial pattern of insects is also essential to the development of accurate sampling methods (Quinn and Hower, 1985). Sampling unit size is an important criterion in the detection of patterns which are aggregated (Hassell, 1985), for the degree of aggregation recorded can depend on the size of the sampling unit (e.g. Heads and Lawton, 1983). Although sampling methodology is beyond the scope of this chapter (but see Southwood, 1978b), an understanding of a species' spatial dynamics is important in terms of the interactions between the herbivore and the plant community.

Virtually all root-feeding insects which have been studied have highly aggregated distributions corresponding to the negative binomial (Elliott, 1979). These distributions generally result from the heterogeneity of the soil and the adaptations of the insects to this environment. Good examples are *Pemphigus trehernei* Foster, the sea aster root aphid, which lives in salt-marshes and is restricted to the pore spaces which are of limited occurrence (Foster, 1975), and the grape rootworm, *Fidia viticida* Walsh, which avoids desiccation by aggregating near moisture (Dennehy and Clark, 1987).

Aggregated distributions may also result from the oviposition behaviour of the female. The selection of an egg-laying site is critical, as the young on hatching are relatively immobile and must immediately locate a food source. As a result, when a suitable position is located, a female commonly lays her entire egg complement (e.g. scarabaeids: Milne, 1963; tipulids: Hartman and Hynes, 1977). Clumped egg deposition, resulting from orientation to a host plant, has been reported in the clover root curculio, *Sitona hispidulus* (Fab.) (Elvin and Yeargan, 1985; Quinn and Hower, 1985) and the sweet potato weevil, *Cylas formicarius* (Wilson *et al.*, 1988). The corn rootworms (*Diabrotica* spp.) overwinter in the egg stage, but larval distribution patterns are still aggregated, due to the laying of large clutches by females (Bergman *et al.*, 1983). Larval aggregation may be maintained by the patchy availability of food in the soil and is particularly seen in specialist feeders (Piedrahita *et al.*, 1985a). If generalist root feeders, such as scarabaeid larvae, occur in crop monocultures, the larval distribution is related to the discrete units provided by individual plants (Zhen-Rong *et al.*, 1986). However, when these insects are presented with a wide range of plant species, such as that occurring in pastures, their distribution is only marginally aggregated (Allsopp and Bull, 1989). In some species, there may be a change from clumped egg distributions to random larval ones. The clover root curculio, *S. hispidulus*, exhibits such a change (Quinn and Hower, 1986b), and it is thought that this may result from nodule availability which is highly correlated with first instar larval survival (Quinn and Hower, 1985). A similar event occurs in the scarabaeid *Phyllopertha horticola* and results from larvae moving through the soil in search of food (Raw, 1951; Milne, 1963).

Aggregated distributions may also result from the interactions of the root-feeding insects with other organisms. Virtually all species of root-feeding aphids, at least in Britain, are attended by ants (Langley, 1986). The association is considered to be a symbiosis (Way, 1963), as ants receive honeydew in return for guarding aphids against predators and keeping colonies clean from fungi, which otherwise grow on the sugar-rich honeydew. Root-feeding aphids have extremely clumped distributions (Pontin, 1978; Langley, 1986). It has been shown that these may be determined by the position of ant nests and that, in the proximity of nests, the distributions of particular aphid species result from the extent of their dependency on the ants (Langley, 1986). Such extreme clumping means that accurate population sampling of these insects is difficult, leading to huge ranges in estimates of density, with the upper range often very high. Some examples are 100–31 400 aphids m^{-2} (Macfadyen, 1953), 100–64 500 aphids m^{-2} (Dunn, 1959a) and 16 500 ± 6500 aphids per ants' nest (Pontin, 1978).

Root-feeding insects also show distinct vertical stratification in the soil. This may occur in response to moisture (Dennehy and Clark, 1987) because, as the soil dries out in summer, larvae tend to move deeper seeking a more humid environment. Alternatively, as plant roots grow and penetrate further into the soil, the insects may follow them (Goldson and French, 1983). Insects which overwinter in the soil may also move deeper in winter, to avoid extremes of cold or moisture (Raw, 1951; Milne, 1963; Alleyne and Morrison, 1978b).

D. Temporal Distribution in Vegetation

Many species of root-feeding insects have long life-cycles. As a result, insects such as scarabaeids and elaterids tend to be denizens of old, undisturbed plant assemblages, such as pasture (Clements *et al.*, 1987). Indeed, insecticide application resulted in significant yield increases in old pasture, but none in recently sown grass (Henderson and Clements, 1974; see Fig. 2a). These increases were attributed mainly to the abundance of elaterid larvae in the older pastures.

However, populations of root-feeding insects do not continue to increase through time, but eventually reach a peak and then decline. Such declines may be caused by mortality, due to food availability, parasites and pathogens or a number of abiotic factors (Turner, 1957). East and Willoughby (1983) described population changes in the scarabaeid beetle, *Costelytra zealandica*, in relation to pasture age. It was found that, in a number of differently aged pastures, populations reached a peak between 3 and 7 years after seeding and then declined (Fig. 2b). The collapse in populations resulted from summer mortality, but was not affected by weather, pasture composition, grazing management, larval food supply, predation or population density. Rather, it was postulated that the decline might be related to pathogen (viral or fungal) abundance. Cranshaw (1985), working with *Sitona hispidulus* in alfalfa (*Medicago sativa* L.), investigated the relationship between larval attack and stand age and found that larval density clearly increased with stand age (Fig. 2c). However, sampling of adults revealed that densities peaked in relatively young stands, suggesting that overall population levels would follow similar trends to those of the scarabaeid larvae described by East and Willoughby (1983). Research in New Zealand, however, suggests that the relationship between age of alfalfa (lucerne) and larval density of a root-feeding weevil, in this case *S. discoideus*, is considerably more complicated than that suggested by Cranshaw (1985). Goldson and French (1983) found significant negative correlations between density and lucerne stand age at a number of different localities (Fig. 2d). At the time, this was attributed to the fact that lucerne root nodules tend to be located deeper in the soil as the plants age (Aeschlimann, 1980). Such nodules are vital to the survival of neonate larvae (Byers and Kendall, 1982), which must move through soil to the nodule from eggs deposited at the soil surface (Goldson and French, 1983). However, Goldson and Muscroft-Taylor (1988) found that nodulation failure occurs in old lucerne plants, thus reducing larval survival, and that older plants actually show an increased sensitivity to damage by the larvae. As a result, the threshold number of larvae required to cause economic damage is 1200 larvae m^{-2} in a 3-year-old stand compared to 2100 m^{-2} in a 1-year-old crop (Goldson *et al.*, 1985).

Root-feeding aphids in natural habitats are also associated with established, permanent plant communities (Foster, 1975; Langley, 1986). Their distribution may often be associated with ants, which also colonize stable plant communities (Langley, 1986). Long term studies of subterranean aphid population biology are apparently lacking, but Foster's (1975) 3-year data

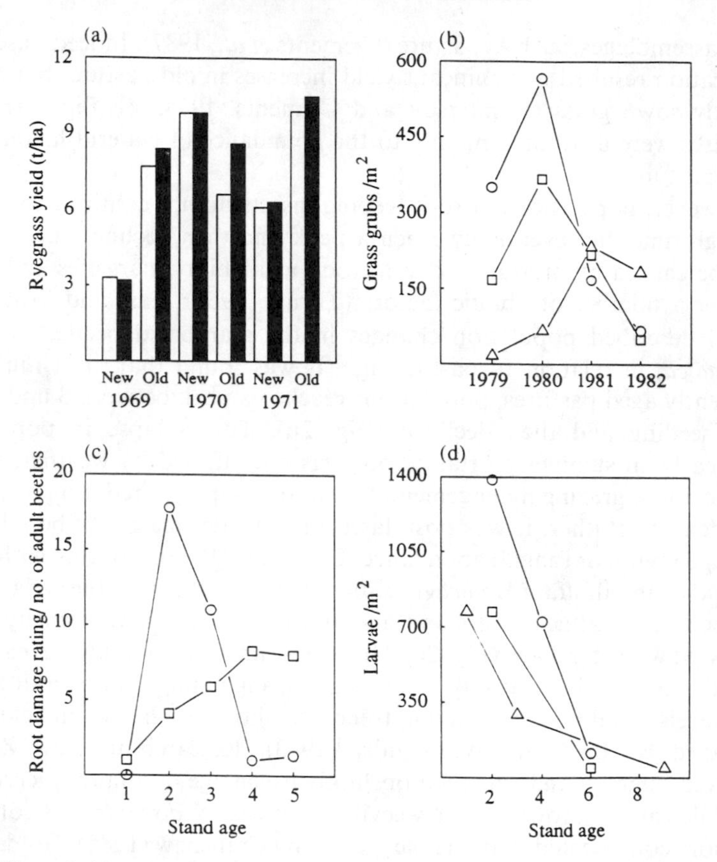

Fig. 2. Effect of plant community age on subterranean insect numbers. (a) Increase in ryegrass yield in old, established pastures as a result of insecticide treatment (■) compared to control, with populations of insect pests (□) (redrawn from Henderson and Clements, 1974). (b) Number of grass grubs per m^2 in three differently aged pastures. Pasture age in 1979: (○) 7 years, (□) 2 years and (△) 1 year (redrawn from East and Willoughby, 1983). (c) Feeding damage (□) and numbers of adult *Sitona hispidulus* (○) in relation to stand age of alfalfa (*Medicago sativa*) (redrawn from Cranshaw, 1985). (d) Numbers of *Sitona discoideus* at three different locations, in relation to stand age of *M. sativa* (redrawn from Goldson and French, 1983).

set suggests that annual population cycles occur, similar to many foliar-feeding species (Dixon, 1985).

E. Habitat Requirements

1. Soil Texture and Compaction

Soil texture describes the relative proportions of mineral particles in a soil and the arrangement of these particles into aggregates defines the soil structure (Brady, 1984). In coarse-textured sandy soils, water retention is lower than in fine-textured soils of silt and clay, because of the abundance of spaces between particles. Such properties have been shown to be very important in determining the survival of soil-dwelling nematodes, because these animals move through the surface film of water between soil particles (Dropkin, 1989). Soil structure is also critical in determining the survival of root-feeding insects, even though they move on a larger scale than nematodes. For larvae of chrysomelids (Lummus *et al.*, 1983; Marrone and Stinner, 1984) and scarabaeids (Régnière *et al.*, 1981b), survival is highest in fine-textured soils and decreases with increasing particle size. Fine-textured soils have a higher water-holding capacity and may reduce the risk of larval desiccation. It has also been suggested that the abrasiveness of large sand particles may reduce the survival of burrowing larvae (Turpin and Peters, 1971).

A similar situation occurs with root-sucking insects. Although Buchanan (1987) found no difference in grape phylloxera infestations over a range of soil types in Australia, survival of this insect is enhanced in fine soils and reduced in sandy ones (Nougaret and Lapham, 1928; de Klerk, 1974; Buchanan, 1978). Apart from clay soils being moister, the cracks which form during drying permit the inter-plant movement of insects; such cracks are not formed in sandy soils (Buchanan, 1984; King and Buchanan, 1986). Large pore spaces resulting from drying also contribute to population increases in root-feeding aphids (Lange *et al.*, 1957; Harper, 1963; Foster, 1975).

It has been known for a century that compacting a soil can reduce the survival of insects (Ormerod, 1890). Compaction presents a physical barrier to the movement of larvae in the soil (Strnad and Bergman, 1987) and has been shown to reduce the damage to seedling sugar beet by Collembola (Dunning and Baker, 1977). Recently, Stewart *et al.* (1988) have recommended rollers to control scarabaeid larvae in pastures when insecticidal treatment is costly.

2. *Soil pH*

The information available on the effect of soil pH on insects is conflicting. Wessel and Polivka (1952) and Polivka (1960a,b) reported that Japanese beetle larvae did not survive as well at high (6·8–7·0) as opposed to low (4·5–5·0) pH. However, Vittum and Tashiro (1980) and Vittum (1984) found no response in this insect or other species of Scarabaeidae over the pH range 4·6–7·6. Edwards (1977) found that Collembola populations were dramatically reduced below a pH of 4·0, but there was no effect between 4·0 and 8·1. Such conflicting results prevent generalizations being made and further research is much needed. Moreover, the nematological literature is equally ambiguous (e.g. Wallace, 1973), although the occurrence of root-feeding nematodes in a range of soil types, as in the insects, tends to suggest that pH is of little importance.

On a much finer scale, the pH at the root surface may be different to that in the open soil and this could be important for microarthropods. The measurement of pH in the rhizosphere is extremely difficult, but has been achieved on a few occasions (e.g. Marschner and Romheld, 1983; Mitsios and Rowell, 1987). It has been shown that roots may secrete HCO_3^- and OH^- ions, thereby tending to increase the pH in the rhizosphere (Rowell, 1988). However, it is also known that the composition of root exudates encompasses a wide variety of acids (Russell, 1977; Vancura *et al.*, 1977) and therefore this might have the overall effect of lowering soil pH in the immediate proximity of the root. It might be hypothesized that rhizosphere pH may be more important to nematodes, which feed on individual cells and which remain longer at the root surface, than to insects, but there appears to be no data on this subject (Dropkin, 1989). However, very local changes in pH may also be important for microarthropods, such as Collembola, which graze on soil microflora or root hairs (Greenslade and Ireson, 1986). Numbers of microorganisms have been shown to be greatly enhanced in the rhizosphere (Russell, 1977), and Wiggins *et al.* (1979) have shown that collembolan numbers may be significantly greater here than in open soil.

3. Soil Temperature

The growth rate of insects is strongly dependent on temperature, although there are threshold temperatures below which no development takes place and above which mortality occurs. Several studies have shown that the growth rate of root-feeding insects increases with increasing temperature of the soil (e.g. Ridsdill Smith *et al.*, 1975; Régnière *et al.*, 1981a; Potter and Gordon, 1984; Jackson and Elliott, 1988). As would be expected, insects from temperate climates can withstand lower temperatures than those from warmer parts of the world. The cabbage root fly pupa can survive temperatures as low −10°C (Block *et al.*, 1987) in Britain, and the lettuce root aphid can survive for 40 weeks at 0°C in Canada (Alleyne and Morrison, 1978a). In contrast, the scarabaeid, *Heteronychus arator* (F.), requires soil temperatures of at least 20°C in New Zealand (King *et al.*, 1981) and the grape phylloxera's optimum range is 16–32°C (Granett and Timper, 1987). Soil temperature is considered to be one of the main reasons why one of the most destructive pests, the Japanese beetle (*Popillia japonica*), is unlikely to become established in Britain (Baker, 1975), as the threshold temperature for larval development is 20°C (Ludwig, 1928).

The soil temperature is important both in determining the size of root-feeding insect larvae and the probability of winter survivorship. Laughlin (1963) found that higher temperatures increased the size of larvae of the garden chafer, *Phyllopertha horticola*, when feeding in pastures. Moreover, it was found that heavier larvae had a better chance of surviving the winter and thus winter survival was related to summer temperatures. A similar situation was reported for the pygmy beetle, *Atomaria linearis* Steph., in sugar beet crops (Cochrane and Thornhill, 1987). The damage to the crop in any year was correlated with the prevailing temperatures and hence larval survival in the previous season.

4. Soil Moisture

Soil moisture is probably the single most important property affecting the survival and abundance of root-feeding insects. Changes in soil moisture have been implicated as being of major importance in population fluctuations of these insects (Régnière *et al.*, 1981c). A number of workers have shown that insect survival is very poor in dry soils for both chewing insects (Jones, 1979; Régnière *et al.*, 1981c; Lummus *et al.*, 1983; Marrone and Stinner, 1984) and sucking species (Bartlett, 1984; Misra *et al.*, 1985; Moran and Whitham, 1988) (Fig. 3). The eggs and young larvae of a number of chewing species are particularly susceptible to desiccation (e.g. Régnière *et al.*, 1981c; Quinn and Hower, 1986a). The eggs of many species of Coleoptera must absorb water before hatching (e.g. Potter, 1983), and Krysan (1976) found that the eggs of a chrysomelid corn rootworm need to be maintained at about 100% R.H. during a critical stage of embryogenesis or desiccation will occur. Subsequent larval survival immediately following egg hatch critically depends on the moisture content of the soil (Régnière *et al.*, 1981c; Quinn and Hower, 1986a). Tipulid larvae, for example, are particularly sensitive and require a soil atmosphere close to saturation (Coulson, 1962; Hartman and Hynes, 1977). As a result, they are more abundant in heavy clay and wet moorland soils (Coulson and Whittaker, 1978). The physiological mechanisms through which soil moisture affects root suckers are unknown (Moran and Whitham, 1988), but mortality may be caused by simple dehydration, caused by a dry soil (Fig. 3c). Foliar-feeding aphids, for example, are very susceptible to desiccation (Dixon, 1985).

In general, increasing soil moisture has a null or enhanced effect on insect survival (Ladd and Buriff, 1979; Régnière *et al.*, 1981c; Lummus *et al.*, 1983; Moran and Whitham, 1988). However, in very wet soils, insect survival may be impaired due to the anaerobic conditions created (Foster and Treherne, 1975; Godfrey and Yeargan, 1985), although in *Pemphigus trehernei* feeding on the roots of sea aster, *Aster tripolium* L., in saltmarshes, flooding provides a dispersal mechanism by first instars floating on sea water (Foster and Treherne, 1978). The application of water to dry soils can have a detrimental effect on the lettuce and sugar beet root aphids (Lange *et al.*, 1957; Harper, 1963), because water causes cracks in the soil to close up, thereby preventing movement of aphids between plants.

Radcliffe (1971a) provided an interesting case of root-feeding larvae

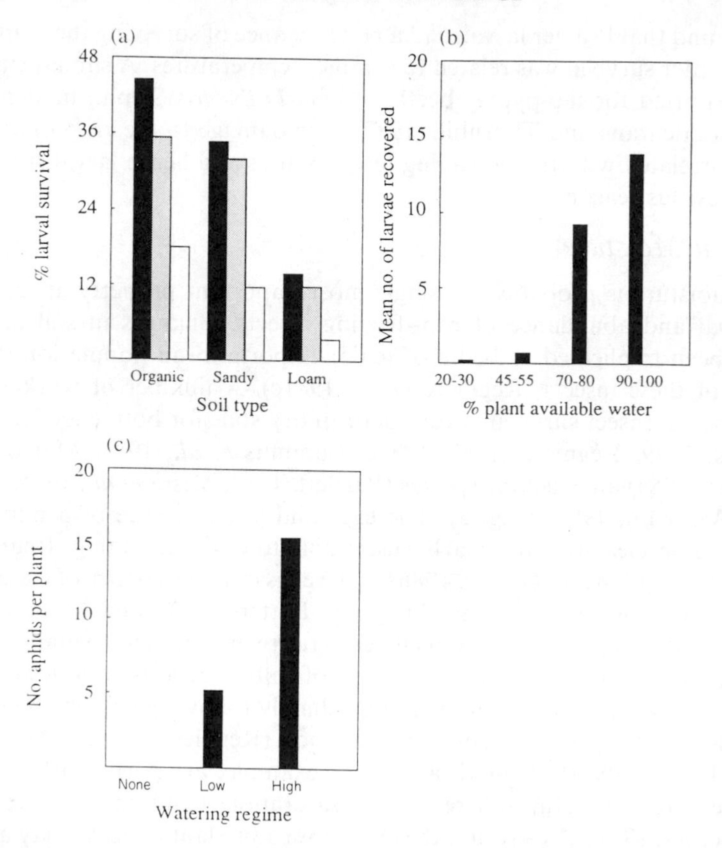

Fig. 3. Effect of soil moisture on subterranean insects. (a) Larval survival of bean leaf beetle in three soils at three watering regimes: low (1000 kPa: □), medium (100 kPa: ▨) and high (10 kPa: ■) (redrawn from Marrone and Stinner, 1984). (b) Number of recovered corn rootworm larvae out of 50 introduced per 15-cm pot after 21 days feeding on maize roots, at different plant available water regimes (redrawn from Lummus *et al.*, 1983). (c) Number of *Pemphigus betae* per root system of *Chenopodium album* at three watering regimes (redrawn from Moran and Whitham, 1988).

modifying the soil moisture. It was suggested that chafer larvae actually reduced the moisture content, by absorption for metabolism or as a result of surface evaporation caused by burrowing. In this way, the environment was made more amenable for the larvae, as high levels of moisture resulted in decreased survival.

5. *Soil Nutrients*

The interaction of root-feeding insects and the nutrient content of a soil is another field in which conflicting results have been obtained. Most authors have found that the addition of nitrogen to the soil, in the form of fertilizers, tends to reduce larval feeding and root damage. Wolfson (1987) found that increased levels of soil nitrogen reduced nodule development in *Medicago sativa* and *Trifolium pratense* L. As a result, the establishment of the larvae of *Sitona hispidulus*, which are dependent on root nodules (Quinn and Hower, 1986b), was impaired and root attack lessened. Radcliffe (1970) added cow dung to soil and found that this resulted in less damage to the roots of grass by chafer larvae. In this case, it was suggested that the larvae ingested the organic matter and reduced their root consumption. Inorganic nitrogen additions have also been shown to reduce root damage by chrysomelid larvae in maize (Hill *et al.*, 1948; Lilly and Gunderson, 1952). These results were attributed to a proliferation of the root system resulting from the added nitrogen and proportionately less root damage occurring. Edwards (1977) demonstrated a clear reduction in wireworm populations as a result of 4 years' consecutive addition of inorganic nitrogen.

In contrast to these results, Foster *et al.* (1986) and Spike and Tollefson (1988) reported root damage by chrysomelid larvae which was positively correlated with the amount of nitrogen applied. Meanwhile, Prestidge *et al.* (1985) found no relation between fertilizer application and feeding by scarabaeid larvae. It appears that the timing of nitrogen application, relative to larval establishment, may be crucial. If the fertilizer is added before establishment, then the proliferation of the root system will result in an increased food supply for the larvae, better larval survival and thus increased levels of damage. If the nitrogen additions occur after establishment, then damage is proportionately less (Spike and Tollefson, 1988). It seems likely that fertilizer applications indirectly affect root herbivores through the root system, although acidification of the soil and the resultant effects on the herbivores should not be discounted (Polivka, 1960a; Edwards, 1977; Curry, 1987).

F. Factors Affecting Root Feeding by Insects

1. Nutritional Quality of Roots

The root system of a plant is the major site of mineral acquisition, and at certain times may represent a place for the storage of carbohydrates (Andersen, 1987). It is generally supposed that roots provide an abundant food supply, but one that is low in nitrogen, compared to the foliar parts of plants (Chapin *et al.*, 1980; Seastedt, 1985). Even so, nitrogen content may vary seasonally and at certain times can actually be higher than in other plant parts (Andersen, 1987). The long life-cycle of some root-feeding insects may be a reflection of the relatively poor nitrogen supply for these insects.

In contrast, there are two genera of Coleoptera and one of Diptera in which the larvae feed on root nodules of the Leguminosae, thereby having access to an extremely rich nitrogen supply (Quinn and Hower, 1986b). The best known of these is the curculionid genus, *Sitona*. For some species, such as the pea and bean weevil, *S. lineatus*, nodules supply nutritional requirements for all the larval stages. In other species, such as *S. hispidulus* (clover root curculio), only the first instar larvae are dependent on nodules (Quinn and Hower, 1986b). The risks of nodule feeding are high (e.g. Goldson and Muscroft-Taylor, 1988), but the rewards are potentially large, because the nodules contain high concentrations of many essential amino acids (Greenwood and Bathurst, 1978).

The storage of carbohydrate in underground plant tissue provides a food supply with a high energy content. Sutherland (1971) reported that sucrose composed 50% of the total soluble sugar content of ryegrass (*Lolium* spp.) roots and demonstrated that this was a major phagostimulant for grass grub (*Costelytra zealandica*) larvae. The high starch and sugar content of sweet potato tubers also represent an energy-rich food supply for the weevil, *Cylas formicarius* (Sutherland, 1986). However, carbohydrate reserves in roots can show rapid temporal changes (e.g. Mooney and Billings, 1960) and this must add to the unpredictability of the root system as a food source.

Because roots absorb minerals from the soil, the concentrations of certain ions may be higher than in foliar parts. For example, sodium, although not required for growth by plants, is sequestered in the roots (Andersen, 1987). However, all animals require sodium and this may represent an important component of nutrition for root-feeding insects.

It is generally assumed that root-sucking insects feed in the phloem. Phloem contents are rich in sugars and amino acids and provide a very good food supply (e.g. Minoranskij, 1971). However, cicada nymphs have been shown to feed only in the xylem (White and Strehl, 1978). It is known that xylem sap is 99·9% water (Cheung and Marshall, 1973), with amino acids and sucrose comprising most of the remainder. Cicadas commonly have long life-cycles, spending several years as a root-feeding nymph, with the extreme of 17 years in the periodical species (*Magicicada* spp.) (Borror *et al.*, 1976). It is not unreasonable to suggest that the long life-cycles in these species may be due, at least in part, to their extremely poor food supply. The cicadas are well adapted for xylem feeding, because they have very well-developed cibarial pumps, in order to overcome the negative pressure of xylem (Snodgrass, 1935). In addition, the excess water which has to be excreted is used to maintain mud cells in which the nymphs live, and to maintain a humid environment (White and Strehl, 1978).

2. *Plant Preferences*

Although many root-feeding insects are restricted to one family of plants, they often show distinct species preferences within these families. A clear example is provided by Wapshere and Helm (1987), working with *Daktulosphaira vitifolii*, the grape phylloxera. This species is restricted to members of the genus *Vitis*. In parts of North America, where phylloxera occurred naturally with *Vitis* spp., a large proportion of the vine species was resistant to attack (very few galls produced by the insect on shoots or roots). However, in areas where the insect was introduced, the proportion of resistant species was very low (Fig. 4a,b) and resulted in the rapid spread of infestations, causing a major problem in the grape industry. The physiological basis of resistance is not clear, but it appears that it is related to the failure of the nymphs to feed or the adults to lay eggs and to healing of the lesions caused by root-feeding (Boubals, 1966). Root-feeding aphids often show a narrow range in their subterranean hosts, in keeping with their foliar-feeding counterparts (Dixon, 1985). For example, the sugar beet root aphid, *Pemphigus betae* Doane, occurs on *Rumex* spp. (Polygonaceae) and in the Chenopodiaceae, on *Chenopodium* (Moran and Whitham, 1988) and sugar beet (*Beta vulgaris* L.) (Lange, 1987); the lettuce root aphid, *P. bursarius*, is restricted to several genera within the Compositae (Dunn, 1959a; Alleyne and Morrison, 1977), while *P. trehernei* occurs only on sea aster (Foster, 1975).

Although root-chewing insects will ingest dead organic matter in the soil, they actively select live root material (Ridsdill Smith, 1975). Radcliffe (1970) grew monocultures and mixtures of white clover (*Trifolium repens* L.) and ryegrass (*Lolium* spp.) and found that clover was heavily attacked by grass grub larvae (*Costelytra zealandica*) when in mixed cultures, although more was eaten in clover monocultures. Prestidge *et al.* (1985) investigated the preference of this species over a range of grasses and legumes. Larval growth rate was greatest on red clover, *T. pratense*, whereas *Dactylis glomerata* L. (cocksfoot grass) and *Medicago sativa* (lucerne) were actively avoided and the growth rate was negative (Fig. 4c).

The density of plants in relation to attack by root-feeding insects has received little attention, but it appears to have little direct effect. Radcliffe (1971c) found no interaction between ryegrass density and grass grub attack, although the higher root productivity of plants at the high density was sufficient to mitigate the effects of larval feeding, when measured in terms of foliar yield. Spike and Tollefson (1988) found no effect of plant density on larval root damage by corn rootworms (*Diabrotica* spp.). However, plants in low-density plots were able to develop more roots subsequent to larval attack, thus aiding their stability. The situation may be more complicated, when considered in terms of plant clumping and distance between clumps (Piedrahita *et al.*, 1985a). These authors found that the relationship between these plant variables and corn rootworm larval density was a three-dimensional surface with highest larval densities occurring at moderate clumping (3 plants per location) and relatively wide spacing (50 cm between locations).

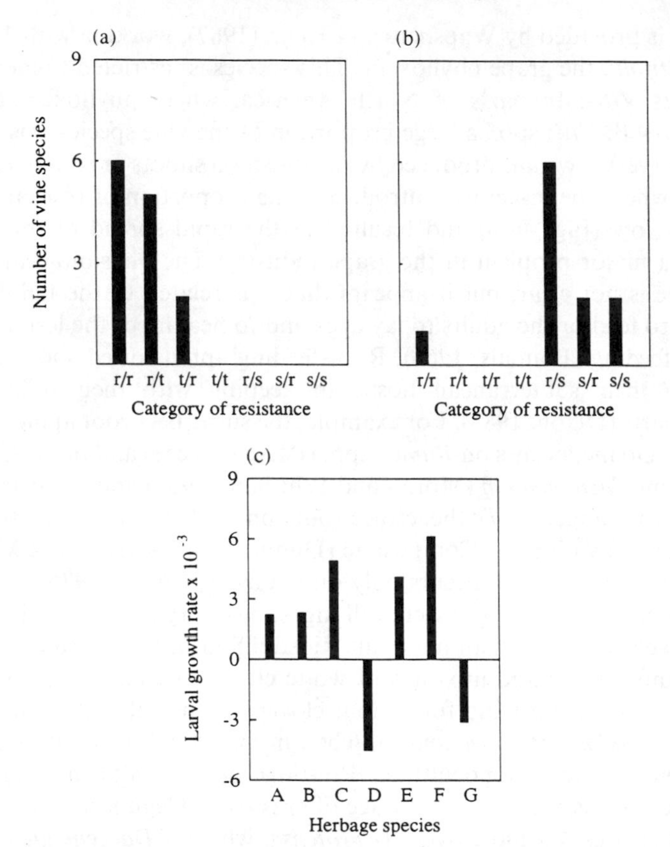

Fig. 4. Host-plant preferences displayed by root-feeding insects. (a) Number of *Vitis* species resistant (r), tolerant (t) or susceptible (s) to galling by *Daktulosphaira vitifolii* on roots/shoots of the vine, in areas of the USA where phylloxera is native. (b) As for (a), but in areas where phylloxera was introduced (redrawn from Wapshere and Helm, 1987). (c) Growth rate of grass grubs on different herbage species. A, *Lolium perenne*; B, *Holcus lanatus*; C, *Festuca arundinacea*; D, *Dactylis glomerata*; E, *Trifolium repens*; F, *T. pratense*; G, *Medicago sativa* (redrawn from Prestidge *et al.*, 1985).

3. *Chemical Attractants and Deterrents in Roots*

The fact that many species of root-feeding insect show distinct host-plant preferences suggests that there are certain chemical compounds, present in root systems, which either enhance or inhibit larval feeding. Chemicals involved in behavioural responses of root-feeding insects may be attractants, phagostimulants or deterrents (Dethier, 1970).

Carbon dioxide appears to be one of the principal chemicals by which root-feeding insects orientate themselves in the soil. This is a primary metabolite which is released from the roots during respiration (Harris and van Bavel, 1957; Koncalova *et al.*, 1989). Strnad *et al.* (1986) and Strnad and Bergman (1987) reported that the larvae of corn rootworms (*Diabrotica* spp.) actively moved through soil to a source of this gas. Many other larvae have also been shown to be actively attracted by this compound (e.g. Jones and Coaker, 1977; Galbreath, 1988). However, as most roots produce CO_2, it is unlikely that this will form the basis for the ability of larvae to distinguish between the roots of different plant species. Indeed, in choice tests in which larvae of *Diabrotica* spp. (Hibbard and Bjostad, 1988) or *Costelytra zealandica* (Sutherland, 1972) were provided with sources of CO_2 or root-volatile extracts plus an equal volume of CO_2, preference was significantly for the latter. Volatile secondary compounds important in the attraction of underground larvae have been identified for several insect species: the carrot fly (*Psila rosae* (F.): Ryan and Guerrin, 1982); the onion maggot (*Delia antiqua* (Meigen): Matsumoto, 1970); other *Delia* spp. (Hibbard and Bjostad, 1988); and corn rootworms (*Diabrotica* spp.: Buttery and Ling, 1985). Volatiles have also been shown to play an important role in host location in other species, but as yet they have not been chemically characterized: the clover root borer (*Hylastinus obscurus* Marsham) (Coleoptera: Scolytidae) is attracted to volatiles from diseased clover roots (*Trifolium pratense*) (Kamm and Buttery, 1984) and western corn rootworms (*Diabrotica virgifera virgifera* Le Conte) are attracted to compounds from maize seedlings (Hibbard and Bjostad, 1988). An excellent discussion of semiochemical insect attractants in roots is provided by Hibbard and Bjostad (1988).

Once at the root, other chemicals serve as phagostimulants. Ladd (1988) reported that five sugars (sucrose, maltose, fructose, glucose and trehalose) were all important phagostimulants for larvae of the Japanese beetle, *Popillia japonica*. Sucrose was also found to be the major phagostimulant in the selection of grass roots by larvae of the grass grub, *C. zealandica* (Sutherland, 1971).

The range of metabolites, which often occur in foliar plant parts and which act as feeding deterrents, may also be found in plant roots (McKey, 1979). Indeed, the degree of specialization recorded among root-feeding insects is likely to be a reflection of the distribution of such deterrents, rather than attractants. Compounds with known insect-feeding deterrent properties which have been isolated from roots include alkaloids, furanocoumarins, cyanogenic glucosides, glucosinolates, isoflavanoids, phenolic acids and saponins (Cole, 1984; Rodman and Louda, 1984; Gaynor *et al.*, 1986; Andersen, 1987). In some cases, the deterrent activity has been related to insect–host selection in the field or laboratory.

Gaynor *et al.* (1986) found that the isoflavanoid, phaseollin was present in high concentrations in bean (*Phaseolus vulgaris* L.) roots. The roots of *Lotus pedunculatus* Cav. were found to have high levels of nitropropanyl esters, while those of white clover (*T. repens*) contained relatively low levels of these two compounds. Correspondingly, it was found that larvae of the grass grub, *C. zealandica*, grew very poorly when fed on *Phaseolus* or *Lotus* roots, but very well on *T. repens*. The susceptibility of clover relative to the other two plants in field conditions was related to the low levels of deterrents in clover roots. Other herbage legumes, such as lucerne (*Medicago sativa*) and sainfoin (*Onobrychis vicifolia* Scop.), have been shown to contain saponins and isoflavanoids in their roots which act as deterrents to grass grub larvae (Sutherland *et al.*, 1975; Russell *et al.*, 1984). Results such as these help to explain the feeding preferences of *C. zealandica*, reported by Prestidge *et al.* (1985) (see Section IV.F.2; Fig. 4c). However, root chemicals which act as feeding deterrents in some species may not affect others. For example, while *C. zealandica* feeds well on white clover roots, these contain a deterrent which repels larvae of another scarabaeid, *Heteronychus arator* (black beetle) (Sutherland and Greenfield, 1978).

Deterrents which affect root-sucking insects are seldom reported, probably as a result of the specialist phloem-feeding habit of these insects and our extremely poor knowledge of the constituents of phloem sap. However, a notable example was provided by Cole (1984), working with the lettuce root aphid, *Pemphigus bursarius*. Certain varieties of lettuce are resistant to this aphid, the resistance being positively associated with isochlorogenic acid and the enzyme phenyl alanine-ammonia lyase levels in the roots. The enzyme converts phenolic acids to quinones, but it is not known at what point in the biochemical pathway "resistance" occurs. It is interesting to note that isochlorogenic acid is the principal cause of "browning" of lettuce roots and stems when these are cut.

4. *Mortality Factors*

Undoubtedly, abiotic factors play a large part in the mortality of root-feeding insects. Soil moisture and temperature are critical for the survival of larvae (see Section IV.E.3 and 4). However, a number of biotic factors have also been shown to be significant causes of mortality, although few have been related to the species' population dynamics. In a number of species, significant larval mortality in the field is attributable to density-dependent effects. High larval densities cause significant reductions in growth rates, through direct competition between larvae for the available food supply (Ridsdill Smith and Roberts, 1976; Régnière *et al.*, 1981c). A reduction of growth rates and larval survival can lead to biased sex ratios in emerging adults, where more of the slower-growing female larvae perish (Weiss *et al.*, 1985). Interspecific larval competition may also occur. Western corn rootworms (WCR: *Diabrotica virgifera virgifera*) outcompete northern corn rootworms (NCR: *D. longicornis barberi* S.&L.) in laboratory and field situations (Piedrahita *et al.*, 1985b) with the result that WCR has spread into areas of the USA, previously occupied by NCR.

Root-feeding insects are generally poorly protected against predators and parasites, presumably relying on their location as a form of protection. However, in those species which have adults above ground, there may be considerable mortality at emergence. It has been proposed that the periodical emergence of vast numbers of cicadas (*Magicicada* spp.) is a method to induce predator satiation and Lloyd (1984) provides an interesting discussion of this phenomenon. In addition to vertebrate predators, such as birds, there are a number of predatory insect species living in the soil or at its surface which may take a heavy toll on subterranean larvae. Examples of predators are carabid beetles, ants, earwigs and empid flies (Jones and Jones, 1984; Sutherland, 1986; Tryon, 1986). In addition to these soil predators attacking root-feeding aphids, the attendant ants may also eat them and this has been estimated to be in the order of 3000 aphids per nest per day (Pontin, 1978). Certain species may also suffer heavy parasitism; for example, Japanese beetle (*Popillia japonica*) populations may be reduced by up to 90% in any one year by tachinid flies (Rikhter, 1975). However, this may be an exception, as Hawkins and Lawton (1987) found relatively few parasitoid species associated with root feeders in Britain.

The effects of nematodes on subterranean insects has been emphasized by the recent interest in the use of entomogenous species as control agents of root-feeding larvae. These nematodes may be present in or added to the soil, where they infect the larvae. The infective stage is the fourth instar nematode. This enters the insect through a natural opening or the soft cuticle and the insect dies rapidly of septicaemia induced by the nematode's symbiotic bacteria. The cadaver then provides a substrate for subsequent generations of nematodes. This method of control has been attempted against the larvae of Scarabaeidae (Kard *et al.*, 1988; Shetlar *et al.*, 1988; Villani and Wright, 1988) and Curculionidae (Rutherford *et al.*, 1987; Jaworska and Wiech, 1988), with considerable success. Nematodes have even been reported to infect the lettuce root aphid, *Pemphigus bursarius*, in Canada (Alleyne and Morrison, 1977).

A number of root-feeding larvae are attacked and killed by entomopathogenic fungi in the field. Japanese beetle larvae frequently die from infections (St. Julian *et al.*, 1982) and fungi have also been reported attacking weevils (Poprawski *et al.*, 1985a), flies (Poprawski *et al.*, 1985b) and cicadas (White and Lloyd, 1983). Bacteria and viruses often provide the most dramatic causes of mortality in root-feeding larval populations. The bacterium *Bacillus popillae* Dutky has caused reductions of 86–94% in successive annual generations of Japanese beetle larvae (Fleming, 1976). More recently, Carter and Green (1988) have shown the effectiveness of viruses against larvae of *Tipula paludosa* Meigen.

V. THE PLANT'S RESPONSE TO ROOT HERBIVORY

A. Vegetative Growth

The large reductions in agricultural yield which are often attributed to root-feeding insects provide evidence of their destructive capabilities (see Section VI.A). However, while there have been many attempts to quantify the effects of various foliar-feeding insects on the growth of individual plants (see Gange, in press), similar studies involving root herbivores are surprisingly scarce. In addition, most of these works involve agricultural pests, where effects on growth are measured in terms of yield (biomass) or herbage nutrient content, rather than the type of measurements used by plant ecologists, such as height, leaf and seed number or seed viability.

Pot trials in the laboratory, often with ranges of larval densities, can provide an indication of the effect of root feeding but often produce overestimates of the amount of damage. In a series of experiments, Radcliffe (1971a,b,c,d) recorded yield losses of up to 70% in pasture grasses and white clover with different densities of grass grub (*Costelytra zealandica*). Ladd and Buriff (1979) found that one Japanese beetle (*Popillia japonica*) larva per 15-cm pot could reduce the yield of *Poa pratensis* L. by 40%, while at 16 larvae per pot the yield reduction was 85%.

There is a serious lack of field experiments which document the effect of root-feeding by insects on individual plants. Two studies do exist, which show what can be achieved by simple field experimentation.

Karban (1980) found that nymphal periodical cicadas, *Magicicada septendecin* (L.), had extensive effects on the growth of the scrub oak, *Quercus ilicifolia* Wang. In eastern North America, this cicada species occurs locally at particularly high densities (Dybas and Davis, 1962). By recording annual growth increments, Karban showed that nymphs were capable of reducing tree growth by as much as 30%, but that the effect was not consistent from year to year. Adult emergence of this cicada species takes place every seventeenth year in the north of its range and every thirteenth year in the south, with the most substantial reduction in growth occurring during the years following the emergence. This can be related to the more mobile young nymphs feeding on a large number of small rootlets (White and Lloyd, 1975), whereas the older nymphs live in permanent earthen cells feeding on larger roots and cause relatively less damage. In this way, below-ground herbivory leads to a periodical pattern of tree growth. Powell and Myers (1988) provide the best example of the effects of root-feeding on the growth of a herbaceous species. They compared plants of knapweed (*Centaurea diffusa*) with roots attacked by larvae of the beetle *Sphenoptera jugoslavica* Ob., with those which were uninfested. It was found that infested plants had a slower growth rate and increased mortality. As a result, uninfested plants grew considerably larger with consequent effects on seed production (see Section V.B). It was concluded that under favourable environmental conditions for the beetle (such as drought or other factors affecting plant stress), attacks could contribute significantly to reductions in knapweed population growth.

Due to the fact that many herbage crops are attacked by root-feeding insects, some attention has been focused on the nutrient contents of this herbage in relation to root attack. Both Godfrey *et al.* (1987) and Goldson *et al.* (1988) reported that *Sitona* larvae feeding on root nodules of *Medicago sativa* (alfalfa) caused reduced nitrogen fixation and a corresponding decrease in crude protein and nitrogen content of the herbage. Such reductions in herbage nitrogen are serious in such an important foliage crop (Hill, 1987). An interesting case of mineral nutrients in maize foliage being altered by corn rootworm (*Diabrotica* spp.) feeding was provided by Kahler *et al.* (1985). It was found that heavily infested plants contained more Na and Fe and less K, Mg and Ca than uninfested plants. The increase in Na and Fe was attributed to increased lateral branching of roots and increased absorption of these elements from the soil. However, it is not clear why concentrations of other elements were decreased. Results such as these have important implications for foliar herbivores, both vertebrate and invertebrate, and more research is needed in this field.

The concept of "economic threshold" is a familiar one to agriculturalists and represents the population density of an organism at which control measures should be started to prevent economic injury to the crop (Hill, 1987). It is somewhat arbitrary, because the threshold will change according to which pest species, crop or situation is being considered. The term is well established in the nematological literature, where plant responses to a range of nematode densities have been described by a function termed the Seinhorst equation (Seinhorst, 1965). Indeed, nematologists have suggested that experiments involving measurements of plant growth in response to root attack should be performed over a range of nematode densities, because at low densities plant growth may be unaffected or even enhanced. Similar relationships appear to exist with insects, although the evidence is mainly from field data. We therefore suggest that experiments involving root herbivores under controlled conditions should always include a range of insect densities. A number of threshold levels of insect abundance have been assessed for major pest species e.g. elaterids (Jones and Jones, 1984), *Sitona* weevils (Goldson *et al.*, 1985; Aeschlimann, 1986) and tipulids (Blackshaw, 1988). In the case of tipulids, farmers and growers may now sample their own crops and, using commercially produced guidelines, predict whether the pest populations are of a sufficient level to justify insecticide application (Stewart and Kozicki, 1987).

However, while threshold population sizes have been described for a number of species in field situations, there is little work involving single plants in laboratory or field conditions. Perhaps the most convincing example of a damage threshold in the laboratory is provided by Davidson (1979). In an experiment with larvae of two scarabaeid species feeding on three grass species, it was found that no measurable reduction in foliage production occurred until 60% of the root system had been removed. There was then a linear relation between the amount of root eaten and the reduction in foliage yield. Ladd and Buriff (1979), working with Japanese beetle (*P. japonica*), found that the number of larvae per 15-cm pot required to produce a measurable drop in foliage yield of *Poa pratensis* depended on soil moisture content. At a low soil moisture (60% of field capacity), one larva per pot was sufficient to cause yield reduction, while at high soil moisture (90% of field capacity) two or sometimes four larvae were needed.

When larval densities are sufficiently high, intraspecific competition for food is important, with the result that little extra damage to the plant is seen. Ridsdill Smith (1977) found that the relations between ryegrass root yields and scarabaeid larvae were "J shaped", with the greatest difference in yield reduction occurring as a result of adding one larva per 15-cm pot. Little difference in yield reduction was seen between treatments of 6, 8 or 10 larvae per pot. Ladd and Buriff (1979) found a similar result; there was no difference in grass yield between treatments with 8 and 16 larvae per pot, irrespective of soil moisture. As described earlier, density-dependent larval mortality is important in these insects (e.g. Régnière *et al.*, 1981c; Weiss *et al.*, 1985).

B. Reproductive Growth

This is an area of research in which there has been very little work, because most agricultural yields lost to root feeders are measured in terms of herbage produced, rather than seed or fruit production. An exception is grain yield in corn, which can be severely reduced by the activities of *Diabrotica* larvae. Kahler *et al.* (1985) recorded grain losses of 14% due to these larvae and linear relationships have been obtained relating yield losses to number of larvae per plant (Chiang, 1973). Dutcher and All (1979) found that root boring by larvae of the moth *Vitacea postiformis* could reduce grape yield by up to 90%.

In natural situations, such studies are even rarer and there is a clear avenue for research on this subject. The results of Powell and Myers (1988) provide an excellent example of what can be achieved. By comparing infested and uninfested knapweed plants, they found that root-feeding by larvae of *Sphenoptera jugoslavica* could significantly reduce the production of seed heads and seeds per head and that the total seed production of individual plants could be reduced by up to 60%. However, individual seed weight and viability were not recorded. Our own work (Brown and Gange, 1989a; see

Section VI.B.2) has shown that the reproductive potential of annual plant communities, in terms of number of flowers produced, may be reduced by root-feeding insects and we are currently seeking to explore the effects on individual plants.

C. Stability and Storage Reserves

The anchorage of a plant in soil and the storage of the products of photosynthesis are two important functions of a root system (Russell, 1977). Although it is intuitively obvious that the severing of parts of a root system by insects would lead to a loss of stability, there are, however, very few recorded instances of this occurring. It may well be that, in natural situations, populations of subterranean insects do not occur in such densities as to impose this constraint. In crop monocultures, however, the number of insects may be very high and a reduction in stability could be a very real problem. Gray and Tollefson (1987) reported that extensive root feeding by corn rootworm larvae (*Diabrotica* spp.) reduced the stability of maize plants, causing lodging to occur, and that extensive lodging reduced the efficiency of grain harvesting from the field. In pasture land, or other large tracts of grassland such as golf courses, extensive feeding by scarabaeid or tipulid larvae can cause turf to become spongy or "weak", with the result that it can be pulled back like a carpet (Ferro, 1976, p. 115).

An example of root-feeding insects increasing plant stability was provided by Simberloff *et al.* (1978). Indeed, this may be one of the very few cases of insect herbivory actually benefiting a plant (cf. Belsky, 1986; see Section V.F.2). These authors found that mangroves produced more branching roots when the main root was attacked by root-feeding larvae. The extra aerial roots produced meant that more found the substrate and therefore aided the stability of the trees. This is important, because many trees get washed away by tropical storms. There was no evidence that root-boring by the insects had any detrimental effect on the trees, although wood and fruit production were not measured.

The root system is an important storage organ for carbohydrates and many plants, such as sweet potato (*Ipomoea batatas* Lam.) and sugar beet (*Beta vulgaris*), contain large amounts of starch and disaccharides such as sucrose. Feeding on these structures by insects can seriously affect storage reserves. In some cases, feeding may directly reduce root sucrose content. For example, Minoranskij (1971) reported that enzymes in the saliva of *Pemphigus fuscicornis* (Koch) split starch and disaccharides into monosaccharides which aphids could easily assimilate. This resulted in up to a 5% loss of sugar content and a 28% loss of crop weight. Dintenfass and Brown (1988a,b) found that feeding by the clover root curculio, *Sitona hispidulus*, on nodules or taproots of alfalfa significantly reduced the carbohydrate reserves in the roots. Goldson *et al.* (1988) working with *Sitona discoideus* feeding on the same plant provided an explanation. Nodule feeding by these insects means that continual nodule regeneration represents a nitrogen sink and increased respiration for such root repair results in a depletion in the carbohydrate reserves of the plant. However, it is likely that simple leakage of carbohydrate from damaged roots also occurs. This phenomenon has been reported in nematode feeding (Wang and Bergeson, 1974), where only individual cells are punctured by the stylets. The damage caused by insects is often more extensive and clearly loss of carbohydrate in this way could be an important problem for the plant.

D. Drought Stress

One of the primary functions of roots is water absorption. A considerable amount of the literature describes the symptoms of drought stress in plants, caused by simple pruning of the root system (Andersen, 1987). In addition, other root feeders, such as nematodes, have been reported to cause drought stress in plants (Dropkin, 1989). It is therefore highly likely that root feeding by insects will have a detrimental effect on the water relations of a plant and there is anecdotal evidence pertaining to the wilting of plants as damage symptoms of chafer grub (French, 1984a), tipulid (French, 1984b) or root aphid attack (Dunn and Kempton, 1974). However, quantification of this in field or laboratory conditions is lacking.

Goldson *et al.* (1987) found that roots of lucerne, when attacked by *Sitona discoideus*, accumulated carbohydrate reserves and that these reserves were not mobilized for the replacement of shoots after harvesting as would normally occur. The accumulation process was attributed as being a response to water stress and it appears that *Sitona* can induce a form of drought-related dormancy in lucerne as a result of larval feeding. Ridsdill Smith (1977) measured the water content of ryegrass leaves when the plant roots were subjected to feeding by larvae of the scarabaeid, *Sericesthis nigrolineata* Boisd. Root feeding significantly reduced leaf water content, thereby inducing water stress and affecting plant growth. More recently, we have found that root feeding by larvae of the garden chafer, *Phyllopertha horticola*, causes drought stress in *Capsella bursa-pastoris* L. (Gange and Brown, 1989).

E. Transmission of Root Pathogens

A potential problem arising from root attack by insects is the subsequent colonization of the root by fungal pathogens. Pathogen entry into roots may be a direct result of transmission by the insect or entry may be via lesions on the root created as a result of insect feeding. In either case, the attack by the fungus on an already weakened plant is usually lethal.

Active transmission of root disease in red clover (*Trifolium pratense*) by the clover root-boring beetle, *Hyalastinus obscurus*, was reported by Kamm and Buttery (1984). The beetle was shown to be attracted to the volatiles released from already diseased roots and the combination of beetle feeding and fungal attack contributed to the decline of clover stands (Leath and Byers, 1973). Witcosky *et al.* (1986) found that three species of weevil acted as vectors of black stain root disease of douglas fir (*Pseudotsuga menziesii* Franco), thereby contributing to the decline of this tree in Oregon, USA.

Most insect larvae in the soil are attracted to living roots (Ridsdill Smith, 1975) and thus a frequent transmission of pathogens is through lesions which are created during feeding. Dickason *et al.* (1968) reported that feeding wounds created by *Sitona hispidulus* on *Medicago sativa* allowed the entry and establishment of fungi. Similar occurrences were described by Rizza *et al.* (1988) working with the syrphid fly, *Cheilosia corydon* (Harris), attacking thistle species in North America.

F. Compensatory Responses

1. *Transport of Photosynthate*

Davidson (1979) considered that the extent to which plants could tolerate root damage depended on the regrowth potential, defined as the quantity of hydrolysable material (mainly carbohydrate) in the plant as a whole. When roots are removed by insect feeding, the plant responds by mobilizing carbohydrate reserves and directing these to the attacked area for use in respiration during the period of new root growth. The larger the plant, the greater the regrowth potential, and thus damage by root feeders will be proportionally less compared to small plants (Davidson *et al.*, 1970). Mobilization of reserves depends on whether only roots or shoots or both are being attacked. For example, Ridsdill Smith (1977) recorded no effect on the foliage yield of ryegrass as a result of scarabaeid larval feeding. However, if above-ground parts were also attacked, then the response to root feeding was clearly seen. Seastedt *et al.* (1988) have shown that foliage removal may alter nitrogen concentrations in roots. Under low or high levels of defoliation, there was little effect, but concentrations were significantly increased under moderate defoliation. Such a change may be one of the main reasons why larval Scarabaeidae are more abundant in grazed than ungrazed pastures (Roberts and Morton, 1985; Seastedt, 1985). Interestingly, a similar response occurs in nematodes, the numbers of which may be increased by up to five times in plants with their foliage clipped compared to unclipped specimens (Stanton, 1983). It is worth remembering that both foliage and root removal is often the normal situation, for example, in pasture grasses attacked by soil-dwelling herbivores and grazed by sheep or cattle (Roberts and Ridsdill Smith, 1979).

The regrowth potential will also depend on the time of year that insect attack occurs. Mooney and Billings (1960) showed that perennial plants that rely on underground food reserves for survival and reproduction during the next year, will often begin allocating photosynthate to the roots prior to the current season's flowering. The loss of stored material late in the growing season would reduce the probability of new reserves being accumulated prior to dormancy. Such a situation may contribute to the fact that the most likely time for a plant to die is during active growth (Harper, 1977; Gange *et al.*, 1989), because reserves would not be sufficient to sustain growth. It is probably more than a coincidence that many temperate zone root feeders have larvae present in the soil during autumn and winter, when underground reserves will be at a maximum.

2. *New Root Growth*

Evidence suggests that when parts of the root system are removed by feeding, the replacement of roots is very rapid (Davidson, 1979; Andersen, 1987). The ability to produce adventitious roots is almost universal among plants and lateral root proliferation in response to moderate feeding damage appears to be the rule; providing a parallel to the results of the artificial root-pruning studies described by Andersen (1987). When lateral root proliferation occurs in response to levels of root herbivory below those affecting foliar production (see Section V.A), root attack may be of possible benefit for a plant. It has already been demonstrated that the boring of mangrove roots by a range of insects aids stability through the production of extra laterals (Simberloff *et al.*, 1978; see Section V.C). A similar situation was reported by Riedell (1989), who found that corn plants with low corn rootworm (*Diabrotica* spp.) attack had larger root systems and, as a result, higher yields than uninfested plants. Lateral root proliferation was associated with root tunnelling by larvae, which at a density of 50 per pot did not reduce yield. However, higher densities (150 per pot) severely reduced yield. Kahler *et al.* (1985) attributed the increased uptake of Na and Fe in corn plants to the fact that infested plants produced extra roots which branched into parts of the soil where Na and Fe could be obtained. The question of whether or not insect herbivory benefits plants has proved to be an emotive one (see Belsky, 1986; McNaughton, 1986), and virtually all reported instances of benefit may be dismissed for one reason or another (Belsky, 1986). However, neither of these authors have considered root herbivores and it is possible that low levels of root herbivory *may* be of *some* benefit to plants. Without more experimental evidence, this cannot be more than speculation, but the compensatory method of lateral root growth could provide a mechanism by which this might occur.

3. *Increased Uptake of Water and Mineral Nutrients*

An obvious way in which a plant may overcome the effects of drought or nutrient stress imposed upon it by root herbivory is by having access to an adequate supply of water or minerals. Experiments (see below) which have sought to address these problems have invariably shown that increased water supplies can mitigate the effects of herbivory. However, the response to nutrients is less clear.

Radcliffe (1971b) found that increasing soil moisture offset the effects of grass grub larvae (*Costelytra zealandica*) feeding on ryegrass and cocksfoot because of reduced grub survival in wet soils. Godfrey and Yeargan (1985) also found that high soil moisture reduced the survival of *Sitona hispidulus* larvae feeding on *Medicago sativa* roots, and that plant yields therefore increased. In general, root-feeding larvae are not adversely affected by high soil moisture levels (see Section IV.E.4) and thus results, such as those of Ladd and Buriff (1979), are typical in that plants can offset the effect of herbivory by having an increased water supply. In this study, one *Popillia japonica* larva at low soil water reduced the yield of *Poa pratensis* by 42%, while at high soil water, the reduction in yield was only 11%.

As described in Section IV.E.5, the application of fertilizers to the soil may have a null, negative or enhanced effect on root feeding by soil insects. However, it is well known that plant roots will proliferate in the presence of a locally abundant supply of nutrients (e.g. Drew, 1975), and therefore it seems reasonable to assume that fertilization may assist plants in overcoming the effects of herbivory. Some studies do indeed show this to be so. Nichols *et al.* (1987) found that recovery of pastureland after application of chlorpyrifos to control tipulid larvae depended on the addition of nitrogen fertilizer. A similar situation was described by Blank and Olson (1988), working with black beetle (*Heteronychus arator*) larval infestations in pastures. Studies which describe no effects of fertilization have been concerned with nutrient addition during larval feeding (e.g. Prestidge *et al.*, 1985) rather than after it. The timing of fertilization is critical (see Section IV.E.5), not only in its possible direct effects on larvae, but also whether root proliferation occurs during larval abundance (and thus gets eaten) or after, thereby aiding plant growth (Spike and Tollefson, 1988).

VI. EFFECTS ON PLANT COMMUNITIES

A. Evidence from Agriculture and Forestry

In the field, the effects of below-ground insect herbivores are seldom obvious and are generally only manifest when insect populations are high. Even then, it is often difficult to attribute visible damage to a particular organism. In the case of arable crops, the effects of species such as carrot fly (*Psila rosae*) or corn rootworms (*Diabrotica* spp.) are generally distinctive and translate into substantial loss in yield. In lucerne (*Medicago sativa*), underground larvae of the weevil, *Sitona discoideus*, are known to cause greater damage than the more apparent leaf-chewing adults. The damage results from larvae feeding on the rhizobial root nodules which can be completely destroyed (Goldson *et al.*, 1985). Even though the duration of the peak larval population may be short, the effects on the crop may be long-term (still being apparent after 3 months) and can result in a loss of yield of up to 43% dry matter. However, Godfrey and Yeargan (1985) report much lower losses due to the clover root curculio, *Sitona hispidulus*, on the same species, and suggest that the main effect is due to feeding lesions providing sites for the entry of pathogenic fungi.

In pastures, mainly because of their long-lived nature, the biomass of invertebrates may be considerable and may even exceed that of the grazing domestic livestock (Clements, 1984). The occurrence of larval insects, such as leatherjackets (*Tipula* spp.), wireworms (*Agriotes* spp.) and chafer grubs (*Phyllopertha* spp.) can often pass unnoticed. This is because grass, unlike other crops, is rarely weighed at harvest, and it is seldom possible to compare infested and uninfested areas. Furthermore, insects often have heterogeneous distributions (see Section IV.C) and their feeding may create bare microsites for the germination of undesirable weed seeds (Clements, 1984). Ueckert (1979) reported densities of scarab larvae, *Phyllophaga crinita* (Burm.), of 46 m^{-2} in patches of shortgrass prairie in Texas which reduced the percentage cover of perennial grasses by as much as 88% and the below-ground biomass by 43% relative to uninfested areas.

In woody species, the effects of subterranean insects are even more difficult to discern. In apple orchards, Hamilton (1961) demonstrated that, in addition to the damage to branches caused by adult female cicadas during oviposition, the prolonged period of root feeding by nymphs imposed a substantial reduction in growth and longevity of the trees and in apple yield. Such effects had previously been attributed to nutrient deficiency. However, digging under trees revealed large populations of nymphs (an average of 72 800 under the spread of each tree), and examination of the root system showed that all fine roots had been destroyed. Tree seedlings are particularly vulnerable. Several species of larval scarabs (white grubs) are destructive pests of red pine seedlings in plantations in central North America (Fowler and Wilson, 1971, 1974). The larvae consume the smaller roots and girdle the larger ones causing a reduction in growth, weakening of seedlings and eventual death. Between 16 and 34% of seedlings can be lost by the second season after planting as a result of white grub feeding.

B. Impact on Natural Plant Communities

1. The Problem of Assessment

The above examples provide some indication of the potential impact of below-ground herbivory in natural plant communities. However, they tell us little about the role of this type of herbivory as a structuring force in plant community dynamics. To influence the latter, herbivory must have an effect on the demography of the species. Such effects are best demonstrated by experimental field studies, well replicated in time and space. Although results from pot trials can provide valuable information on specific aspects of the herbivore/plant interaction (see above), they can seldom be extrapolated to field situations. The variation in abiotic factors, resource levels, plant competition, the effects of above-ground herbivory and the heterogeneity of the subterranean herbivores themselves, combine to make the study of below-ground herbivory in natural communities difficult, and therefore it is rarely attempted.

Glass-faced growth containers have been used to study the phenology of spotted knapweed, *Centaurea maculosa*, and the impact of the root-mining moth, *Agapeta zoegana* (L.) (Müller, 1987, 1989b). On a larger scale, rhizotrons with underground glass-viewing panels can be employed to investigate the activity of various types of herbivores, e.g. when the land is subjected to different management regimes (J. M. Cherrett, unpublished). However, these methods concentrate the root system into two dimensions and may therefore misrepresent what happens in the open soil. An alternative approach is to attempt to simulate herbivory by root pruning. A number of studies have investigated the effects of root pruning on the growth or performance of plants (e.g. Humphries, 1958; Sutton, 1967; Troughton, 1974; Detling *et al.*, 1980; Lovett Doust, 1980). However, the results from such experiments are difficult to synthesize and their interpretation shares the same constraints as the use of simulated foliar herbivory (see Gange, in press). At best, such experiments can test specific responses of single plant species, but have little potential at a community level.

Manipulative field experiments are still in their infancy but, with cautious execution, have great potential in unravelling the complexities of plant/herbivore interactions. The use of such experiments was strongly advocated by Harper (1977). These experiments involve either the artificial increase or decrease in the number of herbivores and the comparison of the vegetation with a control situation with natural levels of herbivores. Experimental studies involving a reduction in herbivory have been successfully employed in a limited number of experiments with foliar-feeding insects (e.g. Cantlon, 1969; Waloff and Richards, 1977; Brown, 1982, 1985, in press; Brown *et al.*, 1988).

Some of the largest-scale, but often least rigorous, manipulative experiments involving an artificial increase in the number of herbivores, are those in which an insect is released in a biological control programme, e.g. the control of diffuse and spotted knapweed, *Centaurea diffusa* and *C. maculosa*, in Canada by the root-feeding beetle, *Sphenoptera jugoslavica* (Harris and Myers, 1981; Powell and Myers, 1988) or of ragwort, *Senecio jacobaea* L., by *Longitarsus jacobaeae* Wat. (Harris *et al.*, 1981). However, these experiments are unusual in that both the herbivore and the plant species were introduced and, as such, were probably free from their natural enemies. They therefore provide relatively little information about the effects of herbivores on plant community dynamics, especially as the control experiment is, understandably, rarely undertaken.

Experiments involving the reduction of below-ground insect herbivores by the use of insecticides have rarely been used. There are a number of potential problems associated with the use of insecticides in field experiments and, although a consideration of these is vital to the critical appraisal of the results, few workers have considered them (but see Brown *et al.*, 1987). Clements and co-workers have explored the use of a wide range of insecticides to control pasture invertebrates in Britain. They found increases in the annual yield of herbage of 16–33%, resulting from the application of the broad-spectrum pesticide, aldicarb (Clements *et al.*, 1987), and that the botanical composition of the resulting sward was modified (Clements and Henderson, 1978). However, these results may not be entirely due to the reduction of below-ground herbivores, as the stem-boring frit fly, *Oscinella frit* (L.), was also affected.

2. *The Silwood Study*

At Silwood Park, Berkshire, UK, a series of long-term, manipulative field experiments was specifically designed to assess and compare the impact of both above- and below-ground insect herbivory on a range of sites of different successional age and therefore floristic composition (e.g. Brown, 1982, 1985, in press; Brown and Gange, 1989a,b, in press). We used a foliar insecticide (Dimethoate-40) and a soil insecticide (Dursban 5G) separately and in combination in a factorial design and compared characteristics of the vegetation in insecticide-treated and control plots. One of the most dramatic effects of reducing below-ground herbivory was an increase in plant species richness. This trend increased with successional age, as can be seen from Fig. 5. Indeed, 10 of the 98 species established during the first 3 years of the colonization of bare ground were only found when a soil insecticide was applied. These species were all forbs (non-graminaceous species) and mainly perennials. The contribution of this life-history grouping to vegetation cover (assessed by touches on point quadrat pins) was also enhanced by a reduction in below-ground herbivory. This was in direct contrast to the perennial

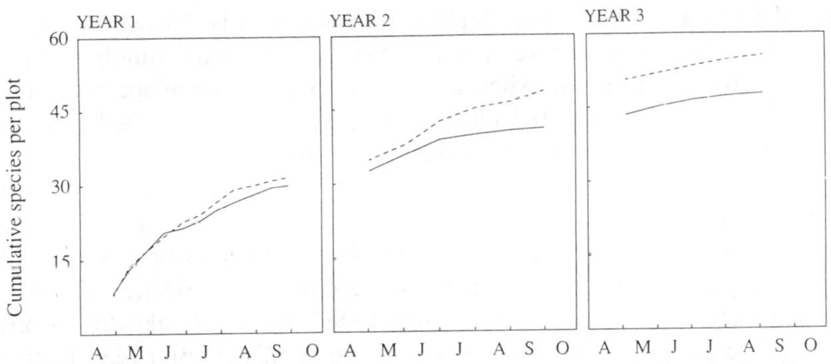

Fig. 5. Accumulated number of plant species per 3 m × 3 m plot during the first 3 years of the colonization of bare ground. ---, Soil insecticide-treated; ——, control (natural levels of insect herbivory).

grasses, which showed their greatest increase in cover abundance when foliar herbivory was reduced by inscticide application (Brown, in press). Within the early successional plant community, the effects of chemical exclusion demonstrated that individual species respond differently to above- and below-ground herbivory (Brown and Gange, 1989a). Species which showed the strongest response to a reduction in below-ground herbivory included *Capsella bursa-pastoris*, *Stellaria media* and *Polygonum persicaria* L. in the first year of treatment and the former species together with *Tripleurospermum inodorum* Schultz Bip. and *Veronica persica* Poiret in the second year. Further field experiments on individual species are currently underway. Although these experiments have focused on the earliest stages of secondary succession, a preliminary analysis of the results from later stages suggest similar results, although the effects maybe manifest more slowly (Brown and Gange, unpublished). It is also encouraging that in a comparable experiment on calcareous soil at Wytham, Oxon, UK, similar results to those obtained from Silwood have been forthcoming.

These results illustrate two features central to plant community dynamics: first, that recruitment is apparently affected by below-ground herbivory and, secondly, that the competitive balance between plant species belonging to different life-history groupings is modified.

3. *Plant Recruitment*

Recruitment is dependent upon the amount of viable seed produced and that reaching and remaining in a microsite until conditions are favourable for germination. The activities of root-feeding insects can reduce plant fecundity, although data are extremely sparse (see Section V.A.2). In the Silwood study, it was found that the number of plant species flowering in soil insecticide-treated plots was significantly higher than in controls (Brown and Gange, 1989a,b). The effect was so strong that it was even visually apparent, particularly in mid-season, when there was a wide range of species in flower in the soil insecticide-treated plots, with many of the species being either uncommon or unrepresented in other treatments.

Normally, predispersal seed predation is an activity of above-ground insect herbivores. Only when a species has a prostrate form or has lodged would it be accessible to a soil-dwelling insect. By contrast, soil-dwelling insects play a major role in post-dispersal seed mortality. Beetles and ants are the main agents, although most recorded seed predation takes place on the soil surface and as such is beyond the scope of this chapter (e.g. Bach and Gross, 1984; Carroll and Risch, 1984; Ohara and Higashi, 1987). However, Janzen (1985) recorded that 95% of the seeds of the tree *Spondias mombin* L. in Costa Rica were eaten by the bruchid beetle, *Amblycerus spondiae* L., living in the litter of the forest floor. Such high levels of seed predation were only circumvented when white-tailed deer dispersed the nuts to the edge of the forest. It is very likely that dormant seeds in the soil will be consumed by insects, either actively in the case of large seeds or passively in very small seeds. However, data are lacking and would at best be very hard to acquire, because of the heterogeneous distribution of most seed banks (Roberts, 1981) and of the herbivores (see Section IV.C). Much more probable though is damage to the seed during and just after germination. If the hypocotyl or radicle is injured or severed, the plant may die. For species rare in the seed bank, this could be critical. The increased species richness and abundance of plants recorded when below-ground herbivory was reduced, must be due at least in part to the successful germination of propagules from the seed bank or immigrant seed (Brown and Gange, 1989a,b, in press). In mature vegetation, this is dependent on the availability of gaps or microsites. However, we found no difference in microsite availability between treatments, which suggests that the higher levels of below-ground herbivory in the control plots may be a major factor determining recruitment. In support of this, there was a higher cover of seedlings in soil insecticide-treated plots. It is not known when the below-ground seed predation occurs or whether it is selective as is that occurring on the soil surface (Thompson, 1985). The work of Rees (1989) on adjacent experimental sites suggests that it may be selective, because the germination of some species was enhanced by the application of soil insecticide whereas in others it was depressed. Because seedling emergence was monitored, mortality at this stage, resulting in differences in species richness, was not a factor (Brown and Gange, unpublished). However, it is very likely that seedling establishment would be significantly influenced by subterranean herbivory. Indeed, Godfrey *et al.* (1986) found that seedling establishment of alfalfa could be reduced by as much as 44% by the larvae of *Sitona hispidulus* and that of *Trifolium repens* increased by the application of soil insecticide when tipulid larvae were abundant (Clements and Bentley, 1983).

4. Competitive Interactions

Herbivory can influence plant competitive interactions in two ways: by modifying growth and morphology and therefore access to resources, and by influencing plant abundance and distribution (Louda *et al.*, 1989). A good example of the former is the diffuse knapweed, in which herbivory by the specialist root-feeding beetle, *Sphenoptera jugoslavica*, causes a decline in rosette size and a slower growth rate (Powell and Myers, 1988). Both of these would serve to reduce the competitive ability of damaged plants. Differential below-ground herbivory, which may ultimately lead to competitive interactions, has been implicated by a number of workers. For example, although Ueckert (1979) found dramatic effects of the larvae of *Phyllophaga crinita* on grass growth, perennial forbs were unaffected. Conversely, White and French (1968) showed that clover was reduced to a greater extent than grasses by tipulid larvae. Clements and Henderson (1983) also found that the proportion of white clover present in clover/ryegrass plots was enhanced by pesticide treatment of the clover weevil (*Sitona* spp.) *inter alia*. Clements and Henderson (1978) noted the capacity of insect pests to bring about subtle changes in sward composition in experiments with Italian and hybrid ryegrasses. They found that the sown species died out within 2 years on untreated plots, but survived for at least another 2 years where pests were controlled. In another experiment, one grass species (*Lolium perenne* L.) survived at the expense of *Phleum* spp. and other grasses. Our manipulative experiments (Brown and Gange, 1989a,b, in press) have enabled an insight into the effects of below-ground herbivory on different plant life-history groupings and species. Although the competitive ability of individual species is notoriously difficult to quantify in natural plant communities, the use of a height index (based on the weighted mean height of the vegetation touches) indicated that many species grew taller and larger when root herbivory was reduced. The perennial forbs were most strongly affected by subterranean herbivory, and this enabled the more vigorously growing grasses to assume a major contribution to vegetation cover. In terms of community processes, such as succession, these effects are of major importance and suggest that below-ground herbivory may actually accelerate the succession to a grass-dominated sward by reducing the competitive ability of the forbs.

In agriculture and forestry, the nature of the background vegetation and its management can also have important implications not only in terms of direct competitive interactions, but on the biology and therefore efficacy of the root-feeding species. Thus, Hruska (1987) in Costa Rica demonstrated a positive relationship between the amount of weed ground cover and the damage caused to cassava by white grubs (*Phyllopertha* spp.). This relationship can be explained by the higher grub populations in weedy ground resulting from increased oviposition by adult beetles, higher resource availability and increased survivorship of eggs and larvae. The latter is most likely due to microclimatic conditions, including the dampening of temperature fluctuations and the maintenance of soil moisture preventing the desiccation of eggs. Finally, weed cover decreases parasitism and predation of the below-ground stages, through difficulty of access. Even though one of the most common hymenopterous parasitoids of scarabaeid larvae has a long ovipositor (Lim *et al.*, 1980), Hruska (1987) considered that parasitoid efficiency may well be reduced by a thick weed cover. Kard and Hain (1988) found a similar relationship between densities of chafer larvae and amount of ground cover in plantations of Fraser fir Christmas trees in North Carolina. However, when the surrounding grass was mowed, the populations of grubs were increased further and the trees sustained the most severe root damage, lowest wood volume and poorest health. The mowed vegetation provided the optimal oviposition habitat for the adult beetles (Chamberlin and Callenbach, 1943).

VII. FUTURE PERSPECTIVES

Insect herbivores can attack both the vegetative and reproductive structures of plants in a variety of ways: defoliating, sap feeding, stem boring, leaf mining, gall forming, flower and seed feeding and root feeding (Brown, in press). A synthesis of the wealth of studies which have demonstrated the impact of above-ground herbivory on a plant's growth and reproduction, and the rather fewer studies describing effects on species' demography, can leave little doubt that plant populations and communities can strongly be influenced by this type of herbivory (Gange, in press).

With few exceptions, the organisms feeding on below-ground plant structures are either different species, or different stages in the life-history of species, to those feeding above ground. Generally, it is the most active feeding stage of the life-cycle which occurs below ground and, furthermore, the bulk of the life-cycle (often spanning several years) may be spent in this situation. It is therefore predictable that such insects will have a significant impact on the root system of a plant and that this will be manifest in a range of plant responses. From this chapter, it is clear that below-ground herbivory can influence the full complement of parameters affected by the feeding activities of herbivores above ground. Moreover, the effects can be just as potent. However, much of the information leading to this conclusion is either anecdotal or stems from the agricultural literature and therefore imparts little substantive information on the role of below-ground herbivory in the dynamics of natural plant communities. In addition, the complexities of interactions in the soil make comparative studies, and therefore the establishment of generalizations, extremely difficult. However, an understanding of the processes such as colonization and succession cannot be complete without more emphasis on subterranean herbivory by insects and indeed by other taxa. We therefore agree with Stanton (1988) that "the foremost challenge in soil ecosystems is experimental manipulation that measures real *in situ* processes".

The study of the plant/herbivore interaction *per se* is now entering an even greater level of complexity, due to the recent awareness and potential importance of the interactions between different herbivore species or guilds mediated through changes in the host plant. One such interaction involving folivorous species, the leaf miners associated with oak, was explored by West (1985). However, to our knowledge, there have been no comparable studies involving the interaction(s) between a root-feeding species and a foliar-feeding one. Intuitively, one might predict some form of interaction, bearing in mind that root-feeding is known to cause water stress in plants (see Section V.D), and that stressed plants often provide a more favourable food resource for herbivores (White, 1969, 1984). To test these ideas, we used root feeding by the garden chafer, *Phyllopertha horticola*, on the annual forb, *Capsella bursa-pastoris*, to assess the nature of the interaction with the polyphagous foliar-feeding aphid, *Aphis fabae* Scop. (Gange and Brown, 1989). By using two watering regimes, we found that a low water regime accompanied by root feeding caused a reduction in plant biomass and a corresponding increase in aphid weight and growth rate, whereas root feeding alone also increased aphid fecundity and adult longevity. The effects of a higher watering regime offset the drought stress caused by root feeding, but did not overcome it completely. We attributed these effects to an improvement in the quality of the phloem sap as a result of amino acid mobilization. The implications of this type of interaction are far-reaching, not least because root herbivory generally goes unnoticed. Thus, the recorded effects of foliar herbivory in natural situations may well be confounded by the activities of subterranean herbivores. In addition, the effects of root herbivory on the population dynamics of foliar-feeding insect species have yet to be explored, although several of the parameters involved in population growth have been shown to be affected (Gange and Brown, 1989). In the case of agricultural pests, such findings may well have important, not to mention undesirable, consequences. Of course, there is potential for the opposite type of interaction to occur every time vegetation is grazed by a herbivore, be it vertebrate or invertebrate. Indeed, Hutchinson and King (1980) reported maximum densities of white grubs under intermediate levels of sheep grazing, while Seastedt (1985) reported that populations of scarabaeids doubled when tall grass prairie was mowed. In addition to relationships involving foliar and root herbivores, there are many possible complex interactions between different root-feeding herbivores. The most likely, and probably most important, involve root-feeding insects and nematodes. Apart from the possibility of these herbivores competing directly for the same food resource, nematodes may significantly alter the physiology of the root system, thus modifying the food supply for the insects. It is documented that nematode feeding may increase mineral uptake, reduce root carbohydrate content, reduce nodulation in legumes and alter the gross morphology of the root system (Dropkin, 1989). Thus nematode feeding may alter the quantity and quality of food available to insects and in some species, such as *Sitona* spp., may reduce larval survival through effects on nodule abundance (cf. Quinn and Hower, 1986b). Clearly, the potential for the study of these interactions is vast, but as yet the subject appears to be virtually unexplored (Dropkin, 1989), although there are some studies of interactions between different nematode species (Yeates, 1987).

The complexity of plant/herbivore interactions enters another dimension when the effects of root-feeding insects on plant symbionts are considered. The benefits of mycorrhizal associations in laboratory conditions are well documented (e.g. Sanders *et al.*, 1975). However, in field situations, mycorrhizas have often appeared to be ineffective (Fitter, 1985). One reason for this may be their removal as an indirect result of root herbivory. This is certainly an area which warrants more attention and one which we are currently exploring at Silwood Park (Gange *et al.*, 1990).

As well as providing a direct challenge to the ecologist, the complex plant/herbivore interactions occurring below ground have much topical as well as practical relevance. In many parts of the world, change in the nature and use of land is a major issue. In parts of Europe, land is coming out of intensive agricultural production and being put to other uses. Farm forestry is one such current development, involving a range of deciduous and coniferous tree species being planted in ex-arable land. However, arable land including pasture has many polyphagous below-ground herbivores associated with it (Clements, 1984; Fowler and Wilson, 1971). It is also generally accepted and intuitively obvious that the establishment phase of woody species is the most vulnerable to pest attack (Fowler and Wilson, 1971). Consequently, the incautious planting of tree seedlings or saplings in unprepared land may well expose them to herbivores not normally encountered in the forest environment. The work of Kard and Hain (1988) is a clear indication of the effects such herbivores may have on Fraser fir Christmas tree plantations in North America.

An alternative use for ex-arable land is "set aside", as a means of recreating or restoring land to a semi-natural habitat. The aim here is to establish areas which are both aesthetically pleasing landscapes and of high conservation value. This is, of course, most readily achieved by promoting perennial forb species at the expense of vigorously growing grasses and pernicious weeds. Because the effects of subterranean insect herbivores have been shown to be strongest on the perennial forbs (Brown and Gange, 1989a), some control measures may be advisable, at least in the early stages of colonization.

Apart from the direct effects of below-ground herbivory in agriculture already discussed, there are subtleties relating to the population dynamics of the pests which may have important effects, particularly in respect of new crops. Blackshaw (1988) has shown that land preparation or management can have a major effect on the populations of subterranean herbivores. In addition, the order of crop rotation can bring about great variability in the size of subterranean herbivore populations. It is hardly surprising that crops following grass are at greatest risk (Blackshaw, 1988; Clements, 1984). However, Blackshaw also found that populations of larval tipulids were particularly high in the year following an oil seed rape crop. Here, the species was *Tipula oleracea* L., rather than the more common pasture species *T. paludosa*. Although Blackshaw does not offer any explanation for this, it does emphasize the need for an awareness of "new", unseen herbivores.

The universal aim to reduce pesticide use highlights the need for more emphasis on biological methods of pest control. There have been several recent and successful attempts at finding and releasing root herbivores as a means of controlling pernicious weeds. The control of diffuse and spotted knapweed, *Centaurea diffusa* and *C. maculosa*, by a buprestrid beetle (Harris and Myers, 1981; Müller *et al.*, 1989; Powell and Myers, 1988) and that of ragwort, *Senecio jacobaea*, by a chrysomelid beetle (Harris *et al.*, 1981) are two such examples. Another extremely successful root-feeding agent, not strictly relevant to this Chapter because the weed is aquatic, is the weevil *Cyrtobagous salviniae* Calder and Sands used to control *Salvinia molesta* (Julien *et al.*, 1987). There is obviously much scope in this field and particularly when the search for possible agents is accompanied by detailed studies of the ecology of the interaction, such as the exemplary studies of Müller (1989a) and Powell and Myers (1988).

These perspectives represent some new and challenging topics in ecology, and it is likely that in the next decade, root feeders will loose their "forgotten status" and be recognized as an important component of plant/herbivore systems.

ACKNOWLEDGEMENTS

We are grateful to many friends and colleagues for discussions on certain parts of the paper. Janet Pryse deserves special mention for her unfailing help in finding references and reading the manuscript. Financial support for our experimental work was provided by the Natural Environment Research Council. Dow Agriculture (Hitchin) generously donated the Dursban for the experimental work.

REFERENCES

Addison, J. A. (1977). Population dynamics and biology of Collembola on Truelove Lowland. In: *Truelove Lowland, Devon Island, Canada: A High Arctic Ecosystem* (Ed. by L. C. Bliss), pp. 363–382. University of Alberta Press, Edmonton.

Aeschlimann, J.-P. (1980). The *Sitona* (Coleoptera: Curculionidae) species occurring on *Medicago* and their natural enemies in the Mediterranean region. *Entomophaga* **25,** 139–153.

Aeschlimann, J.-P. (1986). Rearing and larval development of *Sitona* spp. (Coleoptera: Curculionidae) on the root system of *Medicago* spp. plants (Leguminosae). *J. appl. Entomol.* **101,** 461–469.

Aeschlimann, J.-P. and Vitou, J. (1988). Comparing infestations by *Sitona discoideus* Gyllenhal (Coleoptera: Curculionidae) on Mediterranean and Australian *Medicago* spp. accessions (Leguminosae) in southern France. *J. Aus. entomol. Soc.* **27,** 275–278.

Ahmad, S. (Ed.) (1983). *Herbivorous Insects: Host-seeking Behaviour and Mechanisms.* Academic Press, London.

Alleyne, E. H. and Morrison, F. O. (1977). The lettuce root aphid, *Pemphigus bursarius* (L.) (Homoptera: Aphidoidea) in Quebec, Canada. *Ann. Soc. entomol. Quebec* **22,** 171–180.

Alleyne, E. H. and Morrison, F. O. (1978a). Winter survival and influence of temperature on soil apterae of the lettuce root aphid, *Pemphigus bursarius* (L.) in Quebec, Canada. *Ann. Soc. entomol. Quebec* **23,** 30–38.

Alleyne, E. H. and Morrison, F. O. (1978b). Vertical distribution of overwintering soil apterae of the lettuce root aphid, *Pemphigus bursarius* (L.) in Quebec, Canada. *Ann. Soc. entomol. Quebec* **23,** 155–157.

Allsopp, P. G. and Bull, R. M. (1989). Spatial patterns and sequential sampling plans for melolonthine larvae (Coleoptera: Scarabaeidae) in southern Queensland sugarcane. *Bull. ent. Res.* **79,** 251–258.

Andersen, D. C. (1987). Below-ground herbivory in natural communities: A review emphasizing fossorial animals. *Q. Rev. Biol.* **62,** 261–286.

Anderson, J. M. (1977). The organization of soil animal communities. *Proc. 6th Int. Coll. Soil Zoology, Ecol. Bull. (Stockholm)* **25,** 15–23.

Anon. (1969). Swift moths. *MAFF Advisory Leaflet* No. 160, 5 pp.

Bach, G. G. and Gross, K. L. (1984). Experimental studies of seed predation in old-fields. *Oecologia (Berl.)* **65,** 7–13.

Baker, C. R. B. (1975). Japanese beetle, *Popillia japonica* Newman. Potential for establishment in Britain. *PPL Non-Indigenous Pest Assessment* No. 9, 1 p.

Bartlett, P. W. (1984). Grape phylloxera. *MAFF Plant Health Information Bulletin* No. 3, 5 pp.

Belsky, A. J. (1986). Does herbivory benefit plants? A review of the evidence. *Am. Nat.* **127,** 870–892.

Bergman, M. K., Tollefson, J. J. and Hinz, P. N. (1983). Spatial dispersion of corn rootworm larvae (Coleoptera: Chrysomelidae) in Iowa cornfields. *Environ. Entomol.* **12,** 1443–1446.

Blackshaw, R. P. (1988). Effects of cultivations and previous cropping on leatherjacket populations in spring barley. *Res. Dev. Agric.* **5,** 35–38.

Blank, R. H. and Olson, M. H. (1988). Effect of black beetle, in association with nitrogen and summer spelling, on pasture production on sandy soils. *N.Z.J. Agric. Res.* **31,** 445–453.

Block, W., Turnock, W. J. and Jones, T. H. (1987). Cold resistance and overwintering survival of the cabbage root fly, *Delia radicum* (Anthomyiidae) and its parasitoid, *Trybliographa rapae* (Cynipidae), in England. *Oecologia (Berl.)* **71,** 332–338.

Borror, D. J., DeLong, D. M. and Triplehorn, C. A. (1976). *An Introduction to the Study of Insects*. Holt, Rinehart and Winston, New York, 852 pp.

Boubals, D. (1966). Etude de la distribution et des causes de la résistance au phylloxéra radicole chez les vitacees. *Ann. Amelior. Plant* **16,** 145–184.

Brady, N. C. (1984). *The Nature and Properties of Soils*. Macmillan, New York, 750 pp.

Brown, V. K. (1982). The phytophagous insect community and its impact on early successional habitats. In: *Proceedings of the 5th International Symposium on Insect–Plant Relations, Wageningen 1982* (Ed. by J. H. Visser and A. K. Minks), pp. 205–213. Pudoc, Wageningen.

Brown, V. K. (1985). Insect herbivores and plant succession. *Oikos* **44,** 17–22.

Brown, V. K. (in press). Insect herbivory and its effect on plant succession. In: *Pests, Pathogens and Plant Communities* (Ed. by J. J. Burdon and S. R. Leather). Blackwell, Oxford.

Brown, V. K. and Gange, A. C. (1989a). Differential effects of above- and below-ground insect herbivory during early plant succession. *Oikos* **54,** 67–76.

Brown, V. K. and Gange, A. C. (1989b). Root herbivory by insects depresses plant species richness. *Funct. Ecol.* **3,** 667–671.

Brown, V. K. and Gange, A. C. (in press). Effects of root herbivory on vegetation dynamics. In: *Plant Root Systems: Their Effect on Ecosystem Composition and Structure* (Ed. by D. Atkinson). Blackwell, Oxford.

Brown, V. K., Leijn, M. and Stinson, C. S. A. (1987). The experimental manipulation of insect herbivore load by the use of an insecticide (malathion): The effect of application on plant growth. *Oecologia (Berl.)* **72,** 377–381.

Brown, V. K., Jepsen, M. and Gibson, C. W. D. (1988). Insect herbivory effects on early old field succession demonstrated by chemical exclusion methods. *Oikos* **52,** 293–302.

Buchanan, G. (1978). Grape phylloxera. Old days of rage, new days of caution. *J. Agric.* **76,** 339–340.

Buchanan, G. A. (1984). Factors affecting damage by phylloxera in Victoria. In: *The Biology, Quarantine and Control of Grape Phylloxera in Australia and New Zealand* (Ed. by G. A. Buchanan and T. G. Amos), pp. 24–25. Department of Agriculture, Victoria.

Buchanan, G. A. (1987). The distribution of grape phylloxera, *Daktulosphaira vitifolii* (Fitch), in central and north-eastern Victoria. *Aust. J. exp. Agric.* **27,** 591–595.

Buttery, R. G. and Ling, L. C. (1985). Volatile components of corn roots: possible insect attractants. *J. Agric. Food Chem.* **33,** 772–774.

Byers, R. A. and Kendall, W. A. (1982). Effects of plant genotypes and root nodulation on growth and survival of *Sitona* spp. larvae. *Environ. Entomol.* **11,** 440–443.

Cantlon, J. E. (1969). The stability of natural populations and their sensitivity to technology. In: *Diversity and Stability in Ecological Systems* (Ed. by G. M. Woodwell and H. H. Smith), pp. 197–205. Brookhaven Symposia in Biology No. 22.

Carroll, C. R. and Risch, S. J. (1984). The dynamics of seed harvesting in early successional communities by a tropical ant, *Solenopsis geminata*. *Oecologia (Berl.)* **61,** 388–392.

Carter, J. B. and Green, E. I. (1988). Hemocytes of Baculovirus-infected *Tipula paludosa* larvae (Diptera:Tipulidae). *J. Invertebr. Pathol.* **52,** 393–400.

Chamberlin, T. R. and Callenbach, J. A. (1943). Oviposition of June beetles and the survival of their offspring in grasses and legumes. *J. econ. Ent.* **36,** 681–685.

Chapin, F. S. III, Johnson, D. A. and McKendrick, J. D. (1980). Seasonal movement of nutrients in plants of differing growth form in an Alaskan tundra ecosystem: Implications of herbivory. *J. Ecol.* **68,** 189–209.

Cherry, R. H. (1985). Seasonal phenology of white grubs (Coleoptera: Scarabaeidae) in Florida sugarcane fields. *J. econ. Ent.* **78,** 787–789.

Cheung, W. W. K. and Marshall, A. T. (1973). Water and ion regulation in cicadas in relation to xylem feeding. *J. Insect Physiol.* **19,** 1801–1816.

Chiang, H. C. (1973). Bionomics of the northern and western corn rootworms. *Ann. Rev. Entomol.* **18,** 47–72.

Clements, R. O. (1984). Control of insect pests in grassland. *Span* **27,** 77–80.

Clements, R. O. and Bentley, B. R. (1983). The effect of three pesticide treatments on the establishment of white clover (*Trifolium repens*) sown with a slot-seeder. *Crop Prot.* **2,** 375–378.

Clements, R. O. and Henderson, I. F. (1978). Insects as a cause of botanical change in swards. *J. Br. Grassld Soc., Occ. Symp.* **10,** 157–160.

Clements, R. O. and Henderson, I. F. (1983). An assessment of insidious pest damage to 26 varieties of seven species of herbage legumes. *Crop Prot.* **2,** 491–495.

Clements, R. O., Bentley, B. R. and Nuttall, R. M. (1987). The invertebrate population and response to pesticide treatment of two permanent and two temporary pastures. *Ann. appl. Biol.* **111,** 399–407.

Cochrane, J. and Thornhill, W. A. (1987). Variation in annual and regional damage to sugar beet by pygmy beetle (*Atomaria linearis*). *Ann. appl. Biol.* **110,** 231–238.

Cole, R. A. (1984). Phenolic acids associated with the resistance of lettuce cultivars to the lettuce root aphid. *Ann. appl. Biol.* **105,** 129–145.

Coleman, D. C. (1976). A review of root production processes and their influence on soil biota in terrestrial ecosystems. In: *The Role of Terrestrial and Aquatic Organisms in Decomposition Processes* (Ed. by J. M. Anderson and A. Macfadyen), pp. 417–434. Blackwell, Oxford.

Coleman, D. C., Reid, C. P. P. and Cole, C. V. (1983). Biological strategies of nutrient cycling in soil systems. In: *Advances in Ecological Research* (Ed. by A. Macfadyen and E. D. Ford), Vol. 13, pp. 1–55. Academic Press, London.

Coulson, J. C. (1962). The biology of *Tipula subnodicornis* Zetterstedt, with comparative observations on *Tipula paludosa* Meigen. *J. Anim. Ecol.* **31,** 1–21.

Coulson, J. C. and Whittaker, J. B. (1978). Ecology of moorland animals. In: *The Ecology of Some British Moors and Montane Grasslands* (Ed. by O. W. Heal and D. F. Perkins), pp. 52–93. Springer-Verlag, Berlin.

Cranshaw, W. S. (1985). Clover root curculio (*Sitona hispidulus*) injury and abundance in Minnesota (USA) alfalfa of different stand age. *Grt Lakes Ent.* **18,** 93–95.

Crawley, M. J. (1983). *Herbivory: The Dynamics of Animal–Plant Interaction.* Blackwell, Oxford, 437 pp.

Curry, J. P. (1987). The invertebrate fauna of grassland and its influence on productivity. II. Factors affecting the abundance and composition of the fauna. *Grass For. Sci.* **42,** 197–212.

Davidson, R. L. (1979). Effects of root feeding on foliage yield. In: *Proceedings of the 2nd Australasian Conference on Grassland Invertebrate Ecology* (Ed. by T. K. Crosby and R. F. Pottinger), pp. 117–120. Government Printer, Wellington, NZ.

Davidson, R. L., Wensler, R. J. and Wolfe, V. J. (1970). Damage to ryegrass plants of different size by various densities of the pruinose scarab (*Sericesthis geminata* (Coleoptera)). *Aust. J. exp. Agric. Anim. Husb.* **10,** 166–171.

Dennehy, T. J. and Clark, L. (1987). Biology and control of the grape rootworm, *Fidia viticida* Walsh, in central New York (USA). *J. agric. Entomol.* **4,** 157–166.

Denno, R. F. and McClure, M. S. (Eds) (1983). *Variable Plants and Herbivores in Natural and Managed Systems.* Academic Press, London, 712 pp.

Dethier, V. G. (1970). Some general considerations of insects' responses to the chemicals in food plants. In: *Control of Insect Behaviour by Natural Products* (Ed. by D. L. Wood, R. M. Silverstein and M. Nakajima), pp. 21–28. Academic Press, London.

Detling, J. K., Winn, D. T., Procter-Gregg, C. and Painter, E. L. (1980). Effects of simulated grazing by below-ground herbivores on growth, CO_2 exchange, and carbon allocation patterns of *Bouteloua gracilis. J. appl. Ecol.* **17,** 771–778.

Dickason, E. A., Leach, C. M. and Gross, A. E. (1968). Clover root curculio injury and vascular decay of alfalfa roots. *J. econ. Ent.* **61,** 1163–1168.

Dintenfass, L. P. and Brown, G. C. (1988a). Quantifying effects of clover root curculio (Coleoptera: Curculionidae) larval feeding on biomass and root reserves of alfalfa. *J. econ. Ent.* **81,** 641–648.

Dintenfass, L. P. and Brown, G. C. (1988b). Influence of larval clover root curculio (Coleoptera: Curculionidae) injury on carbohydrate reserves and yield of alfalfa. *J. econ. Ent.* **81,** 1803–1809.

Dixon, A. F. G. (1985). *Aphid Ecology.* Blackie, Glasgow, 157 pp.

Drew, M. C. (1975). Comparison of the effects of a localized supply of phosphate, nitrate, ammonium and potassium on the growth of the seminal root system, and the shoot, in barley. *New Phytol.* **75,** 479–490.

Dropkin, V. H. (1989). *Introduction to Plant Nematology.* John Wiley, New York, 304 pp.

Dunn, J. A. (1959a). The biology of lettuce root aphid. *Ann. appl. Biol.* **47,** 475–491.

Dunn, J. A. (1959b). The survival in soil of apterae of the lettuce root aphid, *Pemphigus bursarius* (L.). *Ann. appl. Biol.* **47,** 766–771.

Dunn, J. A. and Kempton, D. P. H. (1974). Lettuce root aphid control by means of plant resistance. *Plant Path.* **23,** 76–80.

Dunning, R. A. and Baker, A. N. (1977). Some sugar beet cultural practices in relation to incidence and damage by soil-inhabiting pests. *Ann. appl. Biol.* **87,** 528–532.

Dutcher, J. D. and All, J. N. (1979). Damage impact of larval feeding by the grape root borer in a commercial concord grape vineyard. *J. econ. Ent.* **72,** 159–161.

Dybas, H. S. and Davis, D. D. (1962). A population census of seventeen-year periodical cicadas (Homoptera: Cicadidae: *Magicicada*). *Ecology* **43,** 432–444.

East, R. and Willoughby, B. E. (1983). Grass grub (*Costelytra zealandica*) population collapse in the northern North Island. *N.Z.J. agric. Res.* **26,** 381–390.

Edwards, C. A. (1962). Springtail damage to bean seedlings. *Plant Path.* **11,** 67–68.

Edwards, C. A. (1977). Investigations into the influence of agricultural practice on soil invertebrates. *Ann. appl. Biol.* **87,** 515–520.

Elliott, J. M. (1979). *Some Methods for the Statistical Analysis of Benthic Invertebrates.* Freshwater Biological Association Scientific Publication No. 25, 160 pp.

Elvin, M. K. and Yeargan, K. V. (1985). Spatial distribution of clover root curculio, *Sitona hispidulus* (Fabricius) (Coleoptera: Curculionidae), eggs in relation to alfalfa crowns. *J. Kans. ent. Soc.* **58,** 346–348.

Ferro, D. N. (Ed.) (1976). *New Zealand Insect. Pests.* Lincoln University College of Agriculture, Canterbury, 312 pp.

Finch, S., Collier, R. H. and Skinner, G. (1986). Local population differences in emergence of cabbage root flies from south-west Lancashire: Implications for pest forecasting and population divergence. *Ecol. Ent.* **11,** 139–145.

Fitter, A. H. (1985). Functioning of vesicular-arbuscular mycorrhizas under field conditions. *New Phytol.* **99,** 257–265.

Fleming, W. E. (1972). Biology of the Japanese beetle. *USDA Technical Bulletin* No. 1449, 129 pp.

Fleming, W. E. (1976). Integrating control of the Japanese beetle – A historical review. *USDA Technical Bulletin* No. 1545, 65 pp.

Fogel, R. (1985). Roots as primary producers in below-ground ecosystems. In: *Ecological Interactions in the Soil: Plants Microbes and Animals* (Ed. by A. H. Fitter, D. Atkinson, D. J. Read and M. B. Usher), pp. 23–36. Blackwell, Oxford.

Forno, I. W. and Semple, J. L. (1987). Response of *Salvinia molesta* to insect damage: Changes in nitrogen, phosphorous and potassium content. *Oecologia (Berl.)* **73,** 71–74.

Foster, R. E., Tollefson, J. J., Nyrop, J. P. and Hein, G. L. (1986). Value of adult corn rootworm (Coleoptera: Chrysomelidae) population estimates in pest management decision making. *J. econ. Ent.* **79,** 303–310.

Foster, W. A. (1975). The life history and population biology of an intertidal aphid, *Pemphigus trehernei* Foster. *Trans. R. ent. Soc. Lond.* **127,** 193–207.

Foster, W. A. and Treherne, J. E. (1975). The distribution of an intertidal aphid, *Pemphigus trehernei* Foster, on marine salt-marshes. *Oecologia (Berl.)* **21,** 141–155.

Foster, W. A. and Treherne, J. E. (1978). Dispersal mechanisms in an intertidal aphid. *J. Anim. Ecol.* **47,** 205–218.

Fowler, R. F. and Wilson, L. F. (1971). White grub populations, *Phyllophaga* spp., in relation to damaged red pine seedlings in Michigan and Wisconsin plantations (Coleoptera: Scarabaeidae). *Mich. Ent.* **4,** 23–28.

Fowler, R. F. and Wilson, L. F. (1974). Injury to aldrin-treated and untreated red pine by white grubs (Coleoptera: Scarabaeidae) and other agents during first five years after planting. *Grt Lakes Ent.* **7,** 81–88.

French, N. (1984a). Chafer grubs. *MAFF Leaflet* No. 235, 6 pp.

French, N. (1984b). Leatherjackets. *MAFF Leaflet* No. 179, 6 pp.

Galbreath, R. A. (1988). Orientation of grass grub *Costelytra zealandica* (Coleoptera: Scarabaeidae) to a carbon dioxide source. *N.Z. Ent.* **11,** 6–7.

Gange, A. C. (in press). Effects of insect herbivory on herbaceous plants. In: *Pests, Pathogens and Plant Communities* (Ed. by J. J. Burdon and S. R. Leather). Blackwell, Oxford.

Gange, A. C. and Brown, V. K. (1989). Effects of root herbivory by an insect on a foliar-feeding species, mediated through changes in the host plant. *Oecologia (Berl.)* **81,** 38–42.

Gange, A. C., Brown, V. K., Evans, I. M. and Storr, A. L. (1989). Variation in the impact of insect herbivory on *Trifolium pratense* through early plant succession. *J. Ecol.* **77,** 537–551.

Gange, A. C., Brown, V. K. and Farmer, L. M. (1990). A test of mycorrhizal benefit in an early successional plant community. *New Phytol.* **115,** 85–91.

Gaynor, D. L., Lane, G. A., Biggs, D. R. and Sutherland, O. R. W. (1986). Measurement of grass grub resistance of bean in a controlled environment. *N.Z.J. exp. Agric.* **14,** 77–82.

Godfrey, L. D. and Yeargan, K. V. (1985). Influence of soil moisture and weed density on clover root curculio *Sitona hispidulus*, larval stress to alfalfa (*Medicago sativa*). *J. agric. Entomol.* **2,** 370–377.

Godfrey, L. D., Legg, D. E. and Yeargan, K. V. (1986). Effects of soil-borne organisms on spring alfalfa establishment in an alfalfa rotation system. *J. econ. Ent.* **79,** 1055–1063.

Godfrey, L. D., Yeargan, K. V. and Muntifering, R. B. (1987). Digestibility, protein content and nutrient yields of alfalfa stressed by selected early season insect pests and diseases. *J. econ. Ent.* **80,** 257–262.

Goldson, S. L. and French, R. A. (1983). Age-related susceptibility of lucerne to *Sitona* weevil, *Sitona discoideus* Gyllenhal (Coleoptera: Curculionidae), larvae and the associated patterns of adult infestation. *N.Z.J. agric. Res.* **26,** 251–255.

Goldson, S. L. and Muscroft-Taylor, K. E. (1988). Inter-seasonal variation in *Sitona discoideus* Gyllenhal (Coleoptera: Curculionidae) larval damage to lucerne in Canterbury and the economics of insecticidal control. *N.Z.J. agric. Res.* **31,** 339–346.

Goldson, S. L., Dyson, C. B., Proffitt, J. R., Frampton, E. R. and Logan, J. A. (1985). The effect of *Sitona discoideus* Gyllenhal (Coleoptera: Curculionidae) on lucerne yields in New Zealand. *Bull. ent. Res.* **75,** 429–442.

Goldson, S. L., Bourdôt, G. W. and Proffitt, J. R. (1987). A study of the effects of *Sitona discoideus* (Coleoptera: Curculionidae) larval feeding on the growth and development of lucerne (*Medicago sativa*). *J. appl. Ecol.* **24,** 153–161.

Goldson, S. L., Jamieson, P. D. and Bourdôt, G. W. (1988). The response of field-grown lucerne to a manipulated range of insect-induced nitrogen stresses. *Ann. appl. Biol.* **113,** 189–196.

Granett, J. and Timper, P. (1987). Demography of grape phylloxera, *Daktulosphaira vitifoliae* (Homoptera: Phylloxeridae) at different temperatures. *J. econ. Ent.* **80,** 327–329.

Gray, M. E. and Tollefson, J. J. (1987). Influence of tillage and western and northern corn rootworm (Coleoptera: Chrysomelidae) egg populations on larval populations and root damage. *J. econ. Ent.* **80,** 911–915.

Greenslade, P. and Ireson, J. E. (1986). Collembola of the southern Australian culture steppe and urban environments: A review of their pest status and key to identification. *J. Aust. ent. Soc.* **25,** 273–291.

Greenwood, R. M. and Bathurst, N. O. (1978). Effect of rhizobial strain and host on the amino acid patterns in legume root nodules. *N.Z.J. Sci.* **21,** 107–120.

Hamilton, D. W. (1961). Periodical cicadas, *Magicicada* spp., as pests in apple orchards. *Proc. Indiana Acad. Sci.* **71,** 116–121.

Harper, A. M. (1963). Sugar-beet root aphid, *Pemphigus betae* Doane (Homoptera: Aphididae), in southern Alberta. *Can. Ent.* **95,** 863–873.

Harper, J. L. (1977). *Population Biology of Plants.* Academic Press, London, 892 pp.

Harper, J. L., Jones, M. and Sackville-Hamilton, N. R. (in press). The evolution of roots and the problems of analysing their behaviour. In: *Plant Root Systems: Their Effect on Ecosystem Composition and Structure* (Ed. by D. Atkinson). Blackwell, Oxford.

Harris, D. G. and van Bavel, C. H. M. (1957). Root respiration of tobacco, corn and cotton plants. *Agron. J.* **49,** 182–184.

Harris, P. and Myers, J. H. (1981). *Centaurea diffusa* Lam. and *C. maculosa* Lam. *S. lat.*, diffuse and spotted knapweed (Compositae). In: *Biological Control Programmes Against Insects and Weeds in Canada 1969–80* (Ed. by J. S. Kelleher and M. J. Hulme), pp. 127–137. Commonwealth Agricultural Bureau, Slough.

Harris, P., Wilkinson, A. T. S. and Myers, J. H. (1981). *Senecio jacobaea* L., tansy ragwort (Compositae). In: *Biological Control Programmes Against Insects and Weeds in Canada 1969–80* (Ed. by J. S. Kelleher and M. J. Hulme), pp. 195–201. Commonwealth Agricultural Bureau, Slough.

Hartman, M. J. and Hynes, C. D. (1977). Biology of the range crane fly, *Tipula simplex* Doane. *Pan-Pacif. Ent.* **53,** 118–123.

Hassell, M. P. (1985). Insect natural enemies as regulating factors. *J. Anim. Ecol.* **54,** 323–334.
Hawkins, B. A. and Lawton, J. H. (1987). Species richness for parasitoids of British phytophagous insects. *Nature* **326,** 788–790.
Heads, P. A. and Lawton, J. H. (1983). Studies on the natural enemy complex of the holly leaf-miner: The effects of scale on the aggregative responses and the implications for biological control. *Oikos* **40,** 267–276.
Henderson, I. F. and Clements, R. O. (1974). The effect of pesticides on the yield and botanical composition of a newly-sown ryegrass ley and of an old mixed pasture. *J. Br. Grassld Soc.* **29,** 185–190.
Hering, E. M. (1951). *Biology of the Leaf Miners.* W. Junk, The Hague, 420 pp.
Hibbard, B. E. and Bjostad, L. B. (1988). Behavioral responses of western corn rootworm larvae to volatile semiochemicals from corn seedlings. *J. chem. Ecol.* **14,** 1523–1539.
Highland, H. B. and Lummus, P. F. (1986). Use of light traps to monitor flight activity of the burrowing bug, *Pangaeus bilineatus* (Hemiptera: Cydnidae), and associated field infestations in peanuts. *J. econ. Ent.* **79,** 523–526.
Hill, D. S. (1983). *Agricultural Insect Pests of the Tropics and Their Control.* Cambridge University Press, Cambridge, 516 pp.
Hill, D. S. (1987). *Agricultural Insect Pests of Temperate Regions and Their Control.* Cambridge University Press, Cambridge, 659 pp.
Hill, R. E., Hixson, E. and Muma, M. H. (1948). Corn rootworm tests with benzene hexachloride, DDT, nitrogen fertilizers and crop rotations. *J. econ. Ent.* **67,** 748–750.
Hruska, A. J. (1987). Weedy ground-cover increases damage to cassava by white grubs in Costa Rica. *Trop. Agric., Trin.* **64,** 212–216.
Humphries, E. C. (1958). Effect of removal of a part of the root system on the subsequent growth of the root and shoot. *Ann. Bot.* **22,** 251–257.
Hutchinson, K. J. and King, K. L. (1980). The effects of sheep stocking levels on invertebrate abundance, biomass, and energy utilization in a temperate sown grassland. *J. appl. Ecol.* **17,** 369–387.
Ingham, R. E. and Detling, J. K. (1986). Effects of defoliation and nematode consumption on growth and leaf gas exchange in *Bouteloua curtipendula. Oikos* **46,** 23–28.
Jackson, J. J. and Elliott, N. C. (1988). Temperature-dependent development of immature stages of the western corn rootworm, *Diabrotica virgifera virgifera* (Coleoptera: Chrysomelidae). *Environ. Entomol.* **17,** 166–171.
Janzen, D. H. (1985). *Spondias mombin* is culturally deprived in megafauna-free forest. *J. trop. Ecol.* **1,** 131–155.
Jaworska, M. and Wiech, K. (1988). Susceptibility of the clover root weevil, *Sitona hispidulus* F. (Col., Curculionidae) to *Steinernema feltiae*, *S. bibionis*, and *Heterorhabditis bacteriophora. J. appl. Ent.* **106,** 372–376.
Jones, F. G. W. and Jones, M. G. (1984). *Pests of Field Crops.* Edward Arnold, London, 392 pp.
Jones, O. T. (1979). The responses of carrot fly larvae, *Psila rosae*, to components of their physical environment. *Ecol. Entomol.* **4,** 327–334.
Jones, O. T. and Coaker, T. H. (1977). Orientated responses of carrot fly larvae *Psila rosae*, to plant odours, carbon dioxide and carrot root volatiles. *Physiol. Entomol.* **2,** 189–197.
Julien, M. H., Bourne, A. S. and Chan, R. R. (1987). Effects of adult and larval *Cyrtobagous salviniae* on the floating weed *Salvinia molesta. J. appl. Ecol.* **24,** 935–944.

Kahler, A. L., Olness, A. E., Sutter, G. R., Dybing, C. D. and Devine, O. J. (1985). Root damage by western corn rootworm and nutrient content in maize. *Agron. J.* **77,** 769–773.

Kamm, J. A. and Buttery, R. G. (1984). Root volatile components of red clover: Identification and bioassay with the clover root borer. *Environ. Entomol.* **13,** 1427–1430.

Karban, R. (1980). Periodical cicada nymphs impose periodical oak tree wood accumulation. *Nature* **287,** 326–327.

Kard, B. M. and Hain, F. P. (1988). Influence of ground covers on white grub (Coleoptera: Scarabaeidae) populations and their feeding damage to roots of Fraser fir christmas trees in Southern Appalachians. *Environ. Entomol.* **17,** 63–66.

Kard, B. M., Hain, F. P. and Brooks, W. M. (1988). Field suppression of three white grub species (Coleoptera: Scarabaeidae) by the entomogenous nematodes *Steinernema feltiae* and *Heterorhabditis heliothidis*. *J. econ Ent.* **81,** 1033–1039.

Kareiva, P. (1986). Patchiness, dispersal and species interactions: Consequences for communities of herbivorous insects. In: *Community Ecology* (Ed. by J. Diamond and T. J. Case), pp. 192–206. Harper and Row, New York.

Kevan, D. K. McE. (1962). *Soil Animals*. Witherby, London, 237 pp.

King, P. D. and Buchanan, G. A. (1986). The dispersal of phylloxera crawlers and spread of phylloxera infestations in New Zealand and Australian vineyards. *Am. J. Enol. Vitic.* **37,** 26–33.

King, P. D., Mercer, C. F. and Meekings, J. S. (1981). Ecology of black beetle, *Heteronychus arator* (Coleoptera: Scarabaeidae) – influence of temperature, feeding, growth and survival of the larvae. *N.Z.J. Zool.* **8,** 113–117.

de Klerk, C. A. (1974). Biology of *Phylloxera vitifoliae* (Fitch) (Homoptera: Phylloxeridae) in South Africa. *Phytophylactica* **6,** 109–118.

Koncalova, H., Albrecht, G., Pokorny, J. and Wiedenroth, E.-M. (1989). Measurement of respiratory carbon dioxide production of roots in an aquatic medium. *Biol. Plant (Prague)* **31,** 1–7.

Krysan, J. L. (1976). Moisture relationships of the egg of the southern corn rootworm, *Diabrotica undecimpunctata howardi*. *Ent. exp. appl.* **20,** 154–162.

Ladd, T. L. (1988). Japanese beetle (Coleoptera: Scarabaeidae): Influence of sugars on feeding response of larvae. *J. econ. Ent.* **81,** 1390–1393.

Ladd, T. L. and Buriff, C. R. (1979). Japanese beetle: Influence of larval feeding on bluegrass yields at two levels of soil moisture. *J. econ. Ent.* **72,** 311–314.

Lange, W. H. (1987). Insect pests of sugar beet. *Ann. Rev. Entomol.* **32,** 341–360.

Lange, W. H., Benson, L. C., Force, D. C., Grigarick, A. A. and McCalley, N. F. (1957). Lettuce root aphid. *Calif. Agric.* **11,** 7, 8, 15.

Langley, J. M. (1986). The management of subterranean aphids and other soil organisms by the ant *Lasius flavus* (F.). Ph.D. thesis, University of London, 398 pp.

Laughlin, R. (1963). Biology and ecology of the garden chafer, *Phyllopertha horticola* (L.). VIII. Temperature and larval growth. *Bull. ent. Res.* **54,** 745–759.

Lawton, J. H. (1984). Herbivore community organisation: General models and specific tests with phytophagous insects. In: *A New Ecology* (Ed. by P. W. Price, C. N. Slobodchikoff and W. S. Gaud), pp. 329–352. John Wiley, New York.

Leath, K. T. and Byers, R. A. (1973). Attractiveness of diseased red clover roots to the clover root borer. *Phytopathology* **64,** 429–431.

Lewis, T. (1975). *Thrips, Their Biology, Ecology and Economic Importance*. Academic Press, London, 350 pp.

Lilly, J. H. and Gunderson, H. (1952). Fighting the corn rootworm. *Iowa Farm Sci.* **6,** 18–19.

Lim, K. P., Yule, W. N. and Stewart, R. K. (1980). A note on *Pelecinus polyturator* (Hymenoptera: Pelecinidae), a parasite of *Phyllophaga anxia* (Coleoptera: Scarabaeidae). *Can. Ent.* **112,** 219–220.

Lloyd, M. (1984). Periodical Cicadas. *Antenna* **8,** 79–91.

Louda, S. M., Keeler, K. H. and Holt, R. D. (1990). Herbivore influences on plant performance and competitive interactions. In: *Perspectives in Plant Competition* (Ed. by J. B. Grace and D. Tilman) pp. 414–444. Academic Press, London.

Lovett Doust, J. (1980). Experimental manipulation of patterns of resource allocation in the growth cycle and reproduction of *Smyrnium olustratum* L. *Biol. J. Linn. Soc.* **13,** 155–166.

Ludwig, D. (1928). The effects of temperature on the development of an insect (*Popillia japonica* Newman). *Physiol. Zool.* **1,** 358–389.

Lummus, P. F., Smith, J. C. and Powell, N. L. (1983). Soil moisture and texture effects on survival of immature southern corn rootworms, *Diabrotica undecimpunctata howardi* Barber (Coleoptera: Chrysomelidae). *Environ. Entomol.* **12,** 1529–1531.

McCrea, K. D. and Abrahamson, W. G. (1985). Evolutionary impacts on the goldenrod ball gall maker on *Solidago altissima* clones. *Oecologia (Berl.)* **68,** 20–22.

McGonigle, T. P. and Fitter, A. H. (1988). Ecological consequences of arthropod grazing on VA mycorrhizal fungi. *Proc. R. Soc. Edin.* **94B,** 25–32.

McKey, D. (1979). The distribution of secondary compounds within plants. In: *Herbivores – Their Interaction with Secondary Plant Metabolites* (Ed. by G. A. Rosenthal and D. H. Janzen), pp. 55–133. Academic Press, London.

McNaughton, S. J. (1986). On plants and herbivores. *Am. Nat.* **128,** 765–770.

Macfadyen, A. (1953). Notes on methods for the extraction of small soil arthropods. *J. Anim. Ecol.* **22,** 65–77.

Marrone, P. G. and Stinner, R. E. (1984). Influence of soil physical factors on survival and development of the larvae and pupae of the bean leaf beetle *Cerotoma trifurcata* (Coleoptera: Chrysomelidae). *Can. Ent.* **116,** 1015–1023.

Marschner, H. and Romheld, V. (1983). *In vivo* measurement of root-induced changes at the soil-root interface: Effect of plant species and nitrogen source. *Z. Pflanzenphysiol.* **111,** 241–251.

Matsumoto, Y. (1970). Volatile organic sulfur compounds as insect attractants with special reference to host selection. In: *Control of Insect Behaviour by Natural Products* (Ed. by D. L. Wood, R. M. Silverstein and M. Nakajima), pp. 133–160. Academic Press, London.

Miller, G. L., Williams, M. L. and Karr, G. W. (1984). The Japanese beetle: A threat to Alabama agriculture. *Highlts agric. Res.* **31,** 17.

Milne, A. (1963). Biology and ecology of the garden chafer, *Phyllopertha horticola* (L.). IX. Spatial distribution. *Bull. ent. Res.* **54,** 761–795.

Minoranskij, V. A. (1971). Ergebnisse des stadiums von *Pemphigus fuscicornis*, einer in der Sowjet-Union als Zuckerruebenschaedling auftretenden Wurzellaus. *Pedobiologia* **11,** 81–96.

Misra, S. S., Nagia, D. K. and Ram, G. (1985). Efficacy of systemic insecticides against root aphid, *Rhopalosiphum rufiabdominalis* Sasaki on potato crop. *Pesticides (Bombay)* **19,** 63–65.

Mitsios, I. and Rowell, D. L. (1987). Plant uptake of exchangeable and non exchangeable potassium. I. Measurement and modelling for onion roots in a chalky boulder clay soil. *J. Soil Sci.* **38,** 53–63.

Mooney, H. A. and Billings, W. D. (1960). The annual carbohydrate cycle of alpine plants as related to growth. *Am. J. Bot.* **47,** 594–598.

Moran, N. A. and Whitham, T. G. (1988). Population fluctuations in complex life cycles: An example from *Pemphigus* aphids. *Ecology* **69,** 1214–1218.

Müller, H. (1987). Preliminary notes on the use of glass-faced boxes as a tool to study root/herbivore interactions. In: *Insects – Plants* (Ed. by V. Labeyrie, G. Febres and D. Lachaise). p. 405. W. Junk, The Hague.

Müller, H. (1989a). Structural analysis of the phytophagous insect guilds associated with the roots of *Centaurea maculosa* Lam., *C. diffusa* Lam. and *C. vallesiaca* Jordan in Europe. 1. Field observations. *Oecologia (Berl.)* **78,** 41–52.

Müller, H. (1989b). Growth pattern of diploid and tetraploid spotted knapweed, *Centaurea maculosa* Lam. (Compositae), and effects of the root-mining moth *Agapeta zoegana* (L.) (Lep.: Cochylidae). *Weed Res.* **29,** 103–111.

Müller, H., Stinson, C. S. A., Marquardt, K. and Schroeder, D. (1989). The entomofaunas of roots of *Centaurea maculosa* Lam., *C. diffusa* Lam. and *C. vallesiaca* Jordan in Europe. *J. appl. Ent.* **107,** 83–95.

Nichols, D. B. R., Wright, A. J. and French, N. (1987). Yield response of improved upland pasture to the control of leatherjackets under increasing rates of nitrogen. *Ann. appl. Biol.* **110** (Suppl.), 18–19.

Nougaret, R. L. and Lapham, M. H. (1928). A study of phylloxera infestation in California as related to types of soils. *USDA Technical Bulletin* No. 20, pp. 1–38.

Ohara, M. and Higashi, S. (1987). Interference by ground beetles with the dispersal by ants of seeds of *Trillium* species (Liliaceae). *J. Ecol.* **75,** 1091–1098.

Ormerod, E. A. (1890). *A Manual of Injurious Insects and Methods of Prevention.* Simpkin, Marshall, Hamilton, Kent, London, 410 pp.

Parkinson, D. (1988). Linkages between resource availability, micro-organisms and soil invertebrates. *Agric. Ecosystems Env.* **24,** 21–32.

Parkinson, D., Visser, S. and Whittaker, J. B. (1979). Effects of collembolan grazing on fungal colonization of leaf litter. *Soil Biol. Biochem.* **11,** 529–535.

Piedrahita, O., Ellis, C. R. and Allen, O. B. (1985a). Effect of spacing and clumping of corn plants on density of corn rootworm larvae (Coleoptera: Chrysomelidae) per root system. *Can. Ent.* **117,** 139–142.

Piedrahita, O., Ellis, C. R. and Bogart, J. P. (1985b). Interaction of northern and western corn rootworm larvae (Coleoptera: Chrysomelidae) in a controlled environment. *Environ. Entomol.* **14,** 138–141.

Polivka, J. B. (1960a). Effect of lime applications to soil on Japanese beetle larval population. *J. econ. Ent.* **53,** 476–477.

Polivka, J. B. (1960b). Grub population in turf areas with pH levels in Ohio soils. *J. econ. Ent.* **53,** 860–863.

Pontin, A. J. (1978). The numbers and distribution of subterranean aphids and their exploitation by the ant *Lasius flavus* (Fabr.). *Ecol. Entomol.* **3,** 203–207.

Poprawski, T. J., Marchal, M. and Robert, P.-H. (1985a). Comparative susceptibility of *Otiorhynchus sulcatus* and *Sitona lineatus* (Coleoptera: Curculionidae) early stages to five entomopathogenic hyphomycetes. *Environ. Entomol.* **14,** 247–253.

Poprawski, T. J., Robert, P.-H., Majchrowicz, I. and Boivin, G. (1985b). Susceptibility of *Delia antiqua* (Diptera: Anthomyiidae) to eleven isolates of entomopathogenic hyphomycetes. *Environ. Entomol.* **14,** 557–561.

Potter, D. A. (1983). Effect of soil moisture on oviposition, water absorption, and survival of southern masked chafer (Coleoptera: Scarabaeidae) eggs. *Environ. Entomol.* **12,** 1223–1227.

Potter, D. A. and Gordon, F. C. (1984). Susceptibility of *Cyclocephala immaculata* (Coleoptera: Scarabaeidae) eggs and immatures to heat and drought in turf grass. *Environ. Entomol.* **13,** 794–799.

Powell, R. D. and Myers, J. H. (1988). The effect of *Sphenoptera jugoslavica* Obenb.

(Col., Buprestidae) on its host plant *Centaurea diffusa* Lam. (Compositae). *J. appl. Ent.* **106,** 25–45.

Prestidge, R. A., van der Zijpp, S. and Badan, D. (1985). Effects of plant species and fertilisers on grass grub larvae, *Costelytra zealandica. N.Z.J. agric. Res.* **28,** 409–417.

Pritchard, G. (1983). Biology of Tipulidae. *Ann. Rev. Entomol.* **28,** 1–22.

Quinn, M. A. and Hower, A. A. (1985). Changes in the spatial distribution of the clover root curculio (Coleoptera: Curculionidae) in alfalfa field soil. *Environ. Entomol.* **14,** 864–867.

Quinn, M. A. and Hower, A. A. (1986a). Multivariate analysis of the population structure of *Sitona hispidulus* (Coleoptera: Curculionidae) in alfalfa field soil. *Can. Ent.* **118,** 517–524.

Quinn, M. A. and Hower, A. A. (1986b). Effects of root nodules and taproots on survival and abundance of *Sitona hispidulus* (Coleoptera: Curculionidae) on *Medicago sativa. Ecol. Entomol.* **11,** 391–400.

Radcliffe, J. E. (1970). Some effects of grass grub (*Costelytra zealandica* (White)) larvae on pasture plants. *N.Z.J. agric. Res.* **13,** 87–104.

Radcliffe, J. E. (1971a). Effects of grass grub (*Costelytra zealandica* White) larvae on pasture plants. I. Effect of grass grubs and nutrients on perennial ryegrass. *N.Z.J. agric. Res.* **14,** 597–606.

Radcliffe, J. E. (1971b). Effects of grass grub (*Costelytra zealandica* White) larvae on pasture plants. II. Effect of grass grubs and soil moisture on perennial ryegrass and cocksfoot. *N.Z.J. agric. Res.* **14,** 607–617.

Radcliffe, J. E. (1971c). Effects of grass grub (*Costelytra zealandica* White) larvae on pasture plants. III. Effect of grass grubs and plant density on perennial ryegrass and cocksfoot. *N.Z.J. agric. Res.* **14,** 618–624.

Radcliffe, J. E. (1971d). Effects of grass grub (*Costelytra zealandica* White) larvae on pasture plants. IV. Effect of grass grubs on perennial ryegrass and white clover. *N.Z.J. agric. Res.* **14,** 625–632.

Raw, F. (1951). The ecology of the garden chafer, *Phyllopertha horticola* (L.) with preliminary observations on control measures. *Bull. ent. Res.* **42,** 605–646.

Rees, M. (1989). The population ecology of annual crucifers. Ph.D. thesis, University of London, 99 pp.

Régnière, J., Rabb, R. L. and Stinner, R. E. (1981a). *Popillia japonica*: Simulation of temperature-dependent development of the immatures, and prediction of adult emergence. *Environ. Entomol.* **10,** 290–296.

Régnière, J., Rabb, R. L. and Stinner, R. E. (1981b). *Popillia japonica*: Effect of soil moisture and texture on survival and development of eggs and first instar grubs. *Environ. Entomol.* **10,** 654–660.

Régnière, J., Rabb, R. L. and Stinner, R. E. (1981c). *Popillia japonica*: Intraspecific competition among grubs. *Environ. Entomol.* **10,** 661–662.

Ridsdill Smith, T. J. (1975). Selection of living grass roots in the soil by larvae of *Sericesthis nigrolineata* (Coleoptera: Scarabaeidae). *Ent. Exp. Appl.* **18,** 75–86.

Ridsdill Smith, T. J. (1977). Effects of root feeding by scarabaeid larvae on growth of perennial ryegrass plants. *J. appl. Ecol.* **14,** 73–80.

Ridsdill Smith, T. J. and Roberts, R. J. (1976). Insect density effects in root feeding by larvae of *Sericesthis nigrolineata* (Coleoptera: Scarabaeidae). *J. appl. Ecol.* **13,** 423–428.

Ridsdill Smith, T. J., Porter, M. R. and Furnival, M. G. (1975). Effects of temperature and developmental stage on feeding by larvae of *Sericesthis nigrolineata* (Coleoptera: Scarabaeidae). *Ent. exp. appl.* **18,** 244–254.

Riedell, W. E. (1989). Western corn rootworm damage in maize: Greenhouse technique and plant response. *Crop Sci.* **29,** 412–415.

Rikhter, V. A. (1975). Tachinids as parasites of injurious insects. *Zashch. Rast. Vredit.* **2,** 46–48.

Ritcher, P. O. (1958). Biology of the Scarabaeidae. *Ann. Rev. Entomol.* **3,** 311–334.

Rizza, A., Campobasso, G., Dunn, P. H. and Stazi, M. (1988). *Cheilosia corydon* (Diptera: Syrphidae), a candidate for the biological control of musk thistle in North America. *Ann. ent. Soc. Am.* **81,** 225–232.

Roberts, H. A. (1981). Seed banks in soils. In: *Advances in Applied Biology* (Ed. by J. B. Cragg), Vol. 4, pp. 1–55. Academic Press, London.

Roberts, R. J. and Morton, R. (1985). Biomass of larval scarabaeidae (Coleoptera) in relation to grazing pressures in temperate, sown pastures. *J. appl. Ecol.* **22,** 863–874.

Roberts, R. J. and Morton, R. (1985). Biomass of larval Scarabaeidae (Coleoptera) in relation to grazing pressures in temperate, sown pastures. *J. appl. Ecol.* **22,** 863–874.

Roberts, R. J. and Ridsdill Smith, T. J. (1979). Assessing pasture damage and losses in animal production caused by pasture insects. In: *Proceedings of the 2nd Australasian Conference on Grassland Invertebrate Ecology* (Ed. by T. K. Crosby and R. F. Pottinger), pp. 124–125. Government Printer, Wellington, NZ.

Rodman, J. E. and Louda, S. M. (1984). Phenology of glucosinolate concentrations in roots, stems and leaves of *Cardamine cordifolia*. *Biochem. Syst. Ecol.* **12,** 37–46.

Rothschild, G. H. L. (1986). The potato moth – an adaptable pest of short-term cropping systems. In: *The Ecology of Exotic Animals and Plants: Some Australian Case Histories* (Ed. by R. L. Kitching), pp. 150–162. John Wiley, Brisbane.

Rowell, D. L. (1988). Soil acidity and alkalinity. In: *Russell's Soil Conditions and Plant Growth* (Ed. by A. Wild), pp. 844–898. Longman, London.

Russell, G. B., Shaw, G. J., Christmas, P. E., Yates, M. B. and Sutherland, O. R. W. (1984). Two 2-arylbenzofurans as insect feeding deterrents from sainfoin (*Onobrychis vicifolia*). *Phytochemistry* **23,** 1417–1420.

Russell, R. S. (1977). *Plant Root Systems: Their Function and Interaction with the Soil.* McGraw-Hill, London, 298 pp.

Rutherford, T. A., Trotter, D. and Webster, J. M. (1987). The potential of heterorhabditid nematodes as control agents of root weevils. *Can. Ent.* **119,** 67–73. Press, London, 626 pp.

Schowalter, T. D., Hargrove, W. W. and Crossley, D. A. Jr. (1986). Herbivory in forested ecosystems. *Ann. Rev. Entomol.* **31,** 177–196.

Seastedt, T. R. (1985). Maximization of primary and secondary productivity by grazers. *Am. Nat.* **126,** 559–564.

Seastedt, T. R., Ramundo, R. A. and Hayes, D. C. (1988). Maximization of densities of soil animals by foliage herbivory: Empirical evidence, graphical and conceptual models. *Oikos* **51,** 243–248.

Seinhorst, J. W. (1965). The relation between nematode density and damage to plants. *Nematologica* **11,** 159–164.

Shetlar, D. J., Suleman, P. E. and Georgis, R. (1988). Irrigation and use of entomogenous nematodes, *Neoaplectana* spp. and *Heterorhabditis heliothidis* (Rhabdita: Steinernematidae and Heterorhabditidae), for control of Japanese beetle (Coleoptera: Scarabaeidae) grubs in turfgrass. *J. econ. Ent.* **81,** 1318–1322.

Simberloff, D., Brown, B. J. and Lowrie, S. (1978). Isopod and insect root borers may benefit Florida mangroves. *Science* **201,** 630–632.

Skinner, B. (1984). *Colour Identification Guide to Moths of the British Isles*. Penguin, Harmondsworth, 267 pp.

Smith, J. W. and Pitts, J. T. (1971). Pest status of *Pangaeus bilineatus* attacking peanuts in Texas. *J. econ. Ent.* **67,** 111–113.

Snodgrass, R. E. (1935). *Principles of Insect Morphology*. McGraw-Hill, New York, 667 pp.

Southwood, T. R. E. (1962). Migration of terrestrial arthropods in relation to habitat. *Biol. Rev.* **37,** 171–214.

Southwood, T. R. E. (1978a). The components of diversity. In: *Diversity of Insect Faunas* (Ed. by L. A. Mound and N. Waloff): *Symp. R. ent. Soc. Lond.* **9,** 19–40.

Southwood, T. R. E. (1978b). *Ecological Methods with Particular Reference to the Study of Insect Populations*. Chapman and Hall, London, 524 pp.

Spike, B. P. and Tollefson, J. J. (1988). Western corn rootworm (Coleoptera: Chrysomelidae) larval survival and damage potential to corn subjected to nitrogen and plant density treatments. *J. econ. Ent.* **81,** 1450–1455.

Stanton, N. L. (1983). The effect of clipping and phytophagous nematodes on net primary production of blue grama, *Bouteloua gracilis*. *Oikos* **40,** 249–257.

Stanton, N. L. (1988). The underground in grasslands. *Ann. Rev. Ecol. Syst.* **19,** 573–589.

Stewart, K. M., Vantoor, R. F. and Crosbie, S. F. (1988). Control of grass grub (Coleoptera: Scarabaeidae) with rollers of different design. *N.Z.J. exp. Agric.* **16,** 141–150.

Stewart, R. M. and Kozicki, K. K. (1987). DIY assessment of leatherjacket numbers in grassland. In: *Proceedings of Crop Protection in Northern Britain*, pp. 349–353. Scottish Crops Research Institute, Dundee.

St. Julian, G., Toolan, S. C., Detroy, R. W. and Stern, N. (1982). Infectivity of *Nomuraea rileyi* conidia to *Popillia japonica* Newman. *J. Invertebr. Pathol.* **39,** 253–254.

Strnad, S. P. and Bergman, M. K. (1987). Movement of first-instar western corn rootworms (Coleoptera: Chrysomelidae) in soil. *Environ. Entomol.* **16,** 975–978.

Strnad, S. P., Bergman, M. K. and Fulton, W. C. (1986). First instar western corn rootworm (Coleoptera: Chrysomelidae) response to carbon dioxide. *Environ. Entomol.* **15,** 839–842.

Sutherland, J. A. (1986). A review of the biology and control of the sweet potato weevil *Cylas formicarius* (Fab.). *Trop. Pest Mgmt.* **32,** 304–315.

Sutherland, O. R. W. (1971). Feeding behaviour of the grass grub *Costelytra zealandica* (White) (Coleoptera: Melolonthinae). 1. The influence of carbohydrates. *N.Z.J. Sci.* **14,** 18–24.

Sutherland, O. R. W. (1972). Olfactory responses of *Costelytra zealandica* (Coleoptera: Melolonthinae) larvae to grass root odours. *N.Z.J. Sci.* **15,** 165–172.

Sutherland, O. R. W. and Greenfield, W. J. (1978). Effect of root extracts of resistant pasture plants on the feeding and survival of black beetle larvae, *Heteronychus arator* (Scarabaeidae). *N.Z.J. Zool.* **5,** 173–175.

Sutherland, O. R. W., Hood, N. D. and Hillier, J. R. (1975). Lucerne root saponins, a feeding deterrent for the grass grub, *Costelytra zealandica* (Coleoptera: Scarabaeidae). *N.Z.J. Zool.* **2,** 93–100.

Sutton, R. F. (1967). Influence of root pruning on height increment and root development of out-planted spruce. *Can. J. Bot.* **45,** 1671–1682.

Taylor, L. R. (1961). Aggregation, variance and the mean. *Nature* **189,** 732–735.

Thompson, J. N. (1985). Post dispersal seed predation in *Lomatium* spp. (Umbelliferae). Variation among individuals and species. *Ecology* **66,** 1608–1616.

Tolman, D. J. H., McLeod, D. G. R. and Harris, C. R. (1986). Yield losses in potatoes, onions and rutabagas in South-western Ontario, Canada – a case for pest control. *Crop Prot.* **5,** 227–237.

Troughton, A. (1974). The growth and function of the root in relation to the shoot. In: *Structure and Function of Primary Root Tissues* (Ed. by J. Kolek), pp. 153–164. Veda, Bratislava.

Tryon, E. H. (1986). The striped earwig, and ant predators of sugarcane rootstock borer, in Florida citrus. *Florida Entomol.* **69,** 336–343.

Turner, E. C. (1957). Control of the clover root curculio in alfalfa. *J. econ. Ent.* **50,** 645–648.

Turpin, F. T. and Peters, D. C. (1971). Survival of southern and western corn rootworm larvae in relation to soil texture. *J. econ. Ent.* **64,** 1448–1451.

Ueckert, D. N. (1979). Impact of a white grub (*Phyllophaga crinita*) on a short-grass community and evaluation of selected rehabilitation practices. *J. Range Mgmt.* **32,** 445–448.

Vancura, V., Prikryl, Z., Kalachova, L. and Wurst, M. (1977). Some quantitative aspects of root exudation. *Proc. 6th Int. Coll. Soil Zoology, Ecol. Bull. (Stockholm)* **25,** 381–386.

Villani, M. G. and Wright, R. J. (1988). Entomogenous nematodes as biological control agents of European chafer and Japanese beetle (Coleoptera: Scarabaeidae) larvae infesting turfgrass. *J. econ. Ent.* **81,** 484–487.

Vittum, P. J. (1984). Effect of lime applications on Japanese beetle (Coleoptera: Scarabaeidae) grub populations in Massachusetts soils. *J. econ. Ent.* **77,** 687–690.

Vittum, P. J. and Tashiro, H. (1980). Effect of soil pH on survival of Japanese beetle and European chafer larvae. *J. econ. Ent.* **73,** 577–579.

Wallace, H. R. (1973). *Nematode Ecology and Plant Disease*. Edward Arnold, London, 228 pp.

Waloff, N. and Richards, O. W. (1977). The effect of insect fauna on growth mortality and natality of broom, *Sarothamnus scoparius*. *J. appl. Ecol.* **14,** 787–798.

Wang, E. L. H. and Bergeson, G. B. (1974). Biochemical changes in root exudate and xylem sap of tomato plants infected with *Meloidogyne incognita*. *J. Nematol.* **6,** 194–202.

Wapshere, A. J. and Helm, K. F. (1987). Phylloxera and *Vitis*: An experimentally testable coevolutionary hypothesis. *Am. J. Enol. Vitic.* **38,** 216–222.

Warnock, A. J., Fitter, A. H. and Usher, M. B. (1982). The influence of a spring tail *Folsomia candida* (Insecta, Collembola) on the mycorrhizal association of the leek *Allium porrum* and the vesicular-arbuscular mycorrhizal endophyte *Glomus fasciculatus*. *New Phytol.* **90,** 285–292.

Watt, J. C. (1986). Pacific Scarabaeidae and Elateridae (Coleoptera) of agricultural significance. *Agric. Ecosystems Environ.* **15,** 175–187.

Way, M. J. (1963). Mutualism between ants and honeydew producing Homoptera. *Ann. Rev. Entomol.* **8,** 307–344.

Weiss, M. J., Seevers, K. P. and Mayor, Z. B. (1985). Influence of western corn rootworm larval densities and damage on corn rootworm survival, developmental time, size and sex ratio (Coleoptera: Chrysomelidae). *J. Kans. ent. Soc.* **58,** 397–402.

Wessel, R. D. and Polivka, J. B. (1952). Soil pH in relation to Japanese beetle populations. *J. econ. Ent.* **45,** 733–735.

West, C. (1985). Factors underlying the late seasonal appearance of the lepidopterous leaf-mining guild on Oak. *Ecol. Ent.* **10,** 111–120.

White, J. H. and French, N. (1968). Leatherjacket damage to grassland. *J. Br. Grassld Soc.* **23,** 326–329.
White, J. and Lloyd, M. (1975). Growth rates of 17- and 13-year periodical cicadas. *Am. Midl. Nat.* **94,** 127–143.
White, J. and Lloyd, M. (1983). A pathogenic fungus, *Massospora cicadina* Peck (Entomophthorales), in emerging nymphs of periodical cicadas (Homoptera: Cicadidae). *Environ. Entomol.* **12,** 1245–1252.
White, J. and Strehl, C. E. (1978). Xylem feeding by periodical cicada nymphs on tree roots. *Ecol. Ent.* **3,** 323–327.
White, J., Lloyd, M. and Zar, J. H. (1979). Faulty eclosion in crowded suburban periodical cicadas: Populations out of control. *Ecology* **60,** 305–315.
White, T. C. R. (1969). An index to measure weather-induced stress of trees associated with outbreaks of psyllids in Australia. *Ecology* **50,** 905–909.
White, T. C. R. (1984). The abundance of invertebrate herbivores in relation to the availability of nitrogen in stressed foodplants. *Oecologia (Berl.)* **63,** 90–105.
Wiggins, E. A., Curl, E. A. and Harper, J. D. (1979). Effects of soil fertility and cotton rhizosphere on populations of Collembola. *Pedobiologia* **19,** 75–82.
Williams, G. R. (Ed.) (1973). *The Natural History of New Zealand: An Ecological Survey.* A. H. and A. W. Reed, Wellington, NZ, 434 pp.
Wilson, D. D., Severson, R. F., Son, K.-C. and Kays, S. J. (1988). Oviposition stimulant in sweet potato periderm for the sweet potato weevil, *Cylas formicarius elegantulus* (Coleoptera: Curculionidae). *Environ. Entomol.* **17,** 691–693.
Witcosky, J. J., Schowalter, T. D. and Hansen, E. M. (1986). *Hylastes nigrinus* (Coleoptera: Scolytidae), *Pissodes fasciatus*, and *Steremnius carinatus* (Coleoptera: Curculionidae) as vectors of black-stain root disease of Douglas fir (*Pseudotsuga menziesii*). *Environ. Entomol.* **15,** 1090–1095.
Wolfson, J. L. (1987). Impact of *Rhizobium* nodules on *Sitona hispidulus*, the clover root curculio. *Ent. Exp. Appl.* **43,** 237–243.
Yeates, G. W. (1987). How plants affect nematodes. In: *Advances in Ecological Research* (Ed. by A. Macfadyen and E. D. Ford), Vol. 17, pp. 61–113. Academic Press, London.
Zhen-Rong, W., Zhan-Ou, C., Dong-Sheng, Z. and Ji-Kang, X. (1986). Underground distribution pattern of white grubs and sampling method in peanut and soybean fields. [In Chinese; from English summary.] *Act. Ent. Sin.* **29,** 395–400.

Evolutionary and Ecophysiological Responses of Mountain Plants to the Growing Season Environment

A. D. FRIEND and F. I. WOODWARD

ADVANCES IN ECOLOGICAL RESEARCH VOL. 20
ISBN 0-12-013920-0

I. SUMMARY

The responses of mountain plants to their environment are due to a complex mixture of genetic and environmental influences. There is evidence that some mountain plants have evolved in response to their particular altitudinal environment. Instances include the increased ability to fix CO_2 from lower than sea-level concentrations, use high irradiance levels more efficiently, photosynthesize and grow at lower temperatures, and have intrinsically lower growth rates than their lowland counterparts. However, despite such examples, there is also evidence that many of these features could occur in mountain plants without any genetic component. Whether such developmental plasticity has been selected for is difficult to ascertain. Reduced growth temperatures may cause increased amounts of enzymes, such as Rubisco, per unit leaf area, via the differential effect of temperature on growth and photosynthesis, with changes in leaf thickness and the ratio of mesophyll cell area to leaf area (R^{mes}) being of importance. This would result in increased mesophyll conductance, and hence efficiency of carbon uptake (*ECU*) and maximum rate of photosynthesis (A_{max}), which may cause an increased maximum stomatal conductance (g_{max}), under high irradiance conditions, via the control of stomatal density. The above scheme is far from proven but is certainly worthy of further investigation. The evidence for changes in the optimum temperature for photosynthesis (T_o) indicates that amounts of Rubisco and Fru-P_2 phosphate activity per unit leaf area are probably very important. Thus increased *ECU* and A_{max}, and reduced T_o, could all be explained by one scenario, the primal agent being the effect of low temperature on leaf development. The effect of temperature on growth and developmental processes is crucial.

Model simulations of canopy energy balance and CO_2 fixation indicate that canopy structure and leaf area index (*LAI*) strongly influence both photosynthetic rate (*A*) and the ratio of ^{13}C to ^{12}C ($\delta^{13}C$) in leaves. Whereas $\delta^{13}C$ measurements on expanded leaves provide a time integral of CO_2 discrimination during the photosynthetic life of the leaf, they also include some unknown $\delta^{13}C$ contribution from photosynthate exported or remobilized from other leaves and organs. It would be interesting, therefore, to investigate the $\delta^{13}C$ of leaves as they develop during the season. An additional problem is that any mutual shading by other leaves, and any periods of drought, will also influence the ratio of internal leaf to ambient CO_2 partial pressure (p_i/p_a) and therefore $\delta^{13}C$ (Farquhar *et al.*, 1982).

The model simulations, for just the period of peak irradiance during the day, indicate that the energy balance and gas exchange of a leaf are dependent on its aerodynamic coupling with other leaves in the plant canopy, and with the air at some reference height above the canopy. This degree of coupling must be incorporated when individual leaf measurements of gas exchange are taken to indicate plant responses to environmental changes with altitude.

The predictions of canopy and leaf photosynthesis by the model have all been obtained with one A/c_i response, based on interpretations of the data of Körner and Diemer (1987). Therefore, tall trees and short herbs do not differ in terms of A, *ECU* or conductance (g) in the model. Yet this is clearly not so in the real world (e.g. Fitter and Hay, 1987); trees might be expected to exhibit lower A and g. However, Körner *et al.* (1988) showed that $\delta^{13}C$ trends with altitude were broadly similar for trees, shrubs and herbs, indicating that changes in A, *ECU* or g are closely linked, a feature worthy of further study.

Model simulations have indicated that leaf gas exchange and $\delta^{13}C$ are sensitive to both the expected trends in microclimate with altitude, and plant stature and *LAI*. There is a decrease in plant stature with altitude. This will cause an increase in stomatal conductance with altitude because the vapour pressure deficit of the air close to the leaves will tend to be lower due to the decreased boundary layer conductance. The poorly documented effects of temperature on stomatal conductance may prove significant and dependent on canopy stature and energy balance.

II. INTRODUCTION

A. Mountain Mesoclimates

Mountains (see Barry, 1981, for a discussion of the problem in defining mountains) cover about 20% of the land surface of the world (Barry, 1981), extending from Antarctica to the Arctic, throughout the east and west hemispheres and, therefore, through a wide range of climatic provinces. During the growing season, an examination of any mountain of reasonable altitudinal extent, will show a clear gradient of plant stature. There will also be a gradient in air temperature, with lower temperatures at higher altitudes. This phenomenon is striking for mountains which extend up to permanent snow.

This decline in temperature is a fundamental feature of all mountains and is due to the drop in atmospheric pressure (Pa) and air density ρ (kg m^{-3}) with altitude (Barry, 1981). The atmospheric pressure falls with altitude because of the monotonic fall in gravitational attraction towards the earth's surface. The atmospheric pressure P_z, at any altitude z (m), can be approximated by the following equation (from Jones, 1983):

$$P_z = \frac{101325}{e^{((z/29 \cdot 3)(T_k))}} \quad (1)$$

where T_k is air temperature (K). Due to the higher virtual temperatures and humidities in tropical regions, the equation underestimates P_z by about 1000–1500 Pa (Barry, 1981). The virtual temperature is the temperature at which dry air, at the same atmospheric pressure, is equally dense as the sample air.

The fall in air temperature with altitude occurs because a rising parcel of air expands as the atmospheric pressure decreases. Expansion is an endothermic process and heat for the process is extracted from the air itself, causing it to cool. If this cooling occurs at temperatures above the dew point of the air, the (dry) adiabatic lapse rate may be nearly 10°C km^{-1}. Should the air reach the dew point temperature, water vapour condensation occurs. This process of warming reduces the lapse rate to as little as 4–5°C km^{-1} (Barry, 1981).

Altitudinal gradients will therefore be associated with a decline in air temperature, a decline which will be dependent on the temperature, the dew point temperature, the radiation load and local wind patterns. Despite all these possible influences, observed lapse rates of temperature decline for meteorological stations (from Müller, 1982) at a range of altitudes (Fig. 1), are rather similar for mountains from a wide geographical range. Water vapour pressure is strongly linked with temperature (Woodward and Sheehy, 1983), and shows similar trends with altitudes (Fig. 2), reflecting the reduced capacity of the air to hold water vapour as temperature declines. The absolute minimum temperature also declines with altitude (Fig. 3), at a similar rate to the mean annual temperature.

The generality of the temperature, or temperature related lapse rates (Figs 1, 2, 3) contrasts strongly with the altitudinal trends in precipitation (Fig. 4), where no two mountain ranges show the same trend. However, Lauscher (1976) has indicated that some global trends may be found. Over altitudinal gradients in which the air, at some altitude, becomes saturated, precipitation generally increases with altitude, as on equatorial, tropical and polar mountains. In contrast, precipitation appears to decrease with altitude on mountains in the middle latitudes. The non-linear relationship between precipitation and altitude for the Himalayas (Fig. 4) indicates the strong influence of the lowland monsoon climate to about 500 m. At higher altitudes, the cooler air holds less water and precipitation decreases. This simple view may vary between individual mountains through local variations in the patterns and strengths of the mountain winds.

Wind speed generally increases with altitude (Geiger, 1965; Grace, 1977), particularly on mountains of the middle and polar latitudes (Barry, 1981), but with marked local variations depending on wind-funnelling and shelter effects by adjacent mountains. On tropical and equatorial mountains, wind speed may change little or even decrease with altitude. The mean wind speed at 4250 m on Mt Jaya, Papua New Guinea is 2 m s^{-1}, while the mean wind speed is 23 m s^{-1} at 1915 m on Mt Washington, USA (Barry, 1981). The mean trend in wind speed is therefore strongly dependent on the strengths of the global wind belts. However, on any particular mountain there will be

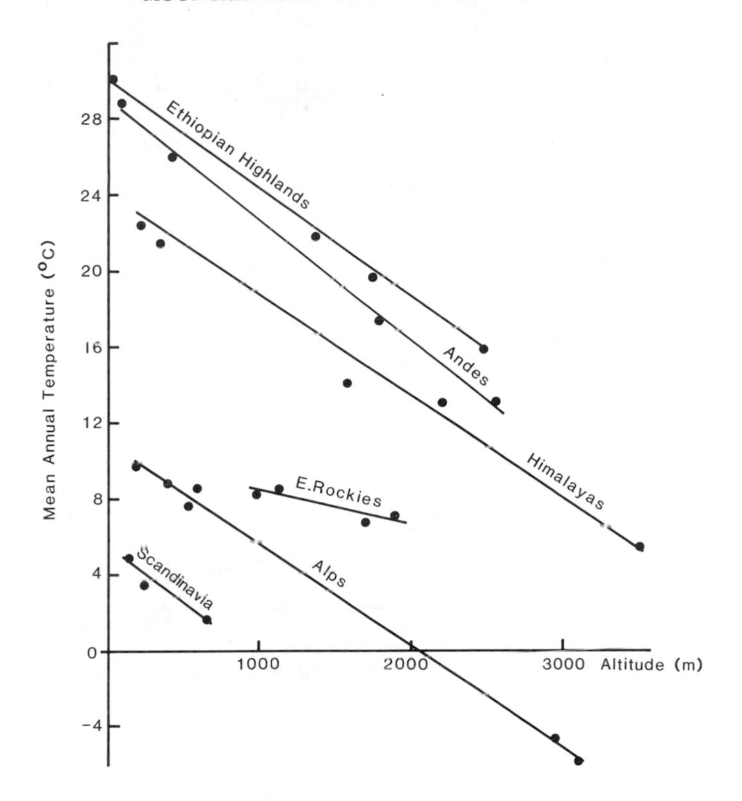

Fig. 1. Trends in mean annual temperature at meteorological stations located on six mountain ranges (from Müller, 1982).

considerable local variations in wind speed dependent on the radiation load, the occurrence of adjacent mountains and topographical variation.

The interaction between the lapse rate in temperature and vapour pressure, plus the characteristics of the lowland regional climate, leads to the complex altitudinal patterns of precipitation (Fig. 4). Associated with these interactions are variations in sunshine duration (Fig. 5). The Himalayas, with a lowland monsoon climate, and mountain ranges with a strong maritime influence such as Scandinavia or Scotland, show a decline in sunshine duration with altitude (Fig. 5; Müller, 1982). The drier continental ranges, such as the European Alps and the American Rockies, show little change in sunshine duration with altitude. Variation in cloudiness strongly influences the receipt of solar and long-wave radiation at ground level. Under both cloudless and cloudy skies, irradiance increases by about 10% km^{-1} (Barry, 1981), due to an altitudinal decline in atmospheric scattering and absorption of the solar beam, as atmospheric pressure and air density decrease.

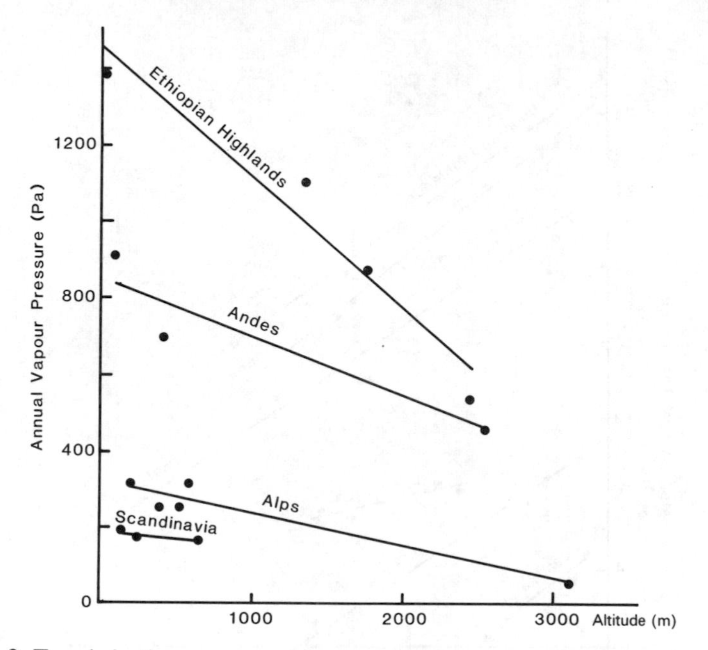

Fig. 2. Trends in the mean annual water vapour pressure (Müller, 1982).

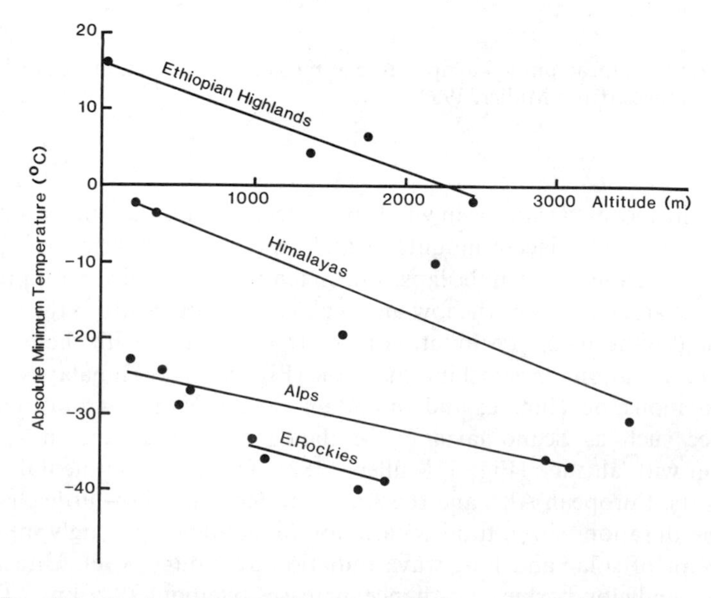

Fig. 3. Trends in the absolute minimum temperature (observed over 30 years) (Müller, 1982).

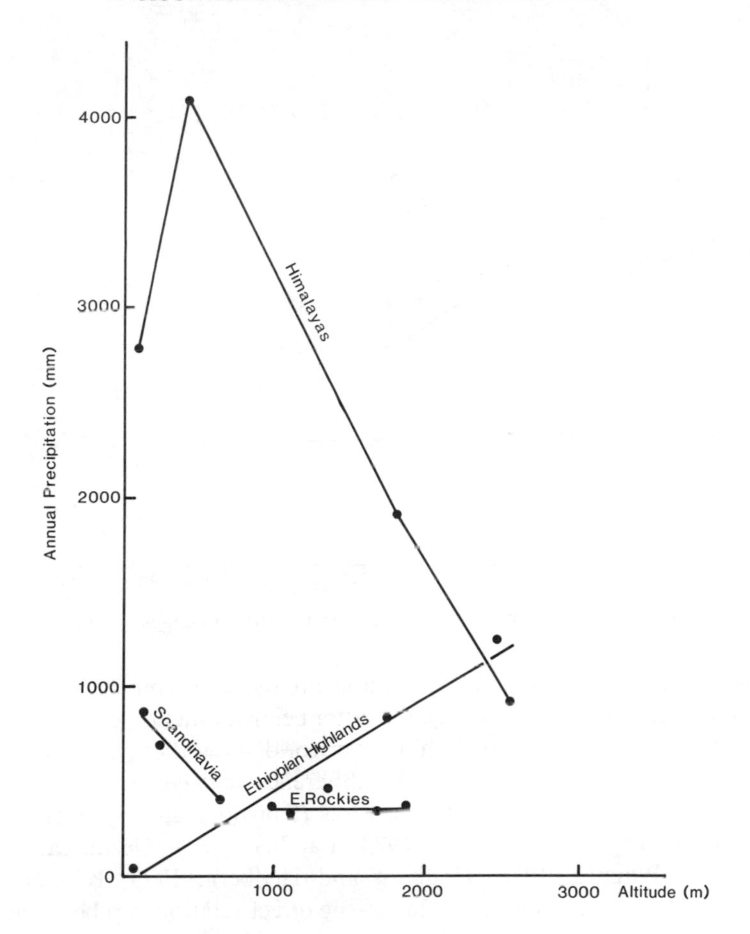

Fig. 4. Annual totals of precipitation along altitudinal gradients (Müller, 1982).

Associated with the increase in solar radiation is an increase in ultraviolet (UV) radiation with altitude. The rate of increase in UV is strongly dependent on cloud cover, but in the Austrian Alps, for example, the increase may be as great as 75% km^{-1} (Barry, 1981).

B. Plant Microclimates

Given the striking differences in some aspects of climate between different mountain ranges, it would be expected that vegetation structure and development also vary, negating any recognizable and repeatable altitudinal trends. However, a study of the range of vegetation heights with altitude (Fig. 6) indicates clear and repeatable reductions in height with altitude. The only

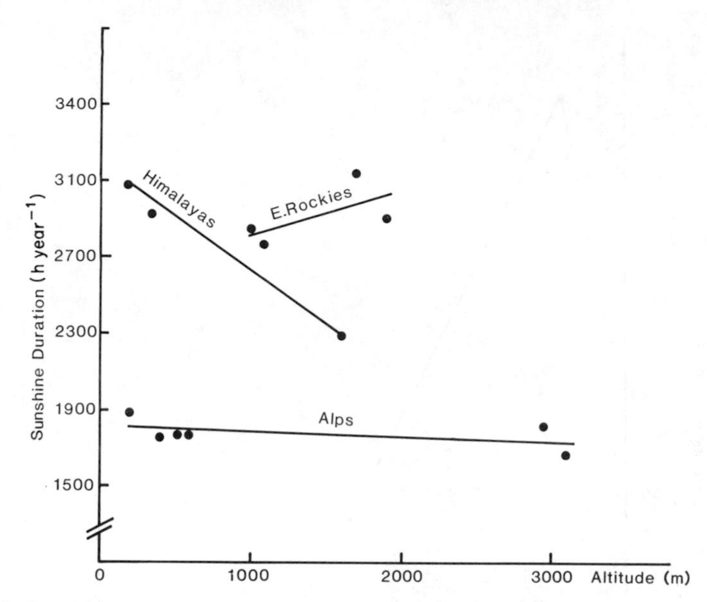

Fig. 5. Sunshine duration for three contrasting mountain ranges (Müller, 1982).

climatic trends which are equally repeatable are those of temperature (Figs 1 and 3) and vapour pressure (Fig. 2), the latter being a function of the former.

Many field observations throughout the world show that as plant height declines with altitude (Fig. 6), so the difference between leaf or canopy temperatures and air temperature increases (Salisbury and Spomer, 1964; Gates and Janke, 1966; Breckle, 1973; Larcher, 1975; Cernusca, 1976; Larcher and Wagner, 1976; Hedberg and Hedberg, 1979; Körner and Cochrane, 1983). This response is due to the direct relationship between the aerodynamic boundary layer conductance to sensible and latent heat transfer and plant height (Grace, 1977), and it can exert a strong influence on leaf temperature. For instance, Körner and Cochrane (1983) found that at an altitude of 2040 m (Australian Snowy Mountains) the leaves of small trees were at air temperature, but adjacent moss cushions were up to 30°C warmer than the air temperature.

Wilson *et al.* (1987) consider that the elevation of leaf and meristem temperatures above air temperatures is a crucial determinant of plant development on mountains. High leaf temperatures will occur only on days of high irradiance, but there will be no differences between leaf and air temperature on days of heavy cloud. Along an altitudinal gradient in the Cairngorm Mountains of Scotland, the lapse rate of air temperature, in sunny conditions, was observed to be 9°C km^{-1}, whereas the lapse rate of leaf temperature, from trees of *Pinus sylvestris* in the lowlands to dwarf-shrubs of *Arctostaphylos uva-ursi* and *Loiseleuria procumbens* in the uplands, was

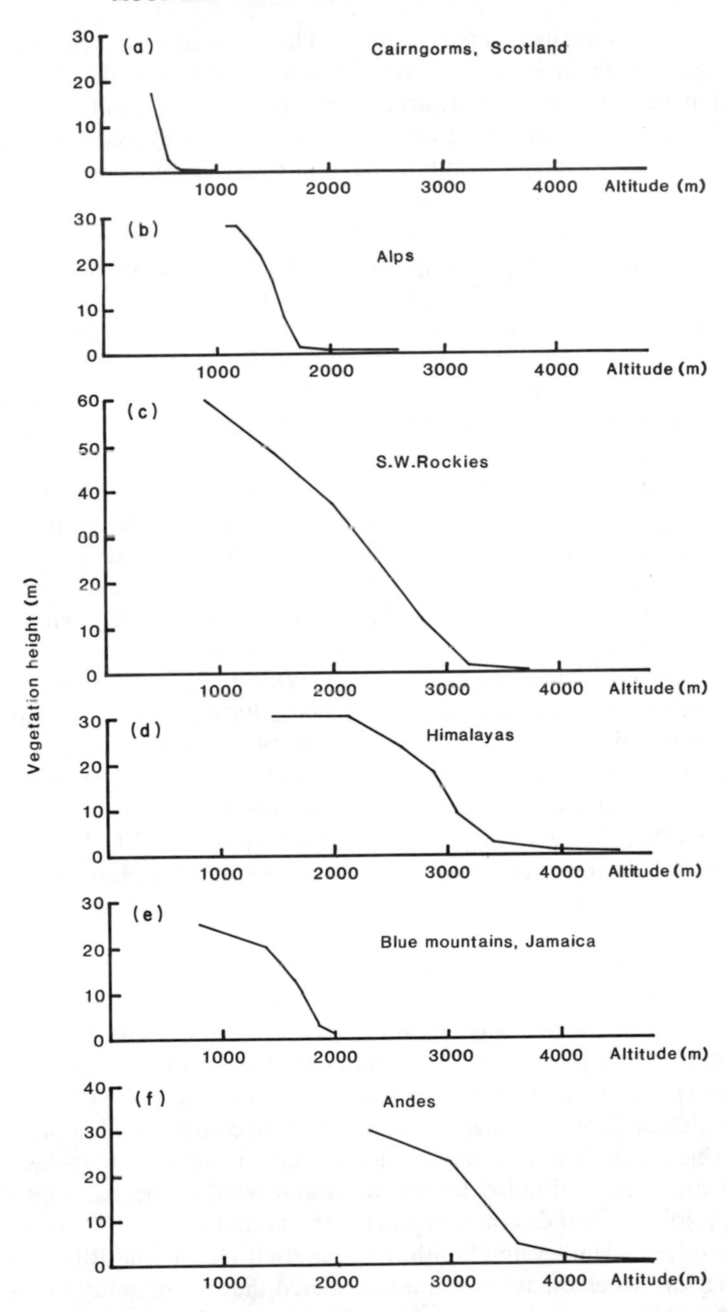

Fig. 6. Trends in vegetation height with altitude. (a) Personal observations; (b) from Cernusca (1976) and Tranquillini (1979); (c) from Barbour (1988); (d) from Osmaston (1922); (e) from J. Healey (pers. comm.); (f) from Beard (1955) and Vareschi (1970).

practically zero (Wilson *et al.*, 1987). This property of vegetation will maximize growth at high altitudes. Because height growth is genetically limited in many species, with dwarf ecotypes often occurring at high altitudes (e.g. Grant and Hunter, 1962; Grant and Mitton, 1977), vegetation cannot grow tall, which would permit higher boundary layer conductances and smaller leaf to air temperature differentials.

The temperature advantage of dwarf vegetation is only realized on days of high irradiance. On very cloudy days, leaf and air temperatures will be similar (Wilson *et al.*, 1987), and so the altitudinal gradient in cloudiness (approximately the inverse of sunshine hours: Fig. 5) will be a crude estimate of the temperature advantage. During nights of clear sky, dwarf vegetation will lose heat by radiation more rapidly than tall vegetation, so that low-temperature tolerance may be an increasing problem with increasing altitude (Larcher and Bauer, 1981; Rada *et al.*, 1985). Woodward *et al.* (1986) demonstrated that leaves of upland species in the Austrian Alps grew both day and night, and in both sunny and cloudy weather, with no measurable low-temperature threshold for growth. In contrast, lowland species, at the same altitude, grew more rapidly at high temperatures but possessed a noticeable low-temperature threshold for leaf growth (3–7°C), with little or no growth on very cloudy days or through the night.

The higher leaf temperatures in dwarf vegetation have a marked effect on photosynthetic rate. At high altitudes (2000–3400 m), for both dwarf trees (Carter and Smith, 1985; Hadley and Smith, 1987; Smith and Carter, 1988) and dwarf shrubs (Grabherr and Cernusca, 1977), the optimum temperature for leaf photosynthesis (T_o) at high irradiance ($> 600\ \mu mol\ m^{-2}\ s^{-1}$) is in the region of 15–25°C, whereas air temperature is often 10°C less than the optimum. Leaf temperatures reach the optimum range for photosynthesis on days of high irradiance, maximizing the photosynthetic rate. However, Körner (1982) has shown that one severe cost of the dwarf life form of the sedge *Carex curvula*, is that the upper-canopy leaves frequently shade lower leaves, reducing the actual photosynthetic rate.

The length of the growing season decreases with altitude (Mooney and Billings, 1961). Any process which enhances the growing season effectiveness, such as by leaf temperature excesses over air temperature (Wilson *et al.*, 1987), will therefore maximize development. One consequence of incomplete leaf development is that immature leaves have incomplete cuticles. These cuticles are readily abraded by the wind and wind-borne particles during winter, leading to leaf desiccation and death (Tranquillini, 1979; Barclay and Crawford, 1982; Hadley and Smith, 1983, 1986; Richards and Bliss, 1986). In the American Rockies, at least, it is considered that the inability of leaves to develop a thick cuticle because of the shortness of the growing season is a crucial factor in determining the treeline (Hadley and Smith, 1986). However, it is also clear that many stages of the life-cycle of a plant are strongly influenced by. the mountain microclimate, with temperature playing a dominant role. Such population effects include vegetative growth and survival (Smith, 1981; Wardle, 1985; Graves and Taylor, 1986), seed germination and seed bank longevity (Miller and Cummins, 1987; Guariguata and Azocar, 1988), seedling emergence and survival (Barclay and Crawford, 1984; Woodward and Jones, 1984), winter survival (Larcher and Bauer, 1981; Larcher, 1983; Woodward and Jones, 1984; Goldstein *et al.*, 1985; Rada *et al.*, 1985; Woodward, 1987a), the frequencies of flower visits by insect pollinators (Arroyo *et al.*, 1985) and seed production (Miller and Cummins, 1982; Young, 1984; Urbanska and Schütz, 1986).

Many processes in the life-cycle of a plant are, therefore, strongly influenced by microclimate, in a manner which exerts its effects on population processes (Woodward and Jones, 1984). Central to many of these processes is the capacity to develop and grow at low temperatures and reduced partial pressures of CO_2. These two features have therefore been selected for the central discussion of this chapter.

III. ECOPHYSIOLOGICAL RESPONSES TO THE MOUNTAIN AERIAL ENVIRONMENT

A. Historical Introduction

The recent use of portable infrared gas analysers (IRGAs) for the *in situ* measurement of photosynthesis, and the application of other sophisticated instruments and techniques, has enabled old ideas about mountain plant ecophysiology to be more thoroughly tested, and has generated new and exciting hypotheses. This chapter concentrates on the above-ground aspects of mountain plant ecology, in particular the interaction of leaf physiology and the environment.

Much of the early work on mountain plants was primarily concerned with the nature of species (e.g. Clausen *et al.*, 1940). In Section IV we briefly discuss evolutionary issues in the light of more recent ecophysiological work. These recent studies have given us a much deeper understanding of the changes in plant ecophysiology with altitude, but comparatively little attention has been given to how they might have arisen.

Mountain plant ecology can be traced back at least as far as the forest provenance work of von Wangenheim (1787, cited in Langlet, 1971). He stressed the importance of the climate at the sites from which seeds are collected, it being correlated with the reponse of the trees when grown under non-native conditions. He found that trees grown from seeds collected from northern sites in Canada grew better at high altitudes in Germany than plants from more southerly populations. For planting at lowland sites, he stated that seeds should be collected from sites of lower latitude. According to Langlet (1971), such knowledge of ecotypic differentiation with respect to altitude was known even prior to this by other German foresters working in mountain districts. Hooker (1853) reported how differences in frost resistance of pine and rhododendron populations were correlated with the altitude of their origin in the Himalayas. Hooker was obviously aware of ecotypic differentiation with respect to altitude in plant populations. Following this, Hoffmann (1886, cited in Langlet, 1971) demonstrated the existence of altitudinal ecotypes in herbs differing in time of flowering. Langlet (1971) cited much of the early work on altitudinal differentiation of genotypes.

Wagner (1892) found that non-woody plants growing at high altitudes in the European Alps, often exhibited structural and physiological modifications, compared to those of low altitudes, and suggested that these were due to the influence of climate on development. Bonnier (1895) showed that these responses could be induced in ramets of the same plant by growing them at different altitudes, in the same soil, for 2–4 decades. Among the structural modifications found by Bonnier to be caused by the high-altitude environment were greater relative below-ground growth, shorter above-ground parts, smaller and thicker leaves, more palisade tissue, more chloroplasts and more stomata (especially adaxial). These are all features commonly found when lowland and upland plants of the same species are compared (e.g. Wagner, 1892; Billings and Mooney, 1968). Bonnier saw these alterations of the plants by the environment as enabling the species to survive at both low and high altitudes. By measuring rates of photosynthesis, Bonnier concluded that at high altitudes the "feeble development" of the aerial parts is compensated for by increased photosynthetic rate (A), thus enabling complete development and carbohydrate reserve accumulation during the short growing season. He also noticed the development of certain "protective tissues" against the "climatic rigours of high altitude". Though he never actually measured any physical environmental variables, he concluded that the morphological and physiological changes were due primarily to the greater irradiance, drier air and lower temperatures at high, compared to low, altitudes. Overall, he implied that genetic differentiation is unnecessary for plant survival at high altitudes, at least in the 81 species that he studied. However, this plasticity may be genetically controlled, perhaps representing an adaptation enabling the species to exploit many habitats. Mountain plants may be very plastic and respond to the environment in particular ways.

The inheritance of acquired characteristics and the universality of the plastic viewpoint were questioned by the Swede, Göte Turesson (1922, 1925, 1927, 1930–31, 1931), the first worker to use the term genecology (Turesson, 1923). He showed that many plant species, which span a range of altitudes, are genetically differentiated in such a way that the high-altitude ecotypes display characteristics seemingly making them fitter under high-altitude conditions, and thus constituting adaptations to the alpine environment. Dobzhansky (1970) has advised that it may be best "to restrict the term ecotype to races that occur mosaic fashion in a quasi-continuously inhabited territory, wherever a certain type of environment (e.g. sand, clay, or calcareous soils) appears". Otherwise, he prefers Mayr's (1963) use of the term "race", such as climatic race. Turesson showed that alpine ecotypes are smaller and flower earlier than their lowland counterparts. Many alpine ecotypes, such as *Myosotis silvatica* Hoffm [*sic*] (Turesson, 1925), also had much thicker leaves. Although Kerner did not publish details of his experiments, he demonstrated genotypic divergence between populations growing at different altitudes, for features such as time of flowering (Langlet, 1971).

Clausen *et al.* (1940) demonstrated the importance of both genetic and environmental factors in the production of plant phenotypes. They discovered genetic races in *Achillea lanulosa* and *Potentilla glandulosa* from sea level, 1402 m and 3048 m in the Colorado (USA) mountains. The main features associated with the high-altitude races were genetically controlled dwarfism, greater frost resistance and phenological differences such as earliness of flowering. Dobzhansky (1970) and Briggs and Walters (1984) gave excellent summaries of this and other works on plant races and ecotypes.

Clements *et al.* (1950) made the most extensive study to that date of the effect of the environment at different altitudes on plant development. They made transplants of many species between three altitudes and assiduously recorded the responses of these plants to the natural environment, as well as to different light, water and nutrient treatments. They regarded the large part played by the environment in determining plant form as being overbearing evidence against the theory for the origin of species by means of natural selection, as proposed by Darwin in the earlier editions of his most famous work (e.g. Darwin, 1859). They went on to include the following statement in their summary: "Natural Selection does not operate upon the forms produced by adaptation [i.e. 'the complete response of plant and species . . . to the habitat complex'], since they are immediately in harmony with the environment that produces them."

Following on from the work of Clements *et al.*, *Oxyria digyna* became a favourite plant for the investigation of plant ecophysiology under arctic and alpine conditions in North America during the 1960s. This very wide-ranging species was shown to display ecotypic differentiation in photosynthetic rate with respect to altitude (Billings *et al.*, 1961). Plants with a high-altitude genotype showed greater photosynthetic rates than those with a low-altitude genotype at all CO_2 concentrations measured. No clear differentiation with respect to temperature was found. However, higher dark respiration rates were found by Mooney and Billings (1961) in alpine ecotypes, and this was thought to allow these plants to continue normal metabolism at low temperatures but limit their ability to thrive south of their distribution.

Three important reviews on mountain plant ecology were also published during the 1960s: Bliss (1962a), Billings and Mooney (1968) and Tranquillini (1964). The first two were about North American mountain ecology. They characterized the properties of angiosperms at high compared to low altitudes by the following features: relatively more root and rhizome reserves; limitation by drought at their upward limit; higher daily productivity during the growing season (see Scott and Billings, 1964); higher calorific and lipid content; growth often limited by soil nitrogen and phosphorus availability; rates of assimilation of nitrogen into organic compounds reduced by low temperatures; less chlorophyll (both on a fresh weight and a leaf area basis); lower temperature optimum for photosynthesis (T_o); greater rates of dark respiration (R_d); higher light levels required for maximal rates of photosynthesis (A_{max}); more fructans; and fewer seeds. The review by Tranquillini (1964) was European-based, being particularly concerned with the eastern Alps. Tranquillini stressed the lack of drought at high altitude, stating that drought becomes less likely with increased elevation. It was concluded that high-altitude plants were adapted photosynthetically to high irradiance, low temperature and low CO_2 concentrations. Soil and leaf temperatures may become very high; leaves have thick palisade layers, high dark respiration rates, with more storage of reserve carbohydrate for insurance against unproductive seasons, and adaptations to high irradiances of ultraviolet (UV-B) radiation.

The leaves of plants from high altitudes were thought to be capable of positive net photosynthesis at lower CO_2 concentrations, and lower temperatures, than those from lower altitudes (Tranquillini, 1964; Billings and Mooney, 1968). It was also suggested that high-altitude leaves achieved maximal photosynthetic rates at higher irradiances. Because CO_2 concentrations and temperature both generally fall with altitude (Fig. 1), and irradiance often increases (Fig. 5), these features might be considered as adaptations.

A tradition of ideas about the process by which plants are seemingly adapted to high altitudes has therefore been established. How do these ideas stand up in the light of subsequent work and what new aspects are of importance? Though the ecophysiology of mountain plants will be discussed under a number of different headings, these various features are all closely related to each other, particularly efficiency of CO_2 uptake (*ECU*), measured by the initial, linear slope of the A/p_i response (Fig. 7), and the rate of photosynthesis at ambient CO_2 concentrations under saturating irradiance (A_{max}). Irradiance, CO_2 concentrations and temperature all interact in a complex way with the biochemistry of photosynthesis. Because of this, a

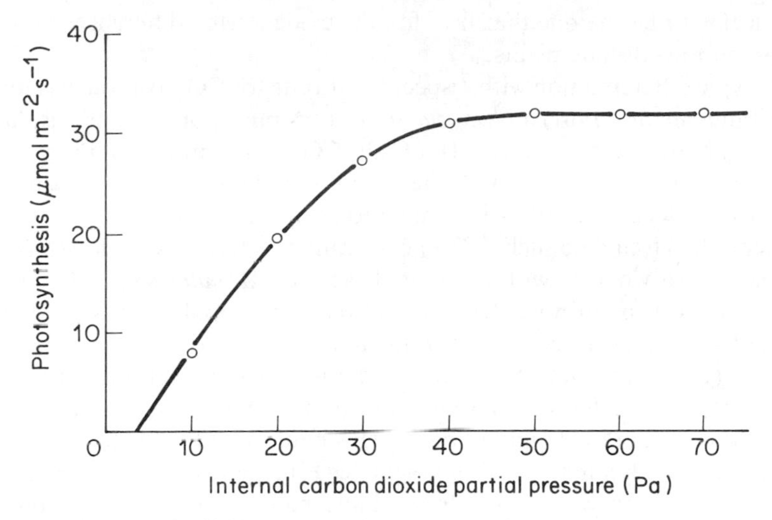

Fig. 7. Example of an A/p_i curve for a C_3 leaf.

reductionist approach is necessary to understand the measured responses of plants at the biochemical level.

B. Photosynthesis at Reduced Carbon Dioxide Levels

1. General

What is the evidence that leaves of plants growing at high altitudes are better at capturing CO_2 from partial pressures lower than the mean at sea-level (34 Pa), than are plants from low altitudes? This question is asked with the possibility of genetic adaptation to low CO_2 levels having occurred. An examination of the available evidence makes such a claim unlikely to be true, except in a very few cases. It seems that any increased ability of high-altitude plants to fix CO_2 from low concentrations could be explained by factors other than adaptation.

At saturating irradiance, the capacity to fix more CO_2 at a given concentration could be due to a greater absolute rate of photon capture, and/or a greater efficiency in converting these captured photons into chemical bond energy. The former could involve more chlorophyll molecules, or those present working more efficiently; the latter could involve greater efficiency of energy conversion to the carboxylase reaction, and/or an increased CO_2 partial pressure at the site of carboxylation, and/or more carboxylation sites per unit leaf area (Sharkey, 1985). This last possibility of more carboxylation sites seems to be the one that best fits the evidence for differences between high- and low-altitude plants.

Ecotypic differentiation with respect to altitude for CO_2 fixation was found by Billings *et al.* (1961) in *Oxyria digyna*. Alpine populations displayed greater photosynthetic rates (A) at all CO_2 concentrations than arctic populations from sea-level, with the alpine population showing positive A down to a CO_2 concentration equivalent to an altitude of 12 200 m. However, Decker (1959) found no such differences within the *Mimulus cardinalis-lewisii* complex. Also working with altitudinal races of *Mimulus* spp., Milner and Hiesey (1964a) found no differences between the populations at a slightly elevated CO_2 mole fraction (425 μmol mol^{-1}); however, at a higher mole fraction (1500 μmol mol^{-1}), the lowland races had higher rates than the upland races. In addition, the increase in A at the higher mole fraction declined most rapidly in upland races. This could well be due to the upland race being more limited by orthophosphate (Pi) regeneration rates (Leegood and Furbank, 1986; Sage *et al.*, 1988). If internal leaf CO_2 mole fraction (c_i), or partial pressure (p_i), limits photosynthesis more at high altitudes than at low, then it is expected that for leaves at high altitudes and saturating irradiance, ribulose-1,5-bisphosphate carboxylase/oxygenase (Rubisco) will be at maximum activation. This is because ribulose bisphosphate (RuP_2) or P-regeneration limitation will not be apparent under such conditions. The absolute concentration of Rubisco will be limiting A at the biochemical level (Sage *et al.*, 1988; Evans, 1989), and photosynthesis will be occurring on the linear part of the A/p_i response (see Fig. 7). It should be stated that a change in the ratio of photosynthesis to photorespiration will not be caused by the change in CO_2 partial pressure with altitude. This is because the partial pressure of CO_2 and O_2 will change at the same rate. However, changes in temperature with altitude will have an effect on this ratio (Brooks and Farquhar, 1985).

McNaughton *et al.* (1974) found that ecotypes of *Typha latifolia* from a range of altitudes, did not differ in their capacity to fix CO_2 at low mole fractions (below 300 μmol mol^{-1}). At higher CO_2 mole fractions (300–550 μmol mol^{-1}), however, the high-altitude ecotypes showed the highest values of A. They inferred higher concentrations of Rubisco in the upland ecotypes, but with identical kinetic properties to the lowland ecotypes. Mooney *et al.* (1964, 1966), working with a large number of species from various altitudes, found no clear relationship between altitude of origin and ability to fix CO_2 from low concentrations; the ability to fix CO_2 at high concentrations was not investigated. Again working with the *Mimulus* complex, Hiesey *et al.* (1971) made extensive investigations into the eco-physiology of *M. cardinalis* and *M. lewisii*, concentrating on comparisons between plants from different altitudes. They grew *M. cardinalis* clones from different altitudes under controlled growth conditions and found that, under increased CO_2 mole fractions (i.e. 1250 μmol mol^{-1} *vs* 300 μmol mol^{-1}), lowland clones responded more favourably, in terms of dry weight accumulation and stem height, than those from high altitudes. The clones from higher altitudes grew faster than the lowland ones at 300 μmol mol^{-1}. McMahon and Bogorad (1966) have shown that Rubisco in the leaves of the upland clones of this species from the same sites used by Hiesey *et al.* (1971) had a lower Michaelis constant ($K_m(CO_2)$) than in those from lower altitudes. In this species it appears that genetic differentiation with respect to altitude has occurred for the ability to fix CO_2 at reduced concentrations, though this was not found in earlier work. Increased CO_2 concentrations caused leaves of all races to be thicker, narrower and shorter, especially in the upland clones.

2. *Stomatal Responses and Efficiency of Carbon Uptake* (ECU)

In order to investigate the increases in stomatal density with altitude found by workers such as Wagner (1892) and Bonnier (1895), it is necessary to examine the literature not directly concerned with altitude. Madsen (1973a, b) investigated the response of young tomato plants to increased CO_2 levels: both dry matter and leaf thickness increased as CO_2 was increased from ambient (350 μmol mol^{-1}) up to 2200 μmol mol^{-1}. In contrast, stomatal number and epidermal cell density decreased. Madsen (1973b) attributed the decrease in stomatal density to increased expansion due to greater carbohydrate concentrations, lower osmotic potentials and higher rates of water flow into enlarging cells. Woodward (1986, 1987b) found that increasing the mole fraction of CO_2 from 200 μmol mol^{-1} caused the stomatal density of leaves in *Vaccinium myrtillus*, *Acer pseudoplatanus*, *Quercus robur*, *Rhamnus catharticus* and *Rumex crispus* to fall, up to a threshold value of about 340 μmol mol^{-1}, beyond which no further change was observed. Subsequently, Woodward and Bazzaz (1988) have shown that CO_2 partial pressure induces a similar response, particularly below ambient partial pressure (from 22.5 to 34 Pa), and that at different atmospheric pressures stomatal density responds to CO_2 partial pressure and not mole fraction. An interpretation of this change in stomatal density may be implicated from the review by Mott *et al.* (1982) on the general significance of amphistomaty. They concluded that a high photosynthetic rate per unit leaf area is often correlated with a high stomatal conductance. Hence stomatal density may be partly controlled by the relationship between CO_2 partial pressure (p_i) and *ECU* for a given leaf. For a developing leaf with a given *ECU*, there will be an optimum p_i for the mature leaf under full irradiance. Hence if p_a (ambient CO_2 partial pressure) is altered, it may be that this is sensed by the leaf and the stomatal density adjusted accordingly. There is, however, no conclusive evidence for such a mechanism. Jarvis and Morison (1981) found that "the evidence for a functional dependence of [stomatal conductance] ... on mesophyll photosynthesis is unconvincing". They pointed out that though stomatal conductance may be independent of the absolute value of p_i, it is sometimes dependent on the CO_2 concentration around the guard cells.

This capacity to control stomatal density and optimise p_i is an extension of the conclusions of Field *et al.* (1983). They found that during photosynthesis, the steady state CO_2 concentration in the leaf air spaces is maintained at a nearly constant value for a given species. In addition, Wong *et al.* (1979) found that in *Gossypium hirsutum*, stomatal conductance is a compromise between water conservation and the maintenance of internal CO_2 concentration at the optimum level for the intrinsic mesophyll capacity; increasing CO_2 concentration caused stomatal conductance to fall. It is possible that stomatal density is somehow controlled during leaf development by the amount of a substance directly related to the balance between mesophyll carboxylation capacity and internal CO_2 concentration. A similar system may operate to control stomatal aperture once the leaf is mature; Wong *et al.* (1979) suggested ATP. Such a mechanism could explain the results of Woodward (1986, 1987b) and Woodward and Bazzaz (1988). However, stomatal density itself may not uniquely determine conductance because, as stomatal density increases, the size of the guard cells often falls; thus the total pore area per unit leaf area may remain almost constant, and consequently stomatal density is only rarely correlated with conductance (Jones, 1987). More work is clearly necessary to establish the causes and physiological implications of changes in stomatal density.

3. *Low Temperature Effects*

Körner and Diemer (1987) investigated *in situ* the responses of various plant species to different partial pressures of CO_2 and irradiances at 600 and 2600 m in the Austrian Alps. Species occurring at 2600 m had both greater efficiency of carbon uptake (i.e. a higher *ECU*, based on the initial A/p_i slope) and greater values of A at CO_2 saturation. They attributed these differences to a greater mesophyll area per unit of leaf area (R^{mes}), and nitrogen content per unit of leaf area. The greater R^{mes} was due to thicker leaves, with a greater volume of palisade mesophyll. Because of the differences in *ECU*, mean A at ambient CO_2 partial pressure did not change with altitude, even though atmospheric and CO_2 partial pressure changed by 21% between the two altitudes. Because these measurements were made on material *in situ*, it is not possible to state whether the genotype, or the environment, or both, gave rise to these anatomical and physiological differences.

To understand these findings fully, it is necessary to examine the general literature on the physiology and biochemistry of C_3 photosynthesis. Farquhar *et al.* (1980) proposed that *ECU* is largely controlled by Rubisco activity per unit leaf area, and that A at saturating CO_2 and irradiance reflects limitations in the regeneration of ribulose bisphosphate (RuP_2) (see also Evans, 1989). Thus Rubisco activity and RuP_2 regeneration rates determine the form of A/p_i curves. Chabot *et al.* (1972) found that Rubisco activity per unit fresh weight of leaf, and per unit weight of protein, increased dramatically in an alpine *Oxyria digyna* population, when grown in a low-temperature regime compared to one at higher temperatures. An arctic population did not show any change. Bunce (1986) noted that leaves which developed at low temperatures had higher photosynthetic capacities (rates of photosynthesis under optimal conditions) than those grown at higher temperatures. This is because the rates of leaf expansion are reduced more by low temperatures than is the manufacture of the photosynthetic system components ("photosynthetic machinery"). Leaves formed at low temperatures will therefore contain greater amounts of photosynthetic machinery per unit area. Also, Chabot *et al.* (1972) found a 58% increase in leaf protein per unit weight in low-temperature grown alpine *O. digyna* compared to high; however, an arctic population showed only a 5% increase. Bunce (1986) suggested that the low temperature effect will only be important for A if stomatal conductance (g_s) also increases. This is possible, given the observations by Wong *et al.* (1979) of the importance of maintaining internal CO_2 partial pressure at the intrinsic mesophyll capacity. Indeed, it seems that the amount of CO_2 within a leaf is often maintained at a point close to the inflexion of the leaf's A/p_i response curve (Long and Hallgren, 1985). This is the point at which the photosynthetic system becomes limited by the regeneration of RuP_2 (Farquhar *et al.*, 1980). Evans (1989) proposed that nitrogen within a leaf is distributed between the Calvin cycle and thylakoid proteins so as to cause a co-limitation to photosynthesis by carboxylation and RuP_2 regeneration (except at high irradiances). Such an optimal distribution would tend to maintain photosynthesis and p_i at the inflexion point.

Any change in *ECU* with altitude might therefore be explained by low temperatures during leaf development, causing an increased mesophyll capacity for CO_2 fixation, with maximum conductance (g_{max}) also increasing. Raven and Glidewell (1981) indicated that the amount of Rubisco per unit leaf area will not affect *ECU* unless R^{mes} also increases. Also, Sharkey (1985) described how mesophyll conductance (conductance of CO_2 through the cell to the sites of carboxylation, per unit leaf area) is higher than stomatal conductance, so that a marked increase in photosynthesis occurs only if stomatal conductance also increases. Körner and Diemer (1987) found only a small increase in g_{max} with altitude, less than the increase in g_{mes}, leading to an increase in the stomatal limitations of photosynthesis. It is probable that g_{max} is controlled by other factors as well as the mesophyll conductance/p_i balance. A caveat with regard to the ideas of Bunce (1986) is that growth at low temperatures does not necessarily lead to an increase in A_{max}, as found for arctic plants (Körner and Diemer, 1987). Interestingly, Körner and Diemer (1987) found that plant species which have a wide altitudinal distribution did not have higher *ECU*s at high altitude. It would be interesting to know the growth rates of these species at different altitudes, and how they differ from specifically upland species under high-altitude conditions.

Some caution should be used in the interpretation of A/p_i relationships. Under conditions of water limitation, leaves possess a non-uniform distribution of open stomata, leading to an overestimation of p_i (Terashima *et al.*, 1988). In such a situation, *ECU* will be underestimated due to an apparent non-stomatal inhibition of *A* (Downton *et al.*, 1988). The generality of leaf patchiness in conductance is unknown but an interesting subject.

4. *Leaf to Air CO_2 Ratios*

If a plant at high altitude has a higher *ECU* than one at low altitude, and/or conductance increasingly limits photosynthesis, the ratio of CO_2 partial pressure inside the leaf air spaces (p_i) to that outside (p_a) will decrease with altitude. Farquhar *et al.* (1982) and Downton *et al.* (1985) showed that the time integral of the p_i/p_a ratio can be obtained for a leaf by measuring the relative proportions of the stable isotopes ^{13}C to ^{12}C. The ratio of these two isotopes is affected by diffusion through the stomata and by the enzyme Rubisco. The isotopic ratio, calculated relative to the Pee Dee Belemnite (PDB) standard, is known as $\delta^{13}C$. Because of its higher mass, $^{13}CO_2$ does not diffuse into leaves as quickly as $^{12}CO_2$, and seemingly binds less readily to Rubisco. Thus plant carbohydrates contain lower ratios of ^{13}C to ^{12}C (i.e. lower values of $\delta^{13}C$) than observed for air, because of these discriminations. The scale of $\delta^{13}C$ is set arbitrarily at zero for the standard. Any substance containing a lower ratio of ^{13}C to ^{12}C than this standard will have a negative $\delta^{13}C$ value. Most values for C_3 plants range from -30 to $-22‰$ (Troughton *et al.*, 1974) and only C_3 species will be considered in this review. $\delta^{13}C$ values are believed to measure the average p_i/p_a ratio for the life of a leaf by the following equation:

$$\delta^{13}C = \delta^{13}C_a - \frac{a(p_a - p_i)}{p_a} - \frac{b(p_i)}{p_a} \tag{2}$$

where $\delta^{13}C$ and $\delta^{13}C_a$ are the ^{13}C to ^{12}C ratios in the leaf and air ($-7{\cdot}8‰$) respectively, and a and b are thought to be 4·4 and 27‰ respectively for C_3 plants (Farquhar *et al.*, 1982).

From this equation it can be seen that if average p_i/p_a falls with altitude, the $^{13}C/^{12}C$ ratio of the leaf would be expected to increase. This is what was indeed found by Körner *et al.* (1988), who measured the $\delta^{13}C$ of 100 C_3 species from different altitudes around the world. They found slopes for the regression lines of 0·78 ($P < 0{\cdot}0001$) for forbs, 0·42 ($P = 0{\cdot}009$) for shrubs and 0·88‰ km^{-1} ($P = 0{\cdot}002$) for trees. It was suggested that the lower slope and significance for shrubs may be due to the effects of intra-canopy carbon recycling. Water stress is also known to cause increased $^{13}C/^{12}C$ ratios, but Körner *et al.* (1988) chose plants from sites reputed not to suffer from drought.

Other workers have also examined the relationship between $\delta^{13}C$ ratios and altitude. Vitousek *et al.* (1988) found that $\delta^{13}C$ increased with altitude in the dominant tree species *Metrosideros polymorpha* on a tropical volcano, and that $\delta^{13}C$ was negatively correlated with *SLA*, but offered no explanation. Friend *et al.* (1989) investigated the relationship between $^{13}C/^{12}C$ ratios and altitude in *Vaccinium myrtillus* and *Nardus stricta* for two growing seasons in Scotland. They found that the slope of the relationship between carbon isotope ratios and altitude varied from positive to near zero, dependent on season, and concluded that, even in the Scottish climate, changes in water relations could account for the differences between the years.

There has been some argument as to whether plants at high altitude really do experience reduced availability of CO_2. Verduin (1953) proposed that an alpine plant at 3500 m would experience CO_2 rarefied by a factor of 1·7 compared with one at sea-level, but that because this will also be true for all the other molecular species, the volume % of CO_2 (mole fraction) will be very similar. The product of concentration and diffusivity proved to be very similar at all altitudes; it was thus concluded that the measured high rates of photosynthesis at high altitude are not as surprising as they might first appear. Similarly, in a theoretical paper, Gale (1973) concluded that plants growing at different altitudes will not experience the extremes of CO_2 supply expected directly from the observations of partial pressures. This was explained in similar terms to Verduin with regard to the diffusion coefficient of CO_2 in air, which increases as atmospheric pressure decreases (i.e. altitude increases). This effect to a large degree compensates for the fall in partial pressure. This is also the case for water vapour, with a potential for increasing transpiration rates at reduced atmospheric pressure. Gale (1973) concluded that the reduction in ambient CO_2 partial pressure with altitude will only have a minor effect on plants. However, the diffusion coefficients decrease with temperature, and so when temperature falls with altitude (as is most frequently the case), the changes in CO_2 partial pressure will become important. Therefore, changes in CO_2 availability to plants with altitude will be dependent on the temperature of the air and the leaf. These issues were debated in a series of letters (Cooper, 1986; Gale, 1986; La Marche *et al.*, 1986) following the publication of a paper purporting to show that subalpine conifers growing at high altitudes have responded more to the increased atmospheric CO_2 levels since 1850 than those from lower altitudes (La Marche *et al.*, 1984). If Gale's (1973) conclusions are valid, then the results of

Billings *et al.* (1961) and Hiesey *et al.* (1971) are incongruous given that no selection for increased ability to fix CO_2 at low partial pressure would have been possible.

Overall, however, plants growing at high altitudes do frequently have higher rates of photosynthesis at lower CO_2 concentrations than plants at low altitudes. Körner and Diemer (1987) found differences in *ECU* between plants at different altitudes, but these differences may have been plastic, rather than genetic. Only Billings *et al.* (1961) and Hiesey *et al.* (1971) found ecotypic differentiation between altitudinal clones for this feature. It seems possible that observations of changes in *ECU* with altitude, as well as in stomatal density and g_{max}, may be explained by developmental responses to the environmental conditions, in particular low temperatures, as shown by Bonnier (1895), Clements *et al.* (1950) and Chabot *et al.* (1972). Controlled environment and reciprocal transplant studies are necessary to separate genetic from environmental influences on the ability of plants to fix CO_2 at low concentrations.

C. A_{max}: The Maximum Rate of Photosynthesis at Saturating Irradiance and Ambient CO_2 Concentration

1. General

A_{max} is the rate of photosynthesis, measured at saturating irradiance, occurring on an A/p_i curve where p_i is that partial pressure occurring within the leaf under ambient CO_2 conditions. Thus any differences in *ECU*, and/or CO_2-saturated rates of photosynthesis, will result in differences in A_{max}. Any mechanism responsible for differences in A/p_i curves between plants at different altitudes, as discussed in the previous section, might also explain differences in A_{max}.

What evidence is there that leaves of plants growing at high altitudes have higher irradiance-saturated rates of photosynthesis (A_{max})? Because irradiance often increases with altitude, at least on sunny days (Barry, 1981, pp. 29–32), one might expect the leaves of plants at high altitudes to have evolved towards a morphology closer to the classic "sun leaf" than those of lower altitudes, enabling full utilization of the high levels of irradiance present during clear summer days. A sun leaf, relative to a shade leaf, is thicker, of lower area, and it has a lower volume of intercellular air space, a higher proportion of palisade cells, longer palisade cells, greater R^{mes} (the ratio of mesophyll to leaf surface area), more and smaller chloroplasts per cell, greater stomatal density and a thicker cuticle (Nobel and Walker, 1985). There are many reports in the literature of leaf thickness increasing with altitude (e.g. Tranquillini, 1964; Hiesey *et al.*, 1971; Woodward, 1979a;

Körner and Cochrane, 1985; Körner *et al.*, 1986; Körner and Diemer, 1987; Körner and Renhardt, 1987). Thicker leaves generally have a lower specific leaf area (SLA), more nitrogen per unit area and a higher A_{max} (e.g. Mooney *et al.*, 1978). Increased leaf thickness can be caused by high irradiance levels during development (see, e.g. Mott *et al.*, 1982) or low temperatures (Wardlaw *et al.*, 1983), and so it may prove possible to find a causal explanation for increased leaf thickness in these factors. Because thicker leaves are likely to have an increased R^{mes}, then they will probably also have a higher A_{max} (Charles-Edwards and Ludwig, 1975; Raven and Glidewell, 1981; Nobel, 1983). However, there are instances in which A_{max} has not been correlated with leaf thickness, such as found by Patton and Jones (1989).

Rabinowitch (1951, p. 997) reported that the highest rates of photosynthesis (with very high quantum yields) found under natural conditions were for alpine plants, in particular those measured by Henrici (1918), Blagowesthchenskij (1935) and Kjär (1937). Verduin (1953) and Zalensky (1954) have made similar claims. Clebsch (1960) found that alpine *Trisetum spicatum* populations had higher A_{max} values than arctic populations. Mooney and Billings (1961) found the same for *Oxyria digyna*, the leaves from the alpine plants possessing a greater internal leaf surface area (Au, 1969). It was suggested by Tranquillini (1964) that plants growing at high altitudes were extreme sun plants able to make good use of high irradiances, and Glagoleva (1962, 1963, cited in Tranquillini, 1964) demonstrated that this may be due to the thickness of the palisade tissue. Milner and Hiesey (1964b) found that light-saturated levels of photosynthesis for six races of *Mimulus cardinalis* increased with their native elevation. These results are particularly interesting as they demonstrated ecotypic differentiation for A_{max} between altitudes. However, Mooney and Johnson (1965) found that alpine clones of *Thalictrum alpinum* had lower A_{max} values and genetically thinner leaves than arctic clones. Also, Godfrey (1969) showed that high-altitude populations of *Geum turbinatum* and *O. digyna* had lower photosynthetic rates than lower-altitude populations at several irradiances.

Hiesey *et al.* (1971) described how upland clones of *M. cardinalis* produced thicker leaves which had higher A_{max} values and irradiance absorptivity than their lowland counterparts. Grabherr (1977) suggested that A_{max} was related to chlorophyll content in *Loiseleuria procumbens*, with two peaks, one in the spring and a higher one in summer. Further evidence for ecotypic differentiation of A_{max} was provided by Mächler and Nösberger (1977), who noted how high-altitude clones of *Trifolium repens* had higher photosynthetic rates than lowland clones when grown under controlled conditions. It should be borne in mind that A_{max} occurs at a higher irradiance level for the canopy than it does for an individual leaf, owing to self-shading of the leaves in the canopy, as found by Grabherr and Cernusca (1977) for *L. procumbens*.

2. *Effects of Leaf Morphology*

Tieszen (1978) found a negative correlation between A_{max} and *SLA* for Alaskan tundra species, while Mooney *et al.* (1978) described a negative correlation between *SLA* and water use efficiency (*WUE*: carbon gained per unit water lost), and a positive correlation between *SLA* and nitrogen use efficiency (*NUE*: carbon gained per unit leaf nitrogen) in *Eucalyptus* spp. Correlations occurred between *A*, *SLA* and leaf nitrogen contents per unit leaf area. Species from drier habitats tended to produce smaller, thicker leaves, with a higher content of leaf nitrogen per unit area, lower *SLA*s, and correspondingly higher levels of A_{max}. This form of leaf, they concluded, must have a higher *WUE*, but lower *NUE*, than the larger and thinner leaves produced by species from wetter habitats. Because leaves are frequently thicker at higher altitudes, one might thus expect them to have higher *WUE*s under identical water vapour pressure deficits. It may also be that any increase in *WUE* is counteracted by an increase in conductance due to the reduced ambient CO_2 concentration. Because the *WUE* of leaves possibly increases with altitude, an increase in potential conductance may be of high selective advantage.

Also of relevance to the relationship between *SLA* and A_{max} is the work of Ledig and Korbobo (1983), in which *Acer saccharum* seedlings were grown from seeds collected along an altitudinal gradient. They found that the population from the highest altitude displayed the highest values of *A*, but, curiously from what has been said above, also had the highest *SLA*s. The mid-altitude population displayed lowest *A* rates but lowest *SLA*s. This pattern was found to repeat itself along another transect. No other data were given which might suggest a possible explanation for these unusual results. The authors did suggest that the differences in *A* could be caused by differences in stomatal density, pore size, mesophyll resistance or carboxylation capacity, but did not provide any data in support of these claims.

It is thought that g_{max} may be an important limitation to A_{max} (Sharkey, 1985). Körner *et al.* (1979) investigated g_{max} for many C_3 vascular plants, and found that plants with a high g_{max} tend to have a high A_{max}, but that a given leaf would not benefit greatly from increasing g_{max} because of the much more important role played by mesophyll conductance. This correlation between g_{max} and A_{max} concurs with the observations by Mott *et al.* (1982) of greater stomatal densities (especially amphistomaty) occurring in leaves with greater values of A_{max}, when developed under high irradiance, as a high mesophyll conductance usually indicates a high carboxylation capacity (e.g. Nobel and Walker, 1985). Körner (1982) highlighted the ability of alpine plants to make full use of high irradiance levels. In a *Carex curvula* canopy, he found that irradiance is by far the major factor limiting photosynthesis over a whole growing season, this limitation being due to clouds, fog and mutual leaf shading. Mott *et al.* (1982) showed that thicker leaves will not necessarily have increased g_{max} unless high irradiance levels also occur during development. It appears that a high A_{max} is only realized by a leaf with a high mesophyll conductance *and* high irradiance levels during development. Thus we might expect to find different relationships between altitude, stomatal density and A_{max} depending on the changes in irradiance with altitude. This idea is supported by the data in Körner *et al.* (1983), which show stomatal density decreasing with altitude in New Guinea, where the climate is maritime and irradiance falls with altitude.

Körner *et al.* (1983) discussed the significance of changes in stomatal density with altitude. As already noted, the most general trend is for an increase in the adaxial density (e.g. Wagner, 1892; Berger-Landefeldt, 1936; Au, 1969; Scheel, 1979, cited in Körner and Mayr, 1981; Körner and Mayr, 1981; Woodward, 1986). This may enhance diffusion, an important feature if the leaves are also thicker (Mott *et al.*, 1982). Körner *et al.* (1983) concluded that as features such as stomatal density, stomatal index and g_{max} do not always change in the same manner with altitude, they must in part be related to specific local environmental factors (such as water availability and/or irradiance levels), and not entirely due to (i.e. selected, or caused by) environmental features which change in the same direction on all mountains, such as temperature and the partial pressure of gases. Increased irradiance frequently has a substantial effect on adaxial stomatal density in dicotyledons (Mott *et al.*, 1982). Bonnier (1895) found increased stomatal density in lowland genotypes when grown at high altitudes, indicating both the plasticity of plants for this feature, and that there were probably increased levels of irradiance and/or reduced growth rates. Clements *et al.* (1950) also found that both increased irradiance and altitude were correlated with increased stomatal density in a wide range of species transplanted between altitudes, and subjected to different shade treatments. The increase in density with altitude was not thought to be due to the increased irradiance levels (irradiance was not found to change with altitude), but rather to reduced leaf expansion at the lower temperatures. However, "[in] some cases, the inverse relation did not hold between growth and number, indicating a more deep-seated modification of the meristem". When transported to the same altitude, plants grown at higher altitudes had equal or greater rates of transpiration than those grown at lower altitudes, reflecting the increased stomatal densities. It may be that both irradiance levels and leaf temperature during development are of importance in determining stomatal density (Körner *et al.*, 1986), as may be CO_2 concentration (see Section III.C.1).

3. Soil Nutrient Effects

Field (1983) showed that A_{max} can be limited by levels of enzymes such as Rubisco, and Field *et al.* (1983) demonstrated a positive relationship between total leaf organic nitrogen content and A_{max} between species. Because levels of leaf nitrogen and enzymes such as Rubisco are closely correlated (e.g. Huffaker, 1982, p. 387; Field, 1983; Evans, 1989; see also Chapin *et al.*, 1980), it is to be expected that leaf nitrogen content and A_{max} will be highly correlated. There is good evidence that leaf nitrogen content generally increases with altitude and latitude (Ehrhardt, 1961; Babb and Whitfield, 1977; Chapin *et al.*, 1980; Nordmeyer, 1980; Haselwandter *et al.*, 1983; Körner and Cochrane, 1985; Körner *et al.*, 1986; Körner and Diemer, 1987; Körner and Renhardt, 1987). This seems surprising, because soil nitrogen availability might be expected to decrease with the altitudinal (and latitudinal) decline in soil temperature, and with the increase in soil moisture, reducing mineralization and nutrient cycling rates (Grubb, 1971; Marrs *et al.*, 1988; Vitousek *et al.*, 1988). Grubb (1989) cited evidence that the growth of montane tropical forest is limited by nitrogen supply, due to decreased rates of organic matter decomposition. However, the overall decline in productivity with altitude is also directly attributable to decreased irradiance and lower temperatures (Grubb, 1989). Vitousek *et al.* (1988) found that concentration of foliar N, P, Mg and K, on a leaf weight basis, declined with altitude in *Metrosideros polymorpha*. However, concentrations per unit leaf area fell less, because *SLA* declined with altitude. Körner and Cochrane (1985), studying *Eucalyptus pauciflora* along an elevational gradient in the Snowy Mountains of south-eastern Australia, found that leaf thickness, number of palisade layers and leaf nitrogen per unit leaf area all increased with altitude. Körner *et al.* (1986) found the same pattern in New Zealand, with a positive correlation between leaf nitrogen and g_{max}, suggesting that high photosynthetic rates are possible in these high-altitude plants. In addition, Körner and Diemer (1987) found that mean total nitrogen per unit leaf area in plants growing at high altitudes was 34% greater compared to those growing at low altitudes. *ECU* in these plants correlated strongly with leaf nitrogen concentration.

Holzmann and Haselwandter (1988) discovered high concentrations of nitrogen in both roots and shoots of plants in an alpine sedge community. Rehder and Schäfer (1978) found that in three alpine dwarf shrub communities of the central Alps, there was greater productivity than expected from the concentrations of mineralizable N in the soil. They concluded that mycorrhizal fungi must contribute to nitrogen uptake, or that some form of root exudation occurs to stimulate N-mineralizing organisms in the rhizosphere. As might be expected, Holzmann and Haselwandter (1988) found that low soil temperatures largely inhibited biological nitrogen fixation in an alpine sedge community. The nitrogen supply per unit ground area by fixation, mineralization and precipitation did not appear to be sufficient to support the measured growth. However, the plants studied were mostly infected with vesicular-arbuscular (VA) mycorrhizas, as are most plants. This, together with evidence for movement of N to the roots for storage towards the end of the growing season for release in the spring, greater relative root length at higher altitudes for absorbing and storage (see also Körner and Renhardt, 1987), and the possibility of large inputs of nitrogen following snow melt (Haselwandter *et al.*, 1983), may help to explain the enigma. Also, it has been found that some commonly occurring ectomycorrhizal fungi can utilize organic nitrogen in peptides and proteins (Abuzinadah and Read, 1986; Abuzinadah *et al.*, 1986), but it is not known how important this may be for alpine plants, most of which are VA mycorrhizal.

As already pointed out, Holzmann and Haselwandter (1988) found no evidence of nitrogen deficiency in the 12 species they examined. Recalling Bunce's (1986) suggestion that low-temperature leaf development can cause increased values of A_{max}, it may be that low temperatures at high altitudes reduce the growth rates of the plant more than the supply of N. Therefore, the concentration of plant nitrogen will increase with altitude, a possible explanation for some of the observed increases in A_{max}. Similarly, the concentration of phosphorus may also increase. Chapin (1980) concluded that plants from infertile habitats may be characterized by the maintenance of high nutrient concentrations as a result of reduced growth rates.

4. *The Importance of Growth Temperature*

To conclude this discussion of increased A_{max} with altitude, it is significant that Bunce (1983) described the dependence of A_{max} not only on irradiance levels during development (as suggested by Chabot *et al.*, 1979), but also on the leaf expansion rate, with increasing expansion causing a dilution of the photosynthetic components. The expansion rate is closely linked to temperature (see below). Bunce (1983) demonstrated that when A_{max} increased with irradiance during development, the increase in the R^{mes} ratio could not on its own explain the increase in A_{max}. It appears that absolute concentrations of critical enzymes such as Rubisco might be involved, or even be of primary importance. Raven and Glidewell (1981), in a review of the literature, concluded that Rubisco activity (i.e. the amount of Rubisco in an activated form per unit leaf area) plays an important role in determining net CO_2 fixation, and that the activity is intimately related to leaf anatomy, especially the R^{mes} ratio. The leaf temperature during development is important for determining A at a given temperature, and hence g_{max}.

D. T_o: The Optimum Temperature for Photosynthesis

1. *General*

As temperature is reduced, A_{max} occurs at lower irradiance levels (e.g. Pisek *et al.*, 1973) and quantum efficiency may fall (Billings, 1974a; Berry and

Björkman, 1980). Wilson (1966) has suggested that this reduced efficiency may be due to a reduction in demand for sugars resulting in product inhibition. There is also evidence that the phosphate status of the chloroplast may be critical for rates of photosynthesis at low temperatures. Such an increase in phosphate limitation may cause a decrease in Rubisco activity, thus altering *ECU* and A_{max}, though Rubisco activity may change in direct response to temperature (Chabot *et al.*, 1972; Labate and Leegood, 1988). If a given plant has a higher photosynthetic rate at a low temperature than another plant, then this may be an adaptation to low temperatures, caused by the same physiological changes which influence A_{max}. If an upland plant has a greater A_{max} at low temperatures than a lowland plant, we may speak of differentiation with respect to altitude for the optimum temperature for photosynthesis (i.e. T_o). Evidence for this and its possible physiological explanation will now be discussed.

2. *Evidence for Genotypic Differentiation*

Björkman *et al.* (1960) found that alpine ecotypes of *Solidago virgaurea* from Sweden had lower T_o values than lowland ecotypes. The alpine ecotypes also had higher rates of dark respiration (Björkman and Holmgren, 1961). Similarly, Mooney *et al.* (1964) found that species from high altitudes in the White Mountains of California, when grown under controlled conditions, displayed lower values of T_o than species from lower altitudes. *Eriogonum ovalifolium* from 3947 m had maximal photosynthetic rates at around 20°C, whereas for *Hymenoclea salsola* from 1372 m they were around 30°C. The photosynthetic rate of lowland ecotypes of *Mimulus cardinalis* increased more rapidly with temperature than upland ecotypes (Milner and Hiesey, 1964b). Tieszen and Helgager (1968) found that an alpine ecotype of *Deschampsia cespitosa* had a higher T_o value for the Hill reaction than an arctic ecotype. Also, alpine populations of *Thalictrum alpinum* have been shown to exhibit higher T_o values than arctic ones (Mooney and Johnson, 1965), but alpine populations of *Oxyria digyna* had lower T_o values than arctic populations, by exhibiting greater acclimation potential (Billings *et al.*, 1971). There are many reports of such acclimation of T_o to growth temperature, for example: *Chamaebatiaria millefolium*, *Artemisia tridentata* and *Haplopappus apargioides* (Mooney and West, 1964); *Erigeron clokeyi*, *H. apargioides*, *C. millefolium* and *Encelia californica* (Mooney *et al.*, 1966); *D. cespitosa* (Pearcy, 1969); *Arenaria nuttallii* (Chabot and Billings, 1972); *Eucalyptus pauciflora* (Slatyer, 1977); and *Geum urbanum* and *G. rivale* (Graves and Taylor, 1988). However, the mechanisms leading to this dependency have not been elucidated.

Even though Hiesey *et al.* (1971) measured minimal reductions in leaf temperature with altitude, they found that growth of *M. cardinalis* clones showed genetic differentiation with respect to temperature for dry weight accumulation. The clones from sea-level had a significantly greater increase in dry weight between 10 and 20°C than clones from higher altitudes. Woodward (1975) reached similar conclusions in comparisons of growth at different temperatures between the lowland *Sedum telephium* and the upland *S. rosea*. This differential growth response to temperature was thought to be responsible for changes in the outcome of competition experiments at different altitudes (Woodward and Pigott, 1975), In addition, Woodward (1979b) found that the growth of the lowland species *Dactylis glomerata* and *Phleum bertolonii* responded much more to an increase in temperature, from 10 to 20°C, than did the growth of the upland species *Sesleria albicans* and *P. alpinum*. However, it is not clear whether these differences were due to differences in T_o or to differences in the temperature response of leaf extension rate, or both (the latter subject is discussed in Section III.D.3).

T_o decreased with the altitude of origin of populations of *Abies balsamea*, when grown from seed under uniform environmental conditions. The change with altitude in T_o was similar to the adiabatic lapse rate, suggesting adaptation (Fryer and Ledig, 1972). Similarly, Slatyer (1977) found that under controlled environmental conditions, T_o for populations of *E. pauciflora* fell as the altitude of the seed source increased. Probable adaptation was also reported by Tranquillini (1979), who found that T_o values for spruce (*Picea abies*) from the timberline were 3°C lower than for spruce from the valley floor, irrespective of irradiance levels. Similar results were reported for larch (*Larix decidua*) and birch (*Betula verrucosa*; *B. pendula*). Thus there is substantial evidence for the evolution of T_o in response to altitude. It should be noted that the optimum temperature for photosynthesis of the whole canopy may be lower than for individual leaves, as has been found for a *Loiseleuria procumbens* canopy (Grabherr and Cernusca, 1977). This may in part be explained by greater self-shading of leaves together with the drop in T_o with a decline in irradiance (e.g. Pisek *et al.*, 1973, p. 113).

Pisek *et al.* (1973) and Billings (1974a) have both written reviews on the T_o values of alpine plants. In the former, it was reported that T_o values of 24°C occurred for *Citrus limon* from 80 m, 19°C for *Taxus baccata* from 550 m, 14°C for *Betula pendula* from 1900 m, and 12·5°C for *O. digyna* from 2500 m. Vegetative propagation of clones of *Trifolium repens* from different altitudes in Switzerland has shown the presence of high-altitude ecotypes which display higher rates of photosynthesis at all temperatures, particularly so at low temperatures (Mächler *et al.*, 1977). It is possible that photosynthetic adaptation to low temperatures may also involve high rates of photorespiration when warm, leading to a drop in net CO_2 fixation. This was confirmed by Neama (1982), who found that growing *Salix herbacea* at low temperatures caused increased rates of both dark respiration and photorespiration, when transferred to higher temperatures.

3. *Mechanisms*

In order to understand better the differences in T_o between plants we must examine the literature not specifically aimed at mountain plant ecology. Björkman and Badger (1979) found *Nerium oleander* to be very plastic for T_o when grown at different temperatures. Until this work, no insight had been made into the possible mechanism by which T_o changes in response to growth temperature (Björkman, 1973), despite its obvious widespread occurrence. Any theories in this connection will have important implications for ecotypic differentiation of T_o with altitude. Björkman and Badger investigated the differences between plants differing in T_o values. Differences in stomata, anatomy, amount of photosynthetic machinery per unit leaf area, electron transport capabilities and general enzyme levels could not explain the differences in photosynthetic rates. However, the maximum activities of the chloroplast enzymes Rubisco and fructose-1,6-bisphosphate phosphatase (Fru-P_2 phosphatase) differed substantially, but only Fru-P_2 phosphatase changed by an amount comparable to the changes in photosynthetic rates at suboptimal temperatures. Photosynthetic capacity of plants grown at 45°C, and then transferred to 20°C, was closely correlated to increases in the levels of Fru-P_2 phosphatase activity. Thus it may be that the levels of this enzyme are critical in photosynthetic acclimation, and adaptation, to low temperatures. Rubisco levels changed in a similar fashion, but not to such an extent. Rubisco does not appear to be particularly temperature-dependent at CO_2 partial pressures typical for C_3 plants. Photosynthetic adaptation to low temperatures may, therefore, lie in the metabolic pathway control mechanisms which regulate the metabolic pools in the Calvin cycle. Fru-P_2 phosphatase catalyses one of the key steps for regulation of the Calvin cycle: the hydrolysis of the phosphate from the C-1 atom of fructose-1,6-bisphosphate, to form fructose-6-phosphate (Portis *et al.*, 1977). It is possible that changes in the storage of phosphate in the vacuole and its release into the cytoplasm may play a significant role in the adaptation of photosynthesis to low temperatures (Labate and Leegood, 1988).

Of particular significance to T_o at different altitudes, Raven and Glidewell (1981) described how a number of workers have found a correlation between low-temperature growth and both net CO_2 fixation and Rubisco activity per unit leaf area. This is known as capacity adaptation and seems to be more dependent on absolute Rubisco amounts than any kinetic changes. Leaf thickness can also be increased by growth at low temperatures (e.g. Wardlaw *et al.*, 1983). Peet *et al.* (1977) found that both Rubisco activity and leaf thickness increased in *Phaseolus vulgaris* when grown at low temperatures, hence R^{mes} probably also increased. Raven and Glidewell (1981) suggested that increases in Rubisco/R^{mes} ratios are not required for increased A at low temperatures, an increase in R^{mes} being sufficient. Klikoff (1969) sampled a number of species from different altitudes and found that the low-temperature activities of isolated chloroplasts were positively correlated with altitude. Similar conclusions were reached by May and Villareal (1974). In Puma rye (*Secale cereale* L. cv. Puma), Rubisco itself changed in response to a low growth temperature by increasing its low-temperature stability (Huner and MacDowell, 1979a). These changes were shown to enhance the activity of Rubisco at low temperatures (Huner and MacDowell, 1979b).

4. *Significance of* T_o

In a previously mentioned study of CO_2 exchange in an alpine *Carex curvula* canopy, it was found that the optimal temperature for photosynthesis, at an altitude of 2310 m in the Austrian Alps, was 22·5°C, close to the mean leaf temperature on clear summer days (Körner, 1982). Thus leaf temperature, on days of high irradiance, does not severely limit photosynthesis in this alpine plant. Körner and Diemer (1987) reached similar conclusions for 8 lowland (600 m) and 11 upland (2600 m) species. At 4°C, the plants at 2600 m maintained 50% of A_{max}, whereas this was achieved at 8°C by the plants at 600 m.

Thus it might appear that low temperatures may not be of great direct significance to carbon gain per unit leaf area because of a combination of changes in T_o and the elevation of leaf temperature above air temperature (Körner and Larcher, 1988). However, it should be noted that the latter will only occur on days of high irradiances, when A is at maximum. On cloudy days, low irradiance and temperatures occur, making rates of A at low temperature crucial. The importance of elevation of leaf temperature above ambient will depend on how much CO_2 is fixed in these two types of climate. Local cloudiness will be very important; the Alps may be very different from Scotland or Scandinavia, for example (Fig. 5).

The effect of low temperatures on A during development can be considerable. In barley, for instance, cold hardening causes a doubling of A_{max} per unit chlorophyll, resulting in a 30% increase per unit leaf area (Sicher *et al.*, 1988). The mechanism whereby a plant developing in the cold has a higher rate of photosynthesis at low temperatures than a plant developing in the warm, is most probably a combination of growth being more reduced than photosynthesis during development (Bunce, 1986), resulting in an increased A_{max}, and some type of change in metabolic pool control as suggested above (Berry and Björkman, 1980). However, the general significance of these processes in the reduction of T_o values is not known.

E. Growth

Whereas low temperatures on mountains may, at least in some cases, be of little significance to carbon gain per unit leaf area, they are crucial to the expansion of leaf area, and hence for the productivity of the whole plant. Plant stature usually declines with altitude (Körner and Renhardt, 1987; Woodward, 1986; Fig. 6), though on some tropical mountains it may increase (e.g. Smith and Young, 1987). That high-altitude ecotypes can be genetically of lower stature has been known for some time (Zederbauer, 1908, cited in Langlet, 1971; Turesson, 1925; Clausen *et al.*, 1940). Plants taken from the lowlands and grown at high altitudes are of lower stature than when they are grown in the same soil at lower altitudes (Bonnier, 1895). Woodward (1986) grew populations of *Vaccinium myrtillus* in a controlled environment and found that the low stature of high-altitude populations is unchanging, and probably has some genetic control. He suggested that this resulted from selection by high wind speeds, and reduced competition for light. The reduction of vegetation stature with altitude may well be due to a combination of environmental and genetic factors. High wind speeds may lead to severe abrasion by ice particles and sand, and to drought during winter. Shorter plants may be protected by snow, except in very exposed sites where the lower boundary layer conductance, associated with the small stature, will reduce the impact of the wind. All these factors will give shorter plants a selective advantage as well as causing dwarfism in otherwise taller plants. In addition, summer growth and metabolism will occur at the higher temperatures experienced closer to the ground (e.g. Körner and Cochrane, 1983). Fitter and Hay (1987) give a full description of the possible selective advantages of dwarfism in arctic and alpine environments.

Ollerenshaw and Baker (1981), Chapin and Oechel (1983), Woodward *et al.* (1986) and Sakai and Larcher (1987), have all demonstrated that plant growth dynamics in response to temperature are genetically controlled and vary between latitudes and altitudes. The earlier work in this regard was discussed in the excellent review of genecology from an historical perspective by Langlet (1971). Cline and Agatep (1970) demonstrated that upland ecotypes of the *Achillea borealis-lanulosa* complex had greater growth at low temperatures than lowland ecotypes under the same conditions. There is also evidence that upland ecotypes and species do not respond to increased temperature as well as their lowland counterparts (e.g. Hiesey *et al.*, 1971; Woodward and Pigott, 1975; Woodward, 1975, 1979a, b).

In situ measurements of leaf growth have demonstrated that upland populations and species are able to grow at lower temperatures than lowland populations and species (Woodward *et al.*, 1986; Körner and Woodward, 1987). Woodward and Friend (1988) found the same for seed-grown material under controlled conditions. It was suggested that the ability to grow at low temperatures was related to adaptation in either, or both, the rate of cell production and the rate of cell expansion in the growing zone of the leaves. An examination of the factors involved in cell expansion suggested that the high-altitude species were able to grow at low temperatures by maintaining significant cell wall extensibility; this capacity was lost by lowland species. Cell wall extensibility is the degree to which the cell walls can irreversibly extend due to turgor pressure. The cell walls of plants have both plastic and elastic properties; it is their plastic properties that are most important to growth. The ability to maintain growth at low temperature was negatively correlated with the rate of growth at higher temperatures, which was more dependent on cell turgor potentials.

Of possible relevance to the decline in stature with altitude, Hiesey *et al.* (1971) found that the stem height of low-altitude clones of *Mimulus cardinalis* increased more with CO_2 mole fraction (from 175 to 1250 μmol mol^{-1}) than the upland clone. However, as photosynthetic rates do not appear to decline with altitude (e.g. Körner and Diemer, 1987; Friend *et al.*, 1989), at least on a leaf area basis, it is unlikely that changes in the partial pressure of CO_2 on mountains are responsible for any plant height differences.

The effect of altitude-related changes in the environment on plant height has attracted much attention in terms of the control of the Alpine treeline. Daubenmire (1954) wrote the first review on treeline research, and later Wardle (1971) concluded that treelines are caused by winter desiccation producing death in trees unable to mature fully because of low rates of production of lignin, cuticle and epicuticular wax. This was supported by Tranquillini (1979), who has suggested that the principal cause of the Alpine treeline is winter desiccation, due to high evaporative demand, low root permeability and frozen soil water. Evergreen conifers are especially prone, with the cuticle of the youngest needles unable to develop fully in one short growing season. Baig and Tranquillini (1976) found that the cuticle and cutinized cell wall thicknesses decreased with elevation for stone (Arolla) pine (*Pinus cembra*) and Norway spruce (*Picea abies*) in Austria, causing needle damage and death at high altitudes. This was also found by DeLucia and Berlyn (1984) for *Abies balsamea*, though the cutinized cell wall thickness increased dramatically at the level of the Krummholz trees, whereas elsewhere it decreased with altitude. Stomatal density also increased with altitude, though not significantly, and measured rates of stomatal and cuticular water losses were very high at high elevations. It was concluded that the treeline of *A. balsamea* is caused by desiccation, a conclusion similar to that reached by Hadley and Smith (1986). Grace (1989) pointed out the correlation between summer temperatures and the treeline. The faster reduction in the temperature of photosynthesizing leaves in a tree canopy compared to short vegetation ultimately causes trees to fail to grow and reproduce. This was thought to occur when the mean temperature of the warmest month is below about 10°C. However, it is still not clear whether one main factor, or a number of interacting factors, limits tree distribution at high altitudes.

Chapin (1983) concluded that the characteristic morphologies, metabolisms and life-histories of alpine and arctic plants are so well adapted to low temperatures that their distribution was limited by the secondary effects of low temperatures on water and nutrient supply. These areas are both very poorly understood for high altitudes and latitudes. Körner and Larcher (1988) suggested that the most important influence of low temperatures on plants is on growth and development, but the results of ecophysiological experiments and observations of plant life in cold climates will not be fully understood until more basic developmental physiology is known. They suggested that selection for short plants at high altitudes may be due to increased warming closer to the ground surface, and that intrinsic low growth rates may be due to an adaptation to long-term low resource availability (see Grime, 1979, p. 30 *et seq.*). Reduced competition at high altitudes may make selection for reduced stature possible (e.g. Körner and Diemer, 1987). Woodward (1979a) found that growing *Phleum alpinum* at low temperatures caused the leaves to be thicker with larger cells. Similarly, Körner and Larcher (1988) found that the smaller leaves produced at low temperatures did not have smaller cells, suggesting that a low rate of cell division was responsible for their size; alternatively, they suggested that low rates of cell expansion might reduce the rate of mitosis by some process of negative feedback.

Important relationships between growth rate, low temperature hardening and tolerance to low temperatures can be inferred from the findings of Pollock *et al.* (1988). They found that the growth rates of *Lolium perenne* were negatively correlated with freezing tolerance and accumulation of soluble carbohydrate. Eagles (1967a, b) came to similar conclusions with regard to *Dactylis glomerata*.

F. Pigments

Bonnier (1895) found that lowland plants, when grown at high altitudes, produced many more chloroplasts in each cell than in their natural habitat, each chloroplast being darker than at lower altitudes. However, there is much evidence that leaves of *in situ* alpine plants contain less chlorophyll on both a leaf area and fresh weight basis than either lowland or arctic leaves (Henrici, 1918; Seybold and Eagle, 1940; Mooney and Billings, 1961; Mooney and Johnson, 1965; Billings and Mooney, 1968; Godfrey, 1969; Tieszen, 1970). There is also evidence that the ratio of chlorophyll a to b is higher for alpine plants than arctic plants (Tieszen, 1970), a characteristic of sun leaves, which also tend to have a greater ratio of chlorophyll a to b than shade leaves (Björkman, 1981, p. 67). Billings and Mooney (1968) and Billings *et al.* (1971) found that the reduced level of chlorophyll in alpine *Oxyria digyna* plants was under partial genetic control, and suggested that this may have been selected for as a mechanism for reducing damage by UV irradiance. Tieszen and Helgager (1968) found that reduction of ferricyanide per unit chlorophyll was much greater for alpine than for near sea-level arctic clones of *Deschampsia cespitosa*. Similarly, Billings *et al.* (1971) found higher rates of photosynthesis per unit chlorophyll in alpine than in arctic ecotypes of *O. digyna*. Chlorophyll may also play a role in cold hardening. As mentioned with respect to T_o, Sicher *et al.* (1988) found that cold hardening in barley caused A_{max} per unit chlorophyll to double, and despite a simultaneous reduction in chlorophyll concentration the photosynthetic rate increased by 30%. Clearly, the photosynthetic impact of a reduction in chlorophyll with altitude may be small because of the ability of plants to increase R^{mes} and A per unit chlorophyll.

Leaves of alpine species frequently contain a high concentration of anthocyanins, causing them to appear very dark (Billings and Mooney, 1968; Klein, 1978). Fitter and Hay (1987) suggested that this may be due to selection for increased absorption of solar radiation, and therefore enhanced leaf temperature. Klein (1978) described how high concentrations of anthocyanins may have been selected for, along with flavenoids and epidermal waxes, because of their ability to reduce damage from UV radiation.

G. Dark Respiration

At the same temperature, alpine species and populations frequently exhibit higher mitochondrial respiration rates (R_d) than those from the lowlands (e.g. Björkman and Holmgren, 1961; Mooney, 1963; Tranquillini, 1964, 1979; Mooney *et al.*, 1964; Klikoff, 1966, 1968; Billings and Mooney, 1968; Billings *et al.*, 1971; Chabot and Billings, 1972; Ledig and Korbobo, 1983). These include examples of genetic differentiation for increased dark respiration at high altitudes. Klikoff (1969) found that the low-temperature activities of isolated mitochondria from a number of widely distributed species were positively correlated with the altitude from which the populations were isolated. High mitochondrial respiration rates are probably an advantage for plants in cold climates because of the short, cool growing season for metabolic activity (Stewart and Bannister, 1974; Klikoff, 1968). There is much evidence that rates of dark respiration in some plants are enhanced if the plants are grown at low temperatures (Billings and Godfrey, 1968; Chatterton *et al.*, 1970; Chabot and Billings, 1972; Neama, 1982), the species with the widest altitudinal range exhibiting the greatest change in dark respiration rates (Chabot and Billings, 1972). However, a high dark respiration rate in the cold will mean a very high rate when warm and a rapid exhaustion of carbohydrate reserves (Crawford and Palin, 1981), hence possibly limiting altitudinal (e.g. Dahl, 1951; Stewart and Bannister, 1974) or latitudinal (e.g. Mooney and Billings, 1961) distribution. Elevated rates of dark respiration enable cold-climate plants to operate with enhanced specific metabolic activities; they are probably a compensatory mechanism in response to low temperatures and will not cause excessive losses of carbon as long as the plants remain cold (Körner and Larcher, 1988).

The importance of dark respiration rates (R_d) for altitudinal distribution was highlighted by Graves and Taylor (1986, 1988). They found that the differential altitudinal distribution of the species *Geum rivale* (upland) and *G. urbanum* (lowland) could not be explained by a climatic limitation of relative growth rate (Graves and Taylor, 1986). They went on to show that the growth rate of *G. urbanum* roots was lower than that of *G. rivale* at low temperatures, and suggested that this may be due to lower rates of root respiration. *G. urbanum* was thought to be unable to use the excess assimilate it was capable of producing at low temperatures because of inherently low R_d, and hence was limited to lower, warmer altitudes (Graves and Taylor, 1988).

Because dark respiration occurs mainly at night, when leaf temperatures are closely tied to the air temperature, the ambient temperature for dark respiration will decrease with altitude. Thus the respiration rates of plants may be similar at all altitudes.

H. Photorespiration

Like dark respiration, photorespiration at high temperatures is enhanced by cold acclimation (e.g. Mächler and Nösberger, 1978; Neama, 1982). The ratio of CO_2 to O_2 remains approximately constant with altitude (Barry, 1981), and hence photorespiration changes will be influenced primarily by temperature (Brooks and Farquhar, 1985). Genotypic adaptation to low temperatures may restrict the distribution of plants in higher-temperature regions because of excessive photorespiration (Zetlitch and Day, 1973). High-altitude ecotypes of some species (e.g. *Trifolium repens*: Mächler *et al.*, 1977; Mächler and Nösberger, 1978) exhibit higher photorespiration at high temperatures than low-altitude ecotypes. The ratio of Rubisco to phosphoenolpyruvate carboxylase has also been observed to increase with altitude in the alpine plant *Selinum vaginatum* (Pandey *et al.*, 1984), reflecting the reduced advantage of C_4 metabolism as temperature is reduced.

J. Non-enzymic Biochemical Changes

Golley (1961) described a significant increase in the calorific content of plants from tropical to temperate to alpine environments. Bliss (1962b) suggested that this increase may result in part from the greater lipid content in alpine tundra plants. He further suggested that the high rates of R_d found in arctic and alpine plants may be provided by lipid-rich reserves as well as carbohydrates, and that the high R_d rates may account for rapid growth and development at low temperatures. The implications of these findings for estimates of primary productivity and yield were discussed by Billings and Mooney (1968). They described how alpine shrubs and herbs have higher calorific values than many tropical and temperate plants, especially in evergreen prostrate shrubs, because of their high lipid content. Zachhuber and Larcher (1978) found that the content of energetically expensive compounds increases with altitude in plant tissue. Baruch (1982) found an increase in calorific energy content of giant rosettes with altitude in Venezuela, and argued that this increase in energy content is consistent with the views of Grime (1979) on stress tolerators. Grime describes stress tolerators as plants with low growth rates and a capacity to store and conserve captured resources. These plants must be able to store energy (and nutrients) for environmentally unfavourable periods, in terms of growth, and hence allow rapid development when a brief favourable period arrives. However, McCown (1978) suggested that increased lipid content is not the result of increased storage but simply of rapid synthesis. Levitt (cited in Grill *et al.*, 1988) suggested that resistance to frost requires increased thiol contents, this being important for the stability of cell structure and frost resistance. Increased levels of thiols were found by Grill *et al.* (1988) in high-altitude grown Norway spruce (*Picea abies*). With regard to the storage and transport of metabolites, Berry and Raison (1981) highlighted how movement of assimilate in plants is highly sensitive to temperature. The importance of this at high altitudes is unexplored.

IV. MODELLING THE EFFECTS OF ALTITUDE ON PLANT GAS EXCHANGE

A. General

The discussion on the effects of altitude-related changes in CO_2 partial pressure and temperature on photosynthetic rate and stomatal conductance indicates considerable degrees of freedom in plant responses. The limited observations that have been obtained in the field (Körner and Diemer, 1987; Friend *et al.*, 1989) indicate that mesophyll and stomatal conductance increase with altitude, while A_{max} either increases or is unchanging. In this work, the vegetation was herbaceous at both low and high altitudes, with measurements of gas exchange usually made at constant leaf-to-air water vapour pressure deficit (*VPD*). However, stomatal conductance is sensitive to *VPD* (Lösch and Tenhunen, 1981), a feature which will influence *in situ* stomatal conductance and photosynthetic capacity. Despite this limitation in the range of environments for analysis, Körner and Diemer (1987) were able to state unequivocally that the increase in mesophyll conductance with altitude exceeded that observed for stomatal conductance. As a consequence, the ratio p_i/p_a decreased with altitude, indicating an increasing stomatal limitation of CO_2 uptake. This conclusion was strengthened by a global survey of $\delta^{13}C$ along altitudinal gradients (Körner *et al.*, 1988), which indicated that the time integrated measure of p_i/p_a, as extracted from the measurements of $\delta^{13}C$ ratios (Farquhar *et al.*, 1982), also decreased with altitude. However, the variation of the measurements of $\delta^{13}C$ were very large. This in part must reflect varying degrees of leaf shading, or perhaps stomatal closure due to high leaf temperatures or short-term drought. Both of these effects will influence $\delta^{13}C$ (Farquhar *et al.*, 1982).

Additional sources of variation which need to be considered are those due to variations in plant height and, therefore, the aerodynamic coupling between the vegetation and the air above. Tall vegetation will be well coupled, with a temperature close to air temperature, whereas dwarf vegetation may be less well coupled, with a temperature often very different from that of the air (McNaughton and Jarvis, 1983). These temperature variations will directly influence the photosynthetic rate (Grabherr and Cernusca, 1977; Carter and Smith, 1985). Stomatal conductance may also be influenced directly by temperature (Jarvis, 1976). In addition, the *VPD* of the leaves, and therefore stomatal conductance, will be strongly influenced by temperature and aerodynamic coupling.

The parallel findings of Körner and Diemer (1987), with *in situ* observations at constant *VPD*, and Körner *et al.* (1988), with *in situ* observations over a range of *VPD*s, encourage the view that these effects of temperature, aerodynamic coupling and plant stature are either unimportant or, perhaps, balance out and have little impact on the relation between CO_2 supply to, and fixation at, the chloroplast. The problem is that neither $\delta^{13}C$ nor the value of p_i/p_a indicate the absolute values of stomatal and mesophyll conductances, or rates of photosynthesis and transpiration. In view of the many aspects of microclimate which can influence these conductances and rates, it seems clear that a model which could identify the sensitivity of $\delta^{13}C$ to microclimate is an important development, for clarification and for considering the large variance shown by Körner *et al.* (1987).

A model of vegetation gas exchange and energy balance has therefore been developed with the specific aim of investigating the relationship between canopy height (Fig. 6), canopy energy balance, leaf area index (*LAI*), photosynthetic rate, stomatal and boundary layer conductances and $\delta^{13}C$.

B. Model Development

1. CO_2 Exchange

Central to the model are the data obtained by Körner and Diemer (1987) for A/p_i curves (Fig. 7, page 73), at saturating irradiance, of a very large selection of herbaceous species at altitudes of 600 and 2600 m. The averaged results from these observations are used in this model, no account being taken of expected differences in photosynthetic capacities between plants of different stature (e.g. trees *vs*. herbs). They indicate that the initial slope of the A/p_i curve (g_m) increases with altitude, as does the transition point from the Rubisco to the RuP_2 limitation of photosynthesis. However the A/c_i curve (from Körner and Diemer, 1987) is virtually unchanging with altitude, as is A_{max}, such that:

$$A = \frac{A_{max}}{(275 - c_{i,r})} \tag{3}$$

where $A_{max} = 18{\cdot}8\ \mu mol\ m^{-2}\ s^{-1}$ and $c_{i,r}$ is the CO_2 compensation point where $A = 0$. Bauer *et al.* (1983) have shown that the CO_2 compensation point is a conservative property of a wide range of species with C_3 metabolism.

However, it is temperature-sensitive according to the following relationship:

$$c_{i,r} = 23{\cdot}71 + (0{\cdot}695 \times T_c) + (0{\cdot}062 \times T_c^{\,2}) \tag{4}$$

where $c_{i,r}$ is measured in $\mu mol\ mol^{-1}$ and T_c is temperature (°C).

Körner and Diemer (1987) showed that the optimum temperature for photosynthesis decreases slightly, by 2·7°C over 2000 m of altitude. In addition, the photosynthetic response to temperature is quadratic, as has been shown elsewhere (Ehleringer and Björkman, 1978). The estimate of A from (3) must therefore be modified in terms of leaf temperature. The following quadratic equation includes, and assumes, a linear decrease in the optimum temperature for photosynthesis with altitude, plus the quadratic response to temperature from the data of Körner and Diemer (1987):

$$A_t = A \times 0{\cdot}01 \times [(-13{\cdot}52 \times (0{\cdot}0136 \times z)) + ((9{\cdot}128 - (0{\cdot}000645 \times z)) \times T_c) - ((0{\cdot}182 + (0{\cdot}0000015 \times z)) \times T_c^{\,2})] \tag{5}$$

where z is altitude (m) and A_t is the temperature-adjusted photosynthetic rate. Given an estimate of the stomatal and boundary layer conductances to CO_2 uptake (from eqns 17, 19, 21 and 34) it then becomes possible to estimate c_i:

$$c_i = \frac{((c_{i,r} \times g_m) + (c_a \times g))}{(g_m + g)} \tag{6}$$

where g_m is the mesophyll conductance ($\mu mol\ m^{-2}\ s^{-1}$), g is the combined stomatal and boundary layer conductance to CO_2, and c_a is the ambient CO_2 mole fraction ($345\ \mu mol\ mol^{-1}$). The mesophyll conductance is determined from the relationship:

$$g_m = \frac{A_t}{(0{\cdot}000275 - c_{i,r})} \tag{7}$$

The predicted photosynthetic rate, A_a, may then be calculated from:

$$A_a = g_m \times (c_i - c_{i,r}) \tag{8}$$

The intercellular mole fraction of CO_2, c_i, may be adjusted for the confounding effect of the opposing mass flow of water vapour through the stomata (estimated by the rate of transpiration ET, mol m^{-2} s^{-1}, which is in turn calculated from eqn 28) according to the equation (von Caemmerer and Farquhar, 1981):

$$c_i = \frac{(((g - ET/2) \times c_a) - P)}{g + ET/2} \tag{9}$$

The δ^{13}C of the leaf may then be predicted from the following equation (from Farquhar *et al.*, 1982):

$$\delta^{13}\mathrm{C} = -7{\cdot}8 - (4{\cdot}4 \times \frac{(c_a - c_i)}{c_a}) - 27 \times \frac{c_i}{c_a}) \tag{10}$$

2. *Energy Balance*

The radiant energy balance of the leaf, or canopy, determines leaf temperature. One aim of this model is to predict the effect of changing environmental conditions, at different altitudes, on some typical and initial stomatal conductance, g_o, to water vapour. For the simulations described here, g_o is equal to 275 mmol m^{-2} s^{-1}, a value typical of those observed by Körner and Diemer (1987) and Friend *et al.* (1989).

A number of physical properties in the energy balance equation are influenced by either altitude or temperature, or both. These are described below, using data provided in Monteith (1973), Jones (1983), Woodward and Sheehy (1983) and Landsberg (1986).

The change of pressure with altitude is described in (1). The change in the saturation vapour pressure (Pa) with temperature, e_{svp} is:

$$e_{svp} = 6{\cdot}108 \times \exp((17{\cdot}269 \times T_c)/(237{\cdot}3 + T_c)) \times 100 \tag{11}$$

The latent heat of vaporization (λ, J g^{-1}) changes with temperature as:

$$\lambda = 2500 - (2{\cdot}367 \times T_c) \tag{12}$$

The psychrometric constant (γ, Pa K^{-1}) changes with pressure as:

$$\gamma = \frac{P \times c_p}{0{\cdot}622 \times \lambda} \tag{13}$$

where c_p is the specific heat capacity of air ($1{\cdot}012\ \mathrm{J\,g^{-1}\,K^{-1}}$). The density of air ($\rho$, $\mathrm{g\,m^{-3}}$) is dependent on temperature and atmospheric pressure (P, Pa):

$$\rho = \frac{P}{101\,325} \times (1288{\cdot}4 - (4{\cdot}103 \times T_c)) \tag{14}$$

The diffusion coefficients of water vapour (D_h, $0{\cdot}0000242\ \mathrm{m^2\,s^{-1}}$ at N.T.P.) and CO_2 (D_c, $0{\cdot}0000147\ \mathrm{m^2\,s^{-1}}$ at N.T.P.) are sensitive to both temperature and pressure. These effects may be incorporated as:

$$D_a = D \times \frac{(T_k)^{1{\cdot}75}}{293} \times \frac{(101\,325)}{P} \tag{15}$$

3. *Sensible Heat Transfer*

Meteorological station data, or some given value of wind speed, are used to determine the boundary layer conductance to momentum ($g_{n,a}$, in non-molar units of $\mathrm{m\,s^{-1}}$) by vegetation differing in height (h, m), and leaf area index (LAI). It is assumed that wind speed (u, $\mathrm{m\,s^{-1}}$) is measured at 10 m (u_{10}) above the ground. This speed is adjusted to a reference height of 50 m (u_{50}), which exceeds the maximum canopy height used in the model simulations, by:

$$(u_{50}) = (1{\cdot}219 + (0{\cdot}038 \times h)) \times (u_{10}) \tag{16}$$

The boundary layer conductance of the vegetation is then calculated as follows – ignoring the effects of buoyancy and the radiative conductance (Jones, 1983); and assuming that the conductance to momentum is equal to the conductance for sensible heat:

$$g_{n,a} = \frac{0{\cdot}168 \times u_{50}}{(\log_e(50 - d))^2/z_0} \tag{17}$$

where d is the displacement height ($0{\cdot}7\ h$) and z_0 is the roughness length. The roughness length may be estimated as $0{\cdot}1\ h$, or the effects of plant spacing on z_0 may be approximated from LAI (this assumes that as LAI decreases plant spacing increases), in a simplified linear-log relationship of Garratt (1977) and Landsberg (1986). Roughness length may then be estimated as:

$$z_0 = \exp(-3{\cdot}51 + (0{\cdot}62 \times \log_e(LAI))) \times h \tag{18}$$

The boundary layer conductance and the stomatal conductance may be converted between molar units (as in eqns 6, 7, 8 and 9) and non-molar units (g_n, as in eqn 17) as follows:

$$g_n = g \times \frac{(R \times T_k)}{P} \quad (19)$$

where g has units of mol m^{-2} s^{-1}, R is the gas constant and T_k is temperature (K). Molar units are used here when predicting rates of photosynthesis, while non-molar units are used in the energy balance predictions.

4. *Canopy Transfer Characteristics*

The initial stomatal conductance, g_0, is likely to have a direct response to temperature, with a peak conductance at some optimum temperature. Observations of this nature are limited, but on the basis of Jarvis (1976) a quadratic relation between conductance and temperature has been assumed, with the same coefficients as for the photosynthetic response to temperature (eqn 5). This stomatal conductance is for a leaf, but given a knowledge of canopy *LAI*, the distribution of *LAI* into strata (*LAD*) within a canopy (Woodward, 1987a), and the penetration of solar radiation through the canopy, the canopy conductance g_c may be calculated as follows:

$$S_i = S_0 \times \exp(-0{\cdot}5 \times \sum LAD) \quad (20)$$

where S_0 is the incident irradiance (W m^{-2}), S_i the irradiance at stratum i within the canopy, $\sum LAD$ is the sum of the *LAD* from the top of the canopy to layer i and $-0{\cdot}5$ is a typical extinction coefficient for radiation within the canopy (Woodward, 1987).

The decline in irradiance through the canopy causes a decrease in stomatal conductance as (from Woodward, 1987a):

$$\frac{1}{g_i} = \frac{1}{g_0} + \frac{710}{S_i} \quad (21)$$

Canopy stomatal conductance is then estimated from:

$$g_c = \sum_{i=1}^{i=5} (g_i \times LAD_i) \quad (22)$$

where LAD_i is the *LAD* of layer i. The canopy has been divided into 5 strata (from Woodward, 1987a) and the following fractions of canopy *LAI* are found in the five layers: LAD_1 (top) 0·11; LAD_2 0·38; LAD_3 0·26; LAD_4 0·21; LAD_5 0·04.

The canopy boundary layer conductance is determined from the wind speed profile (eqn 17). This conductance needs to be apportioned to each stratum of the canopy, so that leaf photosynthesis (eqn 8) can be predicted for leaves in the top (or any) layer of the canopy. The heights of the canopy strata are arranged as follows: $LAD_1 = 0{\cdot}9\ h$; $LAD_2 = 0{\cdot}7\ h$; $LAD_3 = 0{\cdot}5\ h$; $LAD_4 = 0{\cdot}3\ h$; $LAD_5 = 0{\cdot}1\ h$. The fractional height f_i therefore has values from 0·9 to 0·1, moving from the top to the bottom of the canopy.

The wind speed at 50 m above ground surface has been estimated (eqn 16), as has the canopy boundary layer conductance (eqn 17). These measurements may then be used to calculate the eddy velocity, u_*:

$$u_* = \sqrt{u_{50}} \times g_{n,a} \tag{23}$$

The wind speed at the top of the canopy, u_h, may then be estimated from:

$$u_h = \frac{u_*}{0{\cdot}41} \times \log_e \frac{(h - d)}{z_0} \tag{24}$$

It is difficult to predict the wind speed within the canopy, but taking the lead of Landsberg (1986), wind speed at depth i in the canopy is estimated from:

$$u_i = u_h \times \exp(-3 \times (1 - f_i)) \tag{25}$$

The expected boundary layer conductance at a wind speed u_i is then calculated for each level of the canopy (from eqn 17). The inverse of these conductances (resistances) act in parallel for the loss of sensible and latent heat. The parallel sum of these resistances is then added and multiplied by a coefficient so that $1/g_{n,a}$ estimated from eqn 17 equals the parallel sum of these resistances. The boundary layer conductance of any layer i may then be extracted.

The conductances to water vapour (in molar units) are converted to the total canopy conductance to CO_2 by:

$$\frac{1}{g_{CO_2}} = \frac{(D_h/D_c)}{g_c} + \frac{1{\cdot}37}{g_a} \tag{26}$$

5. *Calculating the Energy Balance*

The canopy stomatal and boundary layer conductances are used in the equation of the energy balance of the canopy. In addition, the response of stomatal conductance to *VPD* and the aerodynamic coupling between the canopy and the climate above, as measured at a meteorological station, must also be incorporated. This is achieved using the model described by Choudhury and Monteith (1986). It has not been feasible to model the *VPD* profile through the canopy (McNaughton and Jarvis, 1983) and so the response of the total canopy conductance (g_c) to *VPD* has been modelled. Choudhury and Monteith (1986) use non-molar units in their models and this approach is carried on here, so that:

$$g_{n,c} = g_{n,0}\,(1 - \frac{VPD}{VPD_m}) \tag{27}$$

where *VPD* is the *VPD* of the air, as measured at a meteorological station and VPD_m is the *VPD* at which $g_{n,c}$ is zero, because of the stomatal response to *VPD* (Lösch and Tenhunen, 1981). There are few measurements of VPD_m, and so the value of 7000 Pa, as selected by Choudhury and Monteith (1986) for annual crops and deciduous trees, has been used.

The *VPD*-dependent value of $g_{n,c}$, and the canopy boundary layer conductance $g_{n,a}$, are incorporated in the Penman-Monteith equation (Monteith, 1981) to predict canopy transpiration (ET, W m^{-2}):

$$ET = \frac{s \times R_n + \rho \times C_p \times g_{n,a} \times VPD}{s + \gamma(1 + g_{n,a}/g_{n,c})} \tag{28}$$

where s is the change of saturation vapour pressure with temperature (Pa K^{-1}).

The temperature dependence of s is determined as:

$$s = 48{\cdot}7 \times \exp(0{\cdot}0532 \times T_c) \tag{29}$$

R_n is the net radiant balance of the canopy and is determined by iteration of the following two equations and eqn 28, by varying canopy temperature ($T_{k,c}$, in K):

$$R_n = L_d + (-0{\cdot}95 \times \mathbf{s} \times T_{k,c}^{\ 4}) + ((S_0 \times (1 - \exp(-0{\cdot}5 \times LAI))) - (S_0 \times (r \times (1 - \exp(-0{\cdot}5 \times LAI))))) \tag{30}$$

$$0{\cdot}96 \times R_n = C + ET \tag{31}$$

where L_d is the downward flux of long wave radiation (W m^{-2}) from the atmosphere (Monteith, 1973) calculated from the air temperature ($T_{c,a}$):

$$L_d = 208 + (6 \times T_{c,a}) \tag{32}$$

In eqn 30, *LAI* may either be the canopy leaf area index or a modification of the plant leaf area index (LAI_p). The modification may be incorporated when leaf area index is low and individual plants are somewhat isolated (a feature incorporated in eqn 18). The end result of this situation is that plant leaf area index will exceed the mean canopy leaf area index (Woodward, 1987a). In this situation, the equation $LAI_p = \exp(-0{\cdot}05 \times LAI)$ has been employed.

The constant 0·96 in eqn 31 allows for the flux of radiation into the soil beneath the plant canopy. The net radiant balance of the canopy calculates the input of longwave radiation (L_d), and the input of solar radiation (S_0), minus the fraction of solar radiation penetrating through the canopy ($S_0 \times (1 - \exp(-0{\cdot}5 \times LAI))$), which is dependent on leaf area index (LAI) and the extinction coefficient ($-0{\cdot}5$). In addition, the proportion of reflected solar radiation is calculated from $((S_0 \times (r \times (1 - \exp(-0{\cdot}5 \times LAI))))$, where r is the canopy albedo (typically 0·15). The upward flux of long wave radiation from the canopy is calculated using the Stephen-Boltzmann equation, where **s** is the Stephen-Boltzmann constant ($5{\cdot}67 \times 10^{-8}\ \mathrm{W\ m^{-2}\ K^{-4}}$).

In eqn 31, the radiant balance of the canopy is balanced by transpiration (eqn 28) and by sensible, or convective, heat loss (C, $\mathrm{W\ m^{-2}}$):

$$C = (\rho \times C_p(T_{k,c} - T_{k,a})) \times g_{n,a} \tag{33}$$

The Penman-Montieth equation (eqn 28) treats the plant canopy as a single layer of leaves, a treatment which excludes the necessity of calculating *VPD* profiles within the canopy. However, an important variable in eqn 28 is the *VPD* itself. The data input for *VPD* is the *VPD* of the air, but stomata respond to the *VPD* between the canopy surface temperature and the air, at some reference height. Choudhury and Monteith (1986) assume that the *VPD* of the air at the reference height is independent of the vegetation beneath. However, this may not be so, with differences occurring between aerodynamically rough and smooth surfaces (McNaughton and Jarvis, 1983). Due to uncertainty in this respect (Paw and Gao, 1988), and following Choudhury and Monteith (1986), this aspect of the model is not included. The approach of Choudhury and Monteith (1986) to estimate the "canopy *VPD*", i.e. the *VPD* that all of the stomata in the canopy respond to, VPD_c, has also been employed. The reference height *VPD* is retained in eqn 28 but the effect of *VPD* on stomatal conductance (eqn 27) is modified as:

$$g_{n,c} = g_{n,0}\frac{(1 - VPD_c)}{VPD_{\mathrm{m}}} \tag{34}$$

where VPD_c is estimated from:

$$VPD_c = \{\mathbf{b} - (\mathbf{b}^2 - 4 \times \mathbf{a} \times \mathbf{h})^{0{\cdot}5}\}/2 \times \mathbf{a} \tag{35}$$

where,

$$\mathbf{a} = (1 + s/\gamma)(g_{n,0}/(g_{n,a} \times VPD_m)) \tag{36}$$

and,

$$\mathbf{b} = 1 + (1 + s/\gamma)(g_{n,0}/g_{n,a}) \tag{37}$$

and,

$$\mathbf{h} = VPD + s \times R_n/(\rho \times C_p \times g_{n,a}) \tag{38}$$

C. Model Simulations

The model described above has been applied using climatic data for the Austrian Alps (from Müller, 1982; Körner and Diemer, 1987). The responses of vegetation, ranging in height from 0·031 to 32 m, to the climatic gradient over an altitudinal range of 3000 m, have been investigated. The mean sea-level climate and the lapse rates of the selected and various climatic variables are shown in Table 1.

The responses of stomatal conductance, photosynthetic rate and $\delta^{13}C$ of leaves in the top layer of the canopy and the mean canopy temperature have been investigated. Fig. 8 shows the results of a simulation which assumes that *LAI* decreases with altitude, according to the relation (from data presented by Körner and Mayr, 1981):

$$LAI = 7 \times \exp(-0{\cdot}001056 \times z) \tag{39}$$

where z is altitude (m). Over this altitudinal range, plant height decreases (Fig. 6) from 25 m at an altitude of 800 m, to 0·05 m at 2500 m.

From Fig. 8, it may be seen that canopy temperatures will increasingly

Table 1
Climatic conditions for model simulation

Mean temperature at 0 m	21·9°C
Temperature lapse rate	$-$ 6·5°C km^{-1}
Relative humidity at 0 m	64%
Lapse rate in relative humidity	2·7% km^{-1}
Irradiance at 0 m	800 W m^{-2}
Lapse rate in irradiance	80 W m^{-2} km^{-1}
Wind speed at 0 m	4 m s^{-1}
Lapse rate of wind speed	1 m s^{-1} km^{-1}

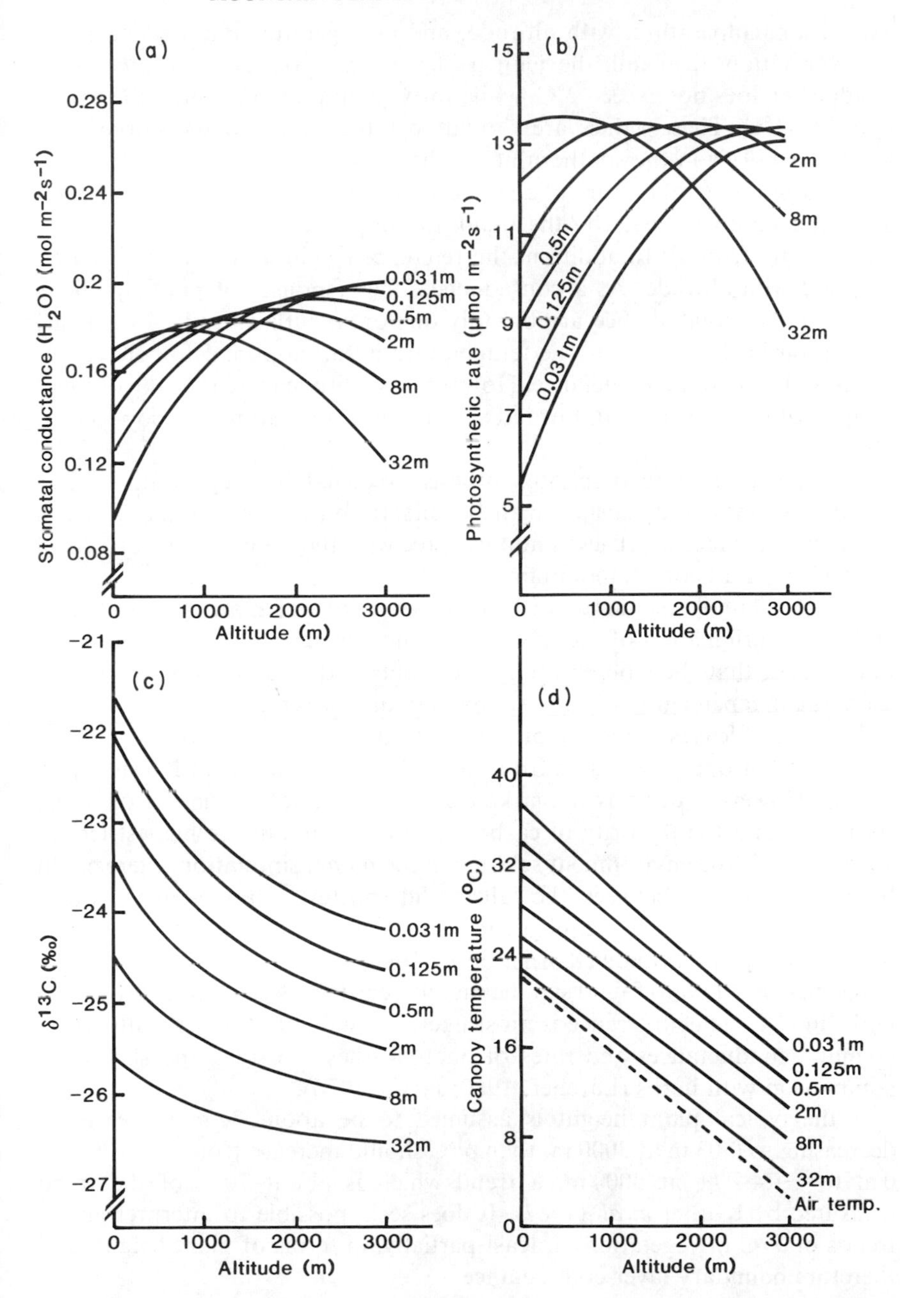

Fig. 8. Model simulations of the responses of (a) stomatal conductance, (b) photosynthetic rate, (c) δ^{13}C and (d) canopy temperature, to the changes in climate with altitude and to vegetation height.

exceed air temperature, with altitude, and as vegetation stature decreases. For vegetation 32 m tall, the temperature excess (for the maximum daily irradiance) does not exceed 2°C, while for vegetation 0·03 m high the excess may be 16°C. These values are similar to those observed by Körner and Cochrane (1983) in the southern hemisphere.

The absolute temperature of the vegetation influences stomatal conductance and photosynthesis by the quadratic responses of these processes to temperature (eqn 5). In addition, the reference height and "canopy" *VPD*s decrease with altitude. As a consequence, the responses of photosynthesis and stomatal conductance are broadly quadratic with altitude. For dwarf vegetation in the lowlands, the temperature is supraoptimal for photosynthesis and stomatal conductance. However, as altitude increases, the optimal temperature is approached. For tall vegetation, temperature is suboptimal at altitude.

The maximum rate of photosynthesis predicted for vegetation of any stature remains about constant with altitude. In contrast, maximum stomatal conductance increases. These findings agree with the observations of Friend *et al.* (1989) on Scottish mountains.

The predictions also agree with the observations of Körner and Diemer (1987), the originators of the raw data in the model. However, it should be pointed out that their observations were obtained at a constant *VPD* and therefore independently of the canopy boundary layer conductance.

The $\delta^{13}C$ of leaves in the top of the plant canopy decreases with altitude, a trend which is opposite in sign from the global observations of Körner *et al.* (1988). However, there is a marked effect of vegetation height on $\delta^{13}C$. Assuming that the majority of carbon fixation occurs in the highest irradiance, which is the environment selected in the model simulation, there could be $\delta^{13}C$ differences between the tallest and shortest canopies in the order of 4‰. This difference is equivalent to the trend in $\delta^{13}C$ associated with a 6000 m change in altitude (Korner *et al.*, 1988).

Körner *et al.* (1988) found significant differences between trees, shrubs and herbs in $\delta^{13}C$. This difference is most likely to be due to the generally lower stomatal conductances and rates of photosynthesis in trees and shrubs, in comparison with herbs (Larcher, 1980; Jarvis, 1976).

If the typical plant height is assumed to be about 32 m at sea-level, decreasing to 0·03 m at 3000 m, then $\delta^{13}C$ should increase from −25·7‰ at 0 m to −24·2‰ at 3000 m, a trend which is about 70% of the trend measured by Körner *et al.* (1987). It does seem possible to interpret global trends of $\delta^{13}C$ in vegetation, at least partially, in terms of plant height, and therefore boundary layer conductance.

The influence of *LAI* alone on the model predictions is shown in Fig. 9. In this case, *LAI* has a constant value of 1, simply simulating the observation on some Scottish mountains where sheep grazing may be intense at all altitudes.

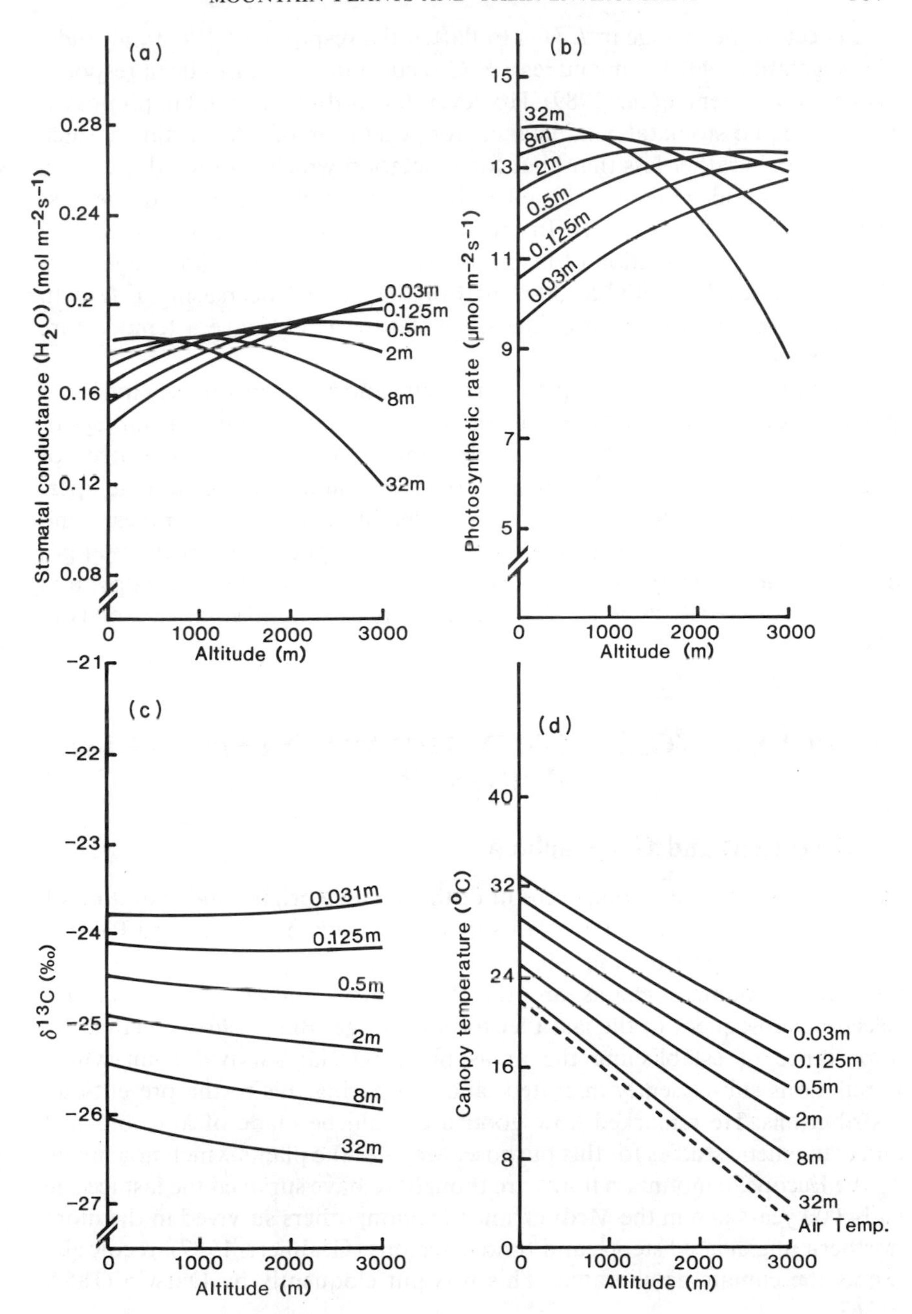

Fig. 9. As Fig. 8, but $LAI = 1$ at all altitudes.

The effect of the change in *LAI* is to flatten the response of $\delta^{13}C$ to altitude. For vegetation of 0·125 m and less, $\delta^{13}C$ is constant with altitude, a response observed by Friend *et al.* (1989). However, the altitudinal trend in photosynthetic rate and stomatal conductance, for vegetation of a fixed stature, such as 0·125 m, is rather less than that for vegetation which shows a decrease in *LAI* with altitude (Fig. 8). This effect is due to the change in canopy excess temperature at constant *LAI*. In this case (Fig. 9), the excess increases with altitude, so that vegetation of height 0·031 m has a temperature range from 33°C at sea-level to 19°C at 3000 m. For the case of decreasing *LAI* with altitude (Fig. 8), vegetation of the same stature would have a temperature range from 37 to 16°C.

It appears therefore that vegetation stature and *LAI* can strongly influence plant $\delta^{13}C$, and presumably lead to significant variation in global surveys of $\delta^{13}C$. The variations of $\delta^{13}C$ in the model simulations result from the effects of vegetation stature and *LAI* on boundary layer conductance, canopy temperature and *VPD*. These responses, in turn, influence photosynthesis and stomatal conductance. Given this strong effect, it appears surprising that gas exchange measurements on leaves isolated from the vegetation canopy, and at a constant *VPD*, show such efficiency in predicting altitudinal trends in plant $\delta^{13}C$.

V. PLANT EVOLUTION IN THE MOUNTAIN AERIAL ENVIRONMENT

A. Historical and Geographical

Billings (1974b) gave a good account of the possible origins and evolution of alpine floras. These mountain plants spread in the late Tertiary and Pleistocene, moving upwards and possibly evolving. The present-day floras of temperate mountain chains are the products of migration and natural selection subsequent to the last Pleistocene ice age. Bliss (1962a) related the importance of establishing the areas of periglacial survival from which populations subsequently migrated, and comparing this to the present-day distributions. He remarked how good use could be made of autecological and cytogenetic studies for this purpose. Many of the plants which now make up the European mountain floras are thought to have survived the last ice age to 14 000 years ago in the Mediterranean region; others survived in the more northern unglaciated steppe and tundra regions (Grabherr, 1987), recolonizing as the climate ameliorated. This was put eloquently by Darwin (1859, p. 367):

> As the warmth returned, the arctic forms would retreat northward, closely followed up in their retreat by the productions of the more temperate regions.

> And as the snow melted from the bases of the mountains, the arctic forms would seize on the cleared and thawed ground, always ascending higher and higher, as the warmth increased, whilst their brethren were pursuing their northern journey. Hence, when the warmth had fully returned, the same arctic species, which had lately lived in a body together on the lowlands of the Old and New Worlds, would be left isolated on distant mountain summits (having been exterminated on all lesser heights) and in the arctic regions of both hemispheres.

During the ice ages, the lowland periglacial plains of Europe were rich in arctic-alpine and boreal species, and some alpine plants survived on nunataks, and re-established in a downward direction (Godwin, 1956, pp. 297, 319–320). It should not be forgotten that vegetation change since the end of the last ice age has not been a simple successional process, as alternating periods of warmer and colder climates have caused fluctuations in species distribution. Because of the ecological and geographical isolation of mountains, vicariance, speciation and genetic drift are thought to have occurred frequently. Alpine regions are floristically diverse, the European Maritime Alps containing more species than the whole of Germany (Grabherr, 1987). The migration of species after the last ice age is thought to have brought formerly isolated but closely related species into contact, with subsequent hybridization and possibly the production of allopolyploids.

B. General

Changes in environment with altitude are important in causing genetic differences within plant populations (e.g. Slatyer, 1977; Ledig and Korbobo, 1983). Indeed, the classic demonstration of ecotypes was made for populations from different altitudes (Turesson, 1931). As described in Section III, many of the genetic differences found between populations at different altitudes appear to be of selective advantage, such as those that cause differences in the optimum temperature for photosynthesis (T_o) (e.g. Fryer and Ledig, 1972). However, Larcher (1983) warned that typical morphological and physiological responses may not be directly adaptive, but incidental and not essential for fitness and survival at high altitudes. Awareness of a panglossian approach in mountain plant ecology is as important as in other areas of biology (Gould and Lewontin, 1979).

It is important to point out that even if there were no genetic differentiation with altitude, the word adaptation may still be applicable when discussing plasticity, such as that found by Bonnier (1895) and Clements *et al.* (1950). Acclimation (in the sense that Bonnier, 1895, used "adaptation" to mean changes in development which occur in response to the environment of a plant during growth) may be of primary interest, but if it is important to establish whether modifications in plants with altitude are selective adaptations, or whether they are simply the passive consequence of the environment, a problem presents itself with regard to plasticity. Does a modification of physiology that can be induced in a leaf by growing it in an appropriate environment, and which increases the plant's ability to pass its genes on to future generations in that environment, constitute an adaptation? This question is not easily resolvable. Indeed, it was the great plasticity of plants, particularly in response to growth at different altitudes, that led Bonnier (1920) and Clements *et al.* (1950) to conclude that there is no reason to invoke natural selection in the formation of new species. It is perhaps best to avoid the use of the word adaptation and concentrate on those works where genetic differentiation with respect to altitude is proven. This includes differences between species which have restricted altitudinal distributions, implying that these differences may be, in some way, responsible for their different distributions.

Perhaps with some surprise we find that it is not at all obvious which environmental or plant processes are important for plant success at high altitudes (Körner and Diemer, 1987). Grabherr (1987) wrote that it is because of the relatively low importance of any one environmental factor that a specific alpine plant type has not evolved. There are, however, examples of convergent evolution, such as unbranched, or little branched, giant rosette plants in the tropics (Smith and Young, 1987), and cushion-type plants in more temperate regions (Bliss, 1962a).

With increased altitude there is probably less inter-plant competition for light, water and nutrients (Körner and Renhardt, 1987). Thus the potential for natural selection driven by competition for these resources is perhaps reduced. This view is reinforced when considering the preponderance of apomixis and vegetative reproduction in alpine areas, with a resultant decrease in genetic variation, although Silander (1985) suggested that clonal plants may not necessarily be less genetically diverse than sexual ones. It seems possible that the upper distributional limits of plants on mountains are not set primarily by competition, with the capacity for leaf and root growth, long-term carbon balance and low-temperature survival being of greater importance (Bunce *et al.*, 1979). K-type selection may therefore be more likely at high than low altitudes (Bliss, 1985), and one might thus expect to find life-histories that correspond to the "stress-tolerators" of Grime (1979). Present-day alpine floras may consist mostly of those genera and species which possessed sufficient plasticity of physiology and morphology to survive the encumbent environment. Adaptation need not be invoked to explain features typically associated with alpine plants.

VI. CONCLUSIONS AND SUGGESTIONS FOR FURTHER WORK

Plants growing on mountains experience reduced temperatures and vapour pressures with altitude, as well as a reduction in the partial pressure of air.

The most obvious general trend in plant morphology is a decrease in stature with increasing altitude. When examined in more detail there are many other morphological, physiological and biochemical features of plants which change with altitude. However, whether these changes are direct consequences of the changing environment, or are the result of genetic differentiation, has only rarely been ascertained. It is possible to explain many of the changes with altitude as being developmental responses to the environment alone, with no genetic component. However, such plasticity may have been selected for if it thus results in greater fitness in many environments.

The observed increased efficiency of carbon uptake with increased altitude might be entirely due to the influence of low temperature on leaf development, as could be the increases in A_{max}. Controlled environment and reciprocal transplant studies could easily test this. Of worthy investigation is the control of stomatal density, and whether the observed increases with altitude have any effect on maximal conductance in the field.

T_o is frequently found to fall with altitude, but again this could be due to developmental responses to low temperature alone. We need to know its biochemical explanation, particularly the role played by phosphate and Fru-P_2 phosphatase levels, and changes in the K_m of critical enzymes by low temperatures. The possibility of alterations in metabolic pool control at different temperatures should also be addressed. We need to have a much more general picture of the relationship between genetically controlled stature and altitude. At low temperatures, is it cell expansion or cell division rate which limits growth, or is there some form of self-regulating system between the two? What is the importance of cell-wall extensibility? Is it really this aspect of plant physiology which changes to allow plants to grow in low-temperature environments?

With regard to plant pigments, we still do not know why there is generally less chlorophyll and more anthocyanin in leaves with increased altitude, at least on temperate mountains. Also, growth at low temperatures tends to produce leaves with greater rates of dark respiration and photorespiration when measured under high-temperature conditions than does growth at higher temperatures. The mechanism for this change is not understood, nor is the effect it may have on the differential mean rate of dark respiration and photorespiration between different altitudes. It may be that intrinsically high rates of dark respiration and photorespiration do not allow high-altitude genotypes to colonize warmer habitats.

We do not know the functional significance of the high calorific contents of alpine plants. We also do not know how important reduced temperatures might be for the movement of assimilates in plants at high altitudes. This may be crucial. From the model proposed, it is clear that vegetation stature and leaf area index are probably very important determinants of leaf $\delta^{13}C$ values. The use of $\delta^{13}C$ gives us a powerful tool with which to characterize plant responses to altitude. The model should be subjected to examination using long-term field measurements, and controlled environment measurements, on the influence of boundary layer conductance, canopy temperature and *VPD* on $\delta^{13}C$ values.

ACKNOWLEDGMENTS

We wish to thank the following for useful discussions and criticism: D. Briggs, A. H. Fitter, A. Herrera, M. D. Morecroft and A. C. Newton.

REFERENCES

Abuzinadah, R. A. and Read, D. J. (1986). The role of proteins in the nitrogen nutrition of ectomycorrhizal plants I. Utilization of peptides and proteins by ectomycorrhizal fungi. *New Phytol.* **103,** 481–493.

Abuzinadah, R. A., Finlay, R. D. and Read, D. J. (1986). The role of proteins in the nitrogen nutrition of ectomycorrhizal plants II. Utilization of proteins by mycorrhizal plants of *Pinus contorta. New Phytol.* **103,** 495–506.

Arroyo, M. T. K., Armesto, J. J. and Primack, R. B. (1985). Community studies in pollination ecology in the high temperate Andes of central Chile. II. Effect of temperature on visitation rates and pollination possibilities. *Plant Syst. Evol.* **149,** 187–203.

Au, S.-F. (1969). Internal leaf surface and stomatal abundance in arctic and alpine populations of *Oxyria digyna. Ecology* **50,** 131–134.

Babb, T. A. and Whitfield, D. W. A. (1977). Mineral nutrient cycling and limitation of plant growth in the Truelove Lowland ecosystem. In: *Truelove Lowland, Devon Island, Canada: A High Arctic Ecosystem* (Ed. by L. C. Bliss), pp. 589–606. University of Alberta Press, Edmonton.

Baig, M. N. and Tranquillini, W. (1976). Studies on upper timberline: Morphology and anatomy of Norway spruce (*Picea abies*) and stone pine (*Pinus cembra*) needles from various habitat conditions. *Can. J. Bot.* **54,** 1622–1632.

Barbour, M. G. (1988). Californian highland forests and woodlands. In: *North American Terrestrial Vegetation* (Ed. by M. G. Barbour and W. D. Billings), pp. 131–164. Cambridge University Press, Cambridge.

Barclay, A. M. and Crawford, R. M. M. (1982). Winter desiccation stress and resting bud viability in relation to high altitude survival in *Sorbus aucuparia* L. *Flora* **173,** 21–34.

Barclay, A. M. and Crawford, R. M. M. (1984). Seedling emergence in the rowan (*Sorbus aucuparia*) from an altitudinal gradient. *J. Ecol.* **72,** 627–636.

Barry, R. G. (1981). *Mountain Weather and Climate.* Methuen, London.

Baruch, Z. (1982). Patterns of energy content in plants from the Venezuelan páramos. *Oecologia (Berl.)* **55,** 47–52.

Bauer, H., Martha, P., Kirchner-Heiss, B. and Mairhofer, I. (1983). The CO_2 compensation point of C_3 plants – a re-examination. II. Intraspecific variability. *Zeitschrift für Pflanzenphysiologie* **109,** 143–154.

Beard, J. S. (1955). The classification of tropical American vegetation types. *Ecology* **36,** 89–100.

Berger-Landefeldt, U. (1936). Der Wasserhaushalt der Alpenpflanzen. *Bibliotheca Botanica* **115.** Schweizerbart'sche Verlagsbuchhandlung, Stuttgart.

Berry, J. A. and Björkman, O. (1980). Photosynthetic response and adaptation to temperature in higher plants. *Ann. Rev. Plant Physiol.* **31,** 491–543.

Berry, J. A. and Raison, J. K. (1981). Responses of macrophytes to temperature. In: *Encyclopeadia of Plant Physiology* (Ed. by O. L. Lange, P. S. Nobel, C. B. Osmond and H. Ziegler), Vol. **12A,** pp. 278–338. Springer-Verlag, Berlin.

Billings, W. D. (1974a). Arctic and alpine vegetation: Plant adaptations to cold summer climates. In: *Arctic and Alpine Environments* (Ed. by J. D. Ives and R. G. Barry), pp. 403–443. Methuen, London.

Billings, W. D. (1974b). Adaptations and origins of alpine plants. *Arctic Alpine Res.* **6,** 129–142.

Billings, W. D. and Godfrey, P. J. (1968). Acclimation effects on metabolic rates of arctic and alpine *Oxyria* populations subjected to temperature stress. *Bull. Ecol. Soc. Am.* **49,** 68–69.

Billings, W. D. and Mooney, H. A. (1968). The ecology of arctic and alpine plants. *Biol. Rev.* **43,** 481–529.

Billings, W. D., Clebsch, E. E. C. and Mooney, H. A. (1961). Effects of low concentrations of carbon dioxide on photosynthetic rates of two races of *Oxyria. Science* **133,** 1834.

Billings, W. D., Godfrey, P. J., Chabot, B. F. and Bourque, D. P. (1971). Metabolic acclimation to temperature in arctic and alpine ecotypes of *Oxyria digyna. Arctic and Alpine Res.* **3,** 277–289.

Björkman, O. (1973). Comparative studies of photosynthesis in higher plants. *Photophysiol.* **8,** 1–63.

Björkman, O. (1981). Responses to different quantum flux densities. In: *Encyclopedia of Plant Physiology* (Ed. by O. Lange, P. S. Nobel, C. B. Osmond and H. Ziegler), Vol. **12A,** pp. 57–107. Springer-Verlag, Berlin.

Björkman, O. and Badger, M. (1979). Time course of thermal acclimation of the photosynthetic apparatus in *Nerium oleander. Carnegie Inst. Wash. Yearbook* **78,** 145–148.

Bjorkman, O. and Holmgren, P. (1961). Studies of climatic ecotypes of higher plants. Leaf respiration in different populations of *Solidago virgaurea. Ann. R. Agr. Coll. Sweden* **27,** 297–304.

Björkman, O., Florell, C. and Holmgren, P. (1960). Studies of climatic ecotypes in higher plants. The temperature dependence of apparent photosynthesis in different populations of *Solidago virgaurea. Ann. R. Agr. Coll. Sweden* **26,** 1–10.

Blagowestschenskij, W. A. (1935). Über den Verlauf der Photosynthese im Hochgebirge des Pamirs. *Planta* **24,** 276–287.

Bliss, L. C. (1962a). Adaptations of arctic and alpine plants to environmental conditions. *Arctic* **15,** 117–144.

Bliss, L. C. (1962b). Caloric and lipid content in alpine tundra plants. *Ecology* **43,** 754–757.

Bliss, L. C. (1985). Alpine. In: *Physiological Ecology of North American Plant Communities* (Ed. by B. F. Chabot and H. A. Mooney), pp. 41–65. Chapman and Hall, London.

Bonnier, G. (1895). Recherches expérimentales sur l'adaptation des plantes au climat alpin. *Ann. des Science Naturelles, Botanique, Sér. VII.* **20,** 217–360.

Bonnier, G. (1920). Nouvelles observations sur les cultures expérimentales à diverse altitudes. *Revue Générale botanique* **32,** 305–326.

Breckle, S. W. (1973). Mikroklimatische Messungen und Okologische Beobachtungen in der alpinen Stufe des afghanischen Hindukusch. *Botanische Jahrbücher* **93,** 25–55.

Briggs, D. and Walters, S. M. (1984). *Plant Variation and Evolution*, 2nd edn. Cambridge University Press, Cambridge.

Brooks, A. and Farquhar, G. D. (1985). Effect of temperature on the CO_2/O_2 specificity of ribulose – 1,5-bisphosphate carboxylase/oxygenase and the rate of respiration in the light. Estimates from gas–exchange measurements of spinach. *Planta* **165,** 397–406.

Bunce, J. A. (1983). Photosynthetic characteristics of leaves developed at different irradiances and temperatures: An extension to the current hypothesis. *Photosyn. Res.* **4,** 87–97.

Bunce, J. A. (1986). Measurements and modeling of photosynthesis in field crops. *CRC Crit. Rev. Plant Sci.* **4,** 47–77.

Bunce, J. A., Chabot, B. F. and Miller, L. N. (1979). Rôle of annual leaf carbon balance in the distribution of plant species along an elevational gradient. *Bot. Gaz.* **140,** 288–294.

Caemmerer, S. von and Farquhar, G. D. (1981). Some relationships between the biochemistry of photosynthesis and the gas exchange of leaves. *Planta* **153,** 376–387.

Carter, G. A. and Smith, W. K. (1985). Influence of shoot structure on light interception and photosynthesis in conifers. *Plant Physiol.* **79,** 1038–1043.

Cernusca, A. (1976). Bestandsstruktur, Bioklima und Energiehaushalt von Alpinen Zwergstrauchbeständen. *Oecologia Plantarum* **11,** 71–102.

Chabot, B. F. and Billings, W. D. (1972). Origins and ecology of the Sierran alpine flora and vegetation. *Ecol. Monog.* **42,** 163–199.

Chabot, B. F., Chabot, J. F. and Billings, W. D. (1972). Ribulose-1,5-diphosphate carboxylase activity in arctic and alpine populations of *Oxyria digyna. Photosynthetica* **6,** 364–369.

Chabot, B. F., Jurik, T. W. and Chabot, J. F. (1979). Influence of instantaneous and integrated light-flux density on leaf anatomy and photosynthesis. *Am. J. Bot.* **66,** 940–945.

Chapin, F. S., III (1980). The mineral nutrition of wild plants. *Ann. Rev. Ecol. Syst.* **11,** 233–260.

Chapin, F. S., III (1983). Direct and indirect effects of temperature on arctic plants. *Polar Biol.* **2,** 47–52.

Chapin, F. S., III, and Oechel, W. (1983). Photosynthesis, respiration, and phosphate absorption by *Carex aquatilis* ecotypes along latitudinal and local environmental gradients. *Ecology* **64,** 743–751.

Chapin, F. S., III, Tieszen, L. L., Lewis, M. C., Miller, P. C. and McCown, E. H. (1980). Control of tundra plant allocation patterns and growth. In: *An Arctic Ecosystem: The Coastal Tundra of Northern Alaska* (Ed. by J. Brown, P. C. Miller, L. L. Tieszen, F. L. Bunnell and S. F. MacLean), pp. 140–185. Hutchinson and Ross, Stroudsberg, Penn.

Charles-Edwards, D. A. and Ludwig, L. J. (1975). The basis of expression of leaf photosynthetic activities. In: *Environmental and Biological Control of Photosynthesis* (Ed. by R. Marcells), pp. 37–44. W. Junk, The Hague.

Chatterton, N. J., McKell, C. M. and Strain, B. R. (1970). Interspecific differences in temperature-induced respiration of desert saltbush. *Ecology* **51,** 545–549.

Choudhury, B. J. and Monteith, J. L. (1986). Implications of stomatal response to saturation deficit for the heat balance of vegetation. *Agric. Forest Meteorol.* **36,** 215–225.

Clausen, J., Keck, D. D. and Hiesey, W. M. (1940). Experimental studies on the nature of species. I. Effect of varied environments on western North American plants. *Carnegie Inst. Wash. Pub.* **520,** 1–452.

Clebsch, E. E. C. (1960). Comparative morphology and physiological variation in arctic and alpine populations of *Trisetum spicatum*. Ph.D. thesis, University of Illinois, Urbana.

Clements, F. E., Martin, E. V. and Long, F. L. (1950). *Adaptation and Origin in the Plant World: The Role of the Environment in Evolution*. Chronica Britanica Co., Waltham, Mass.

Cline, M. G. and Agatep, A. O. (1970). Temperature and photoperiodic control of developmental responses in climatic races of *Achillea*. *Plant Cell Physiol.* **11,** 599–608.

Cooper, C. F. (1986). Carbon dioxide enhancement of tree growth at high elevations. Technical comments. *Science* **231,** 859.

Crawford, R. M. M. and Palin, M. A. (1981). Root respiration and temperature limits to the north–south distribution of four perennial maritime plants. *Flora* **171,** 338–354.

Dahl, E. (1951). On the relation between summer temperature and the distribution of alpine vascular plants in the lowlands of Fennoscandia. *Oikos* **3,** 22–52.

Darwin, C. (1859). *On the Origin of Species by Means of Natural Selection*. John Murray, London.

Daubenmire, R. F. (1954). Alpine timberlines in the Americas and their interpretation. *Butler Univ. Bot. Stud.* **11,** 119–136.

Decker, J. P. (1959). Some effects of temperature and carbon dioxide concentration on photosynthesis of Mimulus. *Plant Physiol.* **34,** 103–106.

DeLucia, E. H. and Berlyn, G. P. (1984). The effect of increasing elevation on leaf cuticle thickness and cuticular transpiration in balsam fir. *Can. J. Bot.* **62,** 2423–2431.

Dobzhansky, T. (1970). *Genetics of the Evolutionary Process*. Columbia University Press, New York.

Downton, W. J. S., Grant, J. R. and Robinson, S. P. (1985). Photosynthetic and stomatal response of spinach leaves to salt stress. *Plant Physiol.* **77,** 85–88.

Downton, W. J. S., Loveys, B. R. and Grant, W. J. R. (1988). Non-uniform stomatal closure induced by water stress causes putative non-stomatal inhibition of photosynthesis. *New Phytol.* **110,** 503–509.

Eagles, C. F. (1967a). The effect of temperature on vegetative growth in climatic races of *Dactylis glomerata* in controlled environments. *Ann. Bot.* **31,** 31–9.

Eagles, C. F. (1967b). Variation in the soluble carbohydrate of climatic races of *Dactylis glomerata* (Cocksfoot) at different temperatures. *Ann. Bot.* **31,** 645–651.

Ehleringer, J. R. and Björkman, O. (1978). A comparison of photosynthetic characteristics of *Encelia* species possessing glabrous and pubescent leaves. *Plant Physiol.* **62,** 185–190.

Ehrhardt, F. (1961). Untersuchungen über den Einfluss des Klimas auf die Stickstoffnachlieferung von Waldhumus in verschiedenen Höhenlegen der Tiroler Alpen. *Forstwiss Zentralbl.* **80,** 193–215.

Evans, J. R. (1989). Photosynthetic and nitrogen relationships in leaves of C_3 plants. *Oecologia (Berl.)* **78,** 9–19.

Farquhar, G. D., Caemmerer, S. von and Berry, J. A. (1980). A biochemical model of photosynthetic CO_2 assimilation in leaves of C_3 species. *Planta* **149,** 78–90.

Farquhar, G. D., O'Leary, M. H. and Berry, J. A. (1982). On the relationship between carbon isotope discrimination and the intercellular carbon dioxide concentration in leaves. *Austr. J. Plant Physiol.* **9,** 121–137.

Field, C. (1983). Allocating leaf nitrogen for the maximization of carbon gain: Leaf age as a control on the allocation program. *Oecologia (Berl.)* **56,** 341–347.

Field, C., Merino, J. and Mooney, H. A. (1983). Comparisons between water-use efficiency and nitrogen-use efficiency in five species of Californian evergreens. *Oecologia (Berl.)* **60,** 384–389.

Fitter, A. H. and Hay, R. K. M. (1987). *Environmental Physiology of Plants*, 2nd edn. Academic Press, London.

Friend, A. D., Woodward, F. I. and Switsur, V. R. (1989). Field measurements of photosynthesis, stomatal conductance, leaf nitrogen and $\delta^{13}C$ along altitudinal gradients in Scotland. *Func. Ecol.* **3,** 117–122.

Fryer, J. H. and Ledig, F. T. (1972). Microevolution of the photosynthetic temperature optimum in relationship to the elevational complex gradient. *Can. J. Bot.* **50,** 1231–1235.

Gale, J. (1973). Experimental evidence for the effect of barometric pressure on photosynthesis and transpiration. *Ecol. Conserv.* **5,** 289–294.

Gale, J. (1986). Carbon dioxide enhancement of tree growth at high elevations. Technical comments. *Science* **231,** 859–860.

Garratt, J. R. (1977). Aerodynamic roughness and mean monthly surface stress over Australia. *Technical Paper* No. 29. CSIRO, Division of Atmospheric Physics.

Gates, D. M. and Janke, R. (1966). The energy environment of the alpine tundra. *Oecologia Plantarum* **1,** 39–61.

Geiger, R. (1965). *The Climate Near the Ground.* Harvard University Press, Cambridge, Mass.

Godfrey, P. J. (1969). Factors influencing the lower limits of alpine plants in the medicine bow mountains of southeast Wyoming. Ph.D. dissertation, Duke University, USA.

Godwin, H. (1956). *The History of the British Flora.* Cambridge University Press, Cambridge.

Goldstein, G., Meinzer, F. and Monastero, M. (1985). Physiological and mechanical factors in relation to size-dependent mortality in an Andean giant rosette species. *Oecologia Plantarum* **6,** 263–275.

Golley, F. B. (1961). Energy values for ecological materials. *Ecology* **42,** 581–584.

Gould, S. J. and Lewontin, R. C. (1979). The spandrels of San Marco and the Panglossian paradigm: A critique of the adaptationist programme. *Proc. R. Soc. Lond.* **B205,** 581–598.

Grabherr, G. (1977). Der CO_2-Gaswechsel des immergrünen Zwergstrauches *Loiseleuria procumbens* (L.) DESV. in Abhängigkeit von Strahlung, Temperatur, Wasserstreß und phänologischen Zustand. *Photosynthetica* **11,** 302–410.

Grabherr, G. (1987). *High Alpine Flora and Vegetation of the Tyrolean Alps* (*W. Austria*). *Guide to Excursion* No. 18. Publ. XIV, International Botanical Congress, Berlin.

Grabherr, G. and Cernusca, A. (1977). Influence of radiation, wind, and temperature on the CO_2 gas exchange of the alpine dwarf shrub community *Loiseleurietum cetrariosum. Photosynthetica* **11,** 22–28.

Grace, J. (1977). *Plant Response to Wind.* Academic Press, London.

Grace, J. (1989). Tree lines. *Phil. Trans. R. Soc. Lond.* **B324,** 233–245.

Grant, M. C. and Mitton, J. B. (1977). Genetic differentiation among growth forms of Engelmann Spruce and sub-alpine fir at tree line. *Arctic Alpine Res.* **9,** 259–263.

Grant, S. A. and Hunter, R. F. (1962). Ecotypic differentiation of *Calluna vulgaris* (L.) in relation to altitude. *New Phytol.* **61,** 44–55.

Graves, J. D. and Taylor, K. (1986). A comparative study of *Geum rivale.* L. and *G. urbanum* L. to determine factors controlling their altitudinal distribution. I. Growth in controlled and natural environments. *New Phytol.* **104,** 681–691.

Graves, J. D. and Taylor, K. (1988). A comparative study of *Geum rivale* L. and *G. urbanum* L. to determine those factors controlling their altitudinal distribution. II. Photosynthesis and respiration. *New Phytol.* **108,** 297–304.

Grill, D., Pfeifhofer, H., Tschulik, A., Hellig, K. and Holzer, K. (1988). Thiol content of spruce needles at forest limits. *Oecologia (Berl.)* **76,** 294–297.

Grime, J. P. (1979). *Plant Strategies and Vegetation Processes*. John Wiley, Chichester, 222 pp.

Grubb, P. J. (1971). Interpretation of the "Massenerhebung" effect on tropical mountains. *Nature* **229,** 44–45.

Grubb, P. J. (1989). A plant ecologist's view. In: *Mineral Nutrients in Tropical Forests and Savannas*. (Ed. by J. Proctor). BES Special Publication 9, pp. 417–439.

Guariguata, M. R. and Azocar, A. (1988). Seed bank dynamics and germination ecology of *Espeletia timotensis* (Compositae), an Andean giant rosette. *Biotropica* **20,** 54–59.

Hadley, J. L. and Smith, W. K. (1983). Influence of wind exposure on needle desiccation and mortality for timberline conifers in Wyoming, U.S.A. *Arctic Alpine Res.* **15,** 127–135.

Hadley, J. L. and Smith, W. K. (1986). Wind effects on needles of timberline conifers. Seasonal influence on mortality. *Ecology* **67,** 12–19.

Hadley, J. L. and Smith, W. K. (1987). Influence of krummholz mat microclimate on needle physiology and survival. *Oecologia (Berl.)* **73,** 82–90.

Haselwandter, K., Hofmann, A., Holzmann, H.-P. and Read, D. J. (1983). Availability of nitrogen and phosphorus in the nival zone of the Alps. *Oecologia (Berl.)* **57,** 266–269.

Hedberg, I. and Hedberg, O. (1979). Tropical-alpine life forms of vascular plants. *Oikos* **33,** 297–307.

Henrici, M. (1918). Chlorophyllgehalt und Kohlensäure-Assimilation bei Alpen und Ebenen Pflanzen. *Verhandl. naturforsch. Ges. Basel* **30,** 43–136.

Hiesey, W. M., Nobs, M. A. and Björkman, O. (1971). Experimental studies on the nature of species. V. Biosystematics, genetics, and physiological ecology of the Eryanthe section of *Mimulus*. *Carnegie Inst. Wash. Pub.* **628**.

Holzmann, H.-P. and Haselwandter, K. (1988). Contribution of nitrogen fixation nutrition in an alpine sedge community (*Caricetum curvulae*). *Oecologia (Berl.)* **76,** 298–302.

Hooker, J. D. (1853). *The Botany of the Antarctic Voyage. II. Flora Novae-Zelandiae. Introductory Essay*. Reeve, London.

Huffaker, R. C. (1982). Biochemistry and physiology of leaf proteins. In: *Encyclopaedia of Plant Physiology* (Ed. by D. Boulter and B. Parthier), Vol. **14A,** pp. 370–400. Springer-Verlag, Berlin.

Huner, N. P. A. and MacDowell, F. D. H. (1979a). Changes in the net charge and subunit properties of ribulose carboxylase-oxygenase during cold hardening of Puma rye. *Can. J. Biochem.* **57,** 155–164.

Huner, N. P. A. and MacDowell, F. D. H. (1979b). The effects of low temperature acclimation of winter rye on catalytic properties of its ribulose bisphosphate carboxylase-oxygenase. *Can. J. Biochem.* **57,** 1036–1041.

Jarvis, P. G. (1976). The interpretation of the variations in leaf water potentials and stomatal conductance found in canopies in the field. *Phil. Trans. R. Soc. Lond.* **B273,** 593–610.

Jarvis, P. G. and Morison, J. I. L. (1981). The control of transpiration and photosynthesis by the stomata. In: *Stomatal Physiology* (Ed. by P. G. Jarvis and T. A. Mansfield), pp. 247–279. Cambridge University Press, Cambridge.

Jones, H. G. (1983). *Plants and Microclimate*. Cambridge University Press, Cambridge.

Jones, H. G. (1987). Breeding for stomatal characters. In: *Stomatal Function* (Ed. by E. Zeiger, G. D. Farquhar and I. R. Cowan), pp. 431–443. Stanford University Press, Stanford, Calif.

Kjär, A. (1937). Die Schwankungen der Assimilationsintensität der Blätter von *Sinapis alba* in Laufe des Tages in Abhängigkeit von Inneren Faktoren. *Planta* **26,** 595–607.

Klein, R. M. (1978). Plants and near-ultraviolet radiation. *Bot. Rev.* **44,** 1–127.

Klikoff, L. G. (1966). Temperature dependence of the oxidative rates of mitochondria in *Danthonia intermedia, Penstemon davidsonii,* and *Sitanion hystrix. Nature* **212,** 529–530.

Klikoff, L. G. (1968). Temperature dependence of mitochondrial oxidative rates of several plant species of the Sierra Nevada. *Bot. Gaz.* **129,** 227–230.

Klikoff, L. G. (1969). Temperature dependence of mitochondrial oxidative rates in relation to plant distribution. In: *Physiological Systems in Semi-arid Environments* (Ed. by C. C. Hoft and M. L. Riedesell), pp. 263–269. University of New Mexico Press, Alberquerque.

Körner, Ch. (1982). CO_2 exchange in the alpine sedge *Carex curvula* as influenced by canopy structure, light and temperature. *Oecologia (Berl.)* **53,** 98–104.

Körner, Ch. and Cochrane, P. (1983). Influence of leaf physiognomy on leaf temperature on clear midsummer days in the Snowy Mountains, south-eastern Australia. *Acta Oecologia* **4,** 117–124.

Körner, Ch. and Cochrane, P. M. (1985). Stomatal responses and water relations of *Eucalyptus pauciflora* in summer along an elevational gradient. *Oecologia (Berl.)* **66,** 443–455.

Körner, Ch. and Diemer, M. (1987). *In situ* photosynthetic responses to light, temperature and carbon dioxide in herbaceous plants from low and high altitude. *Functional Ecol.* **1,** 179–194.

Körner, Ch. and Larcher, W. (1988). Plant life in cold climates. In: *Plants and Temperature* (Ed. by S. P. Long and F. I. Woodward), pp. 25–57. Symposia of the Society for Experimental Biology, No. XXXXII, Company of Biologists, Ltd., Cambridge.

Körner, Ch. and Mayr, R. (1981). Stomatal behaviour in Alpine plant communities between 600 and 2600 m above sea level. In: *Plants and Their Atmospheric Environment* (Ed. by J. Grace, E. D. Ford and P. G. Jarvis), pp. 205–218. Blackwell, Oxford.

Körner, Ch. and Renhardt, U. (1987). Dry matter partitioning and root length/leaf area ratios in herbaceous perennial plants with diverse altitudinal distributions. *Oecologia (Berl.)* **74,** 411–418.

Körner, Ch. and Woodward, F. I. (1987). The dynamics of leaf extension in plants with diverse altitudinal ranges. II. Field studies in *Poa* species between 600 and 3200 m altitude. *Oecologia (Berl.)* **72,** 279–283.

Körner, Ch., Scheel, J. A. and Bauer, H. (1979). Maximum leaf diffusive conductance in vascular plants. *Photosynthetica* **13,** 45–82.

Körner, Ch., Allison, A. and Hilscher, H. (1983). Altitudinal variation in leaf diffusive conductance and leaf anatomy in heliophytes of montane New Guinea and their interrelation with microclimate. *Flora* **174,** 91–135.

Körner, Ch., Bannister, P. and Mark, A. F. (1986). Altitudinal variation in stomatal conductance, nitrogen content and leaf anatomy in different plant life forms in New Zealand. *Oceologia (Berl.)* **69,** 577–588.

Körner, Ch., Farquhar, G. D. and Roksandic, Z. (1988). A global survey of carbon isotope discrimination in plants from high altitude. *Oecologia (Berl.)* **74,** 623–632.

Labate, C. A. and Leegood, R. C. (1988). Limitation of photosynthesis by changes in temperature. *Planta* **173,** 519–527.

La Marche, V. C. Jr, Fraybill, D. A., Fritts, H. C. and Rose, M. R. (1984). Increasing atmospheric carbon dioxide: Tree ring evidence for growth enhancement in natural vegetation. *Science* **225,** 1019–1021.

La Marche, V. C. Jr, Graybill, D. A., Fritts, H. C. and Rose, M. R. (1986). Carbon dioxide enhancement of tree growth at high elevations. Technical comments. *Science* **231,** 860.

Landsberg, J. J. (1986). *Physiological Ecology of Forest Production*. Academic Press, London.

Langlet, O. (1971). Two hundred years' genecology. *Taxon* **20,** 653–722.

Larcher, W. (1975). Pflanzenökologische Beobachtungen in der Paramostufe der venezolanischen Anden. *Anzeiger Math.-naturwiss. Klasse Österr. Akad. Wiss.* **11,** 194–213.

Larcher, W. (1980). *Physiological Plant Ecology*. Springer-Verlag, Berlin.

Larcher, W. (1983). Ökophysiologische Konstitutionseigenschaften von Gebirgspflanzen. *Ber. Deutsch. Ges.* **96,** 73–85.

Larcher, W. and Bauer, H. (1981). Ecological significance of resistance to low temperature. In: *Encyclopedia of Plant Physiology, Vol. 12A: Physiological Plant Ecology I* (Ed. by O. L. Lange, P. S. Nobel, C. B. Osmond and H. Ziegler), pp. 403–437. Springer-Verlag, Berlin.

Larcher, W. and Wagner, J. (1976). Temperaturgrenzen der CO_2-Augnahme und Temperaturresistenz der Blätter von Gebirgspflanzen im vegetationsaktiven Zustand. *Oecologia Plantarum*, **11,** 361–374.

Lauscher, W. (1976). Weltweite Typen der Höhenabhängigkeit des Niederschlags. *Wetter und Leben*, **28,** 80–90.

Ledig, F. T. and Korbobo, D. R. (1983). Adaptation of sugar maple along altitudinal gradients: Photosynthesis, respiration, and specific leaf weight. *Am. J. Bot.* **70,** 256–265.

Leegood, R. C. and Furbank, R. T. (1986). Stimulation of photosynthesis by 2% oxygen at low temperatures is restored by phosphate. *Planta* **168,** 84–93.

Long, S. P. and Hallgren, J.-E. (1985). Measurement of CO_2 assimilation by plants in the field and the laboratory. In: *Techniques in Bioproductivity and Photosynthesis* (Ed. by J. Coombs, D. O. Hall, S. P. Long and J. M. O. Scurlock), pp. 62–94. Pergamon Press, Oxford.

Lösch, R. and Tenhunen, J. D. (1981). Stomatal responses to humidity – phenomenon and mechanism. In: *Stomatal Physiology*. (Ed. by P. G. Jarvis and T. A. Mansfield), pp. 137–161. Cambridge University Press, Cambridge.

McCown, B. H. (1978). The interactions of organic nutrients, soil nitrogen, and soil temperature and plant growth and survival in the arctic environment. In: *Vegetation and Production of an Alaskan Arctic Tundra* (Ed. by L. L. Tieszen), pp. 435–456. Springer-Verlag, New York.

McMahon, D. and Bogorad, L. (1966). Some kinetic studies of ribulose-1,5-diphosphate carboxylase (carboxydismutase) from races of *Mimulus cardinalis*. *Carnegie Inst. Wash. Yearbook* **65,** 459–461.

McNaughton, K. G. and Jarvis, P. G. (1983). Predicting the effects of vegetation changes on transpiration and evaporation. In: *Water Deficits and Plant Growth* (Ed. by T. T. Kozlowski), Vol. VII, pp. 1–47. Academic Press, London.

McNaughton, S. J., Campbell, R. S., Freyer, R. A., Mylroie, J. E. and Rodland, K. D. (1974). Photosynthetic properties and root chilling responses of altitudinal ecotypes of *Typha latifolia* L. *Ecology* **55,** 168–172.

Mächler, F. and Nösberger, J. (1977). Effect of light intensity and temperature on apparent photosynthesis of altitudinal ecotypes of *Trifolium repens* L. *Oecologia (Berl.)* **31,** 73–78.

Mächler, F. and Nösberger, J. (1978). The adaptation to temperature of photorespiration and of the photosynthetic carbon metabolism of altitudinal ecotypes of *Trifolium repens*. *Oecologia (Berl.)* **35,** 267–276.

Mächler, F., Nösberger, J. and Erismann, K. H. (1977). Photosynthetic $^{14}CO_2$ fixation products in altitudinal ecotypes of *Trifolium repens* L. with different temperature requirements. *Oecologia (Berl.)* **31,** 79–84.

Madsen, E. (1973a). The effect of CO_2-concentration on development and dry matter production in young tomato plants. *Acta Agriculturae Scandinavica* **23,** 235–240.

Madsen, E. (1973b). Effect of CO_2 concentration on the morphological, histological and cytological changes in tomato plants. *Acta Agriculturae Scandinavica* **23,** 241–246.

Marrs, R. H., Proctor, J., Heaney, A. and Mountford, M. D. (1988). Changes in soil nitrogen-mineralization and nitrification along an altitudinal transect in tropical rain forest in Costa Rica. *J. Ecol.* **76,** 466–482.

May, D. S. and Villareal, H. M. (1974). Altitudinal differentiation of the Hill Reaction in populations of *Taraxacum officinale* in Colorado. *Photosynthetica* **8,** 73–77.

Mayr, E. (1963). *Animal Species and Evolution*. Belknap, Cambridge.

Miller, G. R. and Cummins, R. P. (1982). Regeneration of Scots pine *Pinus sylvestris* at a natural tree-line in the Cairngorm Mountains, Scotland. *Holarctic Ecol.* **5,** 27–34.

Miller, G. R. And Cummins, R. P. (1987). Rôle of buried viable seeds in the recolonisation of disturbed ground by heather (*Calluna vulgaris* [L.] Hull) in the Cairngorm mountains, Scotland, U.K. *Arctic Alpine Res.* **19,** 396–401.

Milner, H. W. and Hiesey, W. M. (1964a). Photosynthesis in climatic races of *Mimulus*. I. Effect of light intensity and temperature on rate. *Plant Physiol.* **39,** 208–213.

Milner, H. W. and Hiesey, W. M. (1964b). Photosynthesis in climatic races of *Mimulus*. II. Effect of time and CO_2 concentration on rate. *Plant Physiol.* **39,** 746–750.

Monteith, J. L. (1973). *Principles of Environmental Physics*. Edward Arnold, London.

Monteith, J. L. (1981). Evaporation and environment. In: *Symposium of the Society for Environmental Biology, No. 19, The State and Movement of Water in Living Organisms* (Ed. by C. E. Fogg), pp. 205–234. Cambridge University Press, Cambridge.

Mooney, H. A. (1963). Physiological ecology of coastal, subalpine and alpine populations of *Polygonum bistortoides*. *Ecology* **44,** 812–816.

Mooney, H. A. and Billings, W. D. (1961). Comparative physiological ecology of arctic and alpine populations of *Oxyria digyna*. *Ecol. Monogr.* **31,** 1–29.

Mooney, H. A. and Johnson, A. W. (1965). Comparative physiological ecology of an arctic and an alpine population of *Thalictrum alpinum* L. *Ecology* **46,** 721–727.

Mooney, H. A. and West, M. (1964). Photosynthetic acclimation of plants of diverse origin. *Am. J. Bot.* **51,** 825–827.

Mooney, H. A., Wright, R. D. and Strain, B. R. (1964). The gas exchange capacity of plants in relation to vegetation zonation in the White Mountains of California. *Am. Midl. Nat.* **72,** 281–297.

Mooney, H. A., Strain, B. R. and West, M. (1966). Photosynthetic efficiency at reduced carbon dioxide tensions. *Ecology* **47,** 490–491.

Mooney, H. A., Ferrar, P. J. and Slatyer, R. O. (1978). Photosynthetic capacity and carbon allocation patterns in diverse growth forms of *Eucalyptus*. *Oecologia (Berl.)* **36,** 103–111.

Mott, K. A., Gibson, A. C. and O'Leary, J. W. (1982). The adaptive significance of amphistomatic leaves. *Plant Cell Environ.* **5,** 455–460.
Müller, M. J. (1982). *Selected Climatic Data for a Global Set of Standard Stations for Vegetation Science.* W. Junk, The Hague.
Neama, J. D. (1982). Physiological and ecological studies on altitudinal distribution in the genus *Salix*. Ph.D. thesis, University College, Cardiff.
Nobel, P. S. (1983). *Biophysical Plant Physiology and Ecology*. W. H. Freeman, San Francisco.
Nobel, P. S. and Walker, D. B. (1985). Structure of leaf photosynthetic tissue. In: *Photosynthetic Mechanisms and the Environment* (Ed. by J. Barber and N. R. Baker), pp. 502–536. Elsevier, London.
Nordmeyer, A. H. (1980). Tree nutrient concentrations near timberline, Craigieburn Range, New Zealand. In: *Mountain Environments and Subalpine Tree Growth* (Ed. by U. Benecke and M. R. Davis), pp. 83–94. *N.Z. Forest Service Technical Paper* No. 70.
Ollerenshaw, J. H. and Baker, R. H. (1981). Low temperature growth in a controlled environment of *Trifolium repens* from northern latitudes. *J. Appl. Ecol.* **18,** 229–239.
Osmaston, A. E. (1922). Notes on the forest communities of the Garhwal Himalaya. *J. Ecol.* **10,** 129–167.
Pandey, O. P., Bhadula, S. K. and Purohit, A. N. (1984). Changes in the activity of some photosynthetic and photorespiratory enzymes in *Selinum vaginatum* Clarke grown at two altitudes. *Photosynthetica* **18,** 153–155.
Patton, L. and Jones, M. B. (1989). Some relationships between leaf anatomy and photosynthetic characteristics of willows. *New Phytol.* **111,** 657–661.
Paw, U. K. T. and Gao, W. (1988). Applications of solutes to non-linear energy budget equations. *Agric. Forest Meteorol.* **43,** 121–145.
Pearcy, R. W. (1969). Physiological and varied environment studies of ecotypes of *Deschampsia caespitosa* (L.) Beauv. Ph.D. thesis, Colorado State University, Fort Collins.
Peet, M. M., Ozbun, J. L. and Wallace, D. H. (1977). Physiological and anatomical effects of growth on *Phaseolus vulgaris* L. cultivars. *J. Exp. Bot.* **28,** 57–69.
Pisek, A., Larcher, W., Vegis, A. and Napp-Zin, K. (1973). The normal temperature range. In: *Temperature and Life* (Ed. by H. Precht, J. Christopherson, H. Hensel and W. Larcher), pp. 102–194. Springer-Verlag, Berlin.
Pollock, C. J., Eagles, C. F. and Sims, I. M. (1988). Effect of photoperiod and irradiance changes upon development of freezing tolerance and accumulation of soluble carbohydrate in seedlings of *Lolium perenne* grown at 2°C. *Ann. Bot.* **62,** 95–100.
Portis, A. R., Chon, C. J., Mosbach, A. and Heldt, H. W. (1977). Fructose and sedoheptulose phosphatase. The sites of a possible control of CO_2 fixation by light dependent changes of the stromal Mg^{2+} concentration. *Biochim. Biophys. Acta* **461,** 313–325.
Rabinowitch, E. I. (1951). *Photosynthesis and Related Processes*, **Vol. II,** Part I. John Wiley, New York.
Rada, F., Goldstein, G., Azocar, A. and Meinzer, F. (1985). Freezing avoidance in Andean giant rosette plants. *Plant Cell Environ.* **8,** 501–507.
Raven, J. A. and Glidewell, S. M. (1981). Processes limiting photosynthetic conductance. In: *Physiological Processes Limiting Plant Productivity* (Ed. by C. B. Johnson,), pp. 109–136. Butterworths, London.
Rehder, H. and Schäfer, A. (1978). Nutrient turnover studies in alpine ecotypes. IV.

Communities of the central Alps and comparative survey. *Oecologia (Berl.)* **34,** 309–327.

Richards, J. H. and Bliss, L. C. (1986). Winter water relations of a deciduous timberline conifer, *Larix lyalli* Parl. *Oecologia (Berl.)* **69,** 16–24.

Sage, R. F., Sharkey, T. D. and Seemann, J. R. (1988). The *in-vivo* response of ribulose-1,5-bisphosphate carboxylase activation state and the pool sizes of photosynthetic metabolites to elevated CO_2 in *Phaseolus vulgaris* L. *Planta* **174,** 407–416.

Sakai, A. and Larcher, W. (1987). Frost survival of plants. In: *Ecological Studies* (Ed. by W. D. Billings, F. Golley, O. L. Lange, J. S. Olsen and H. Remmert), Vol. 62, Springer-Verlag, Berlin.

Salisbury, F. B. and Spomer, G. G. (1964). Leaf temperatures of alpine plants in the field. *Planta,* **60,** 497–505.

Scott, D. and Billings, W. D. (1964). Effects of environmental factors on standing crop and productivity of an alpine tundra. *Ecol. Monogr.* **34,** 243–270.

Seybold, A. and Eagle, K. (1940). Über die Blettpigmente der Alpenpflanzen. *Bot. Arch.* **40,** 560–570.

Sharkey, T. D. (1985). Photosynthesis in intact leaves of C_3 plants: Physics, physiology and rate limitations. *Bot. Rev.* **51,** 53–105.

Sicher, R. C., Sundblad, L.-G. and Öquist, G. (1988). Effects of low temperature acclimation upon photosynthetic induction in barley primary leaves. *Physiologia Plantarum* **73**, 206–210.

Silander, J. A. Jr (1985). Microevolution in clonal plants. In: *Population Biology and Evolution of Clonal Organisms* (Ed. by J. B. C. Jackson, L. W. Buss and R. E. Cook), pp. 107–152. Yale University Press, New Haven, Conn.

Slatyer, R. O. (1977). Altitudinal variation in the photosynthetic characteristics of Snow Gum, *Eucalyptus pauciflora.* Sieb. ex Spreng. IV. Temperature response of four populations grown at different temperatures. *Austr. J. Plant Physiol.* **4,** 583–594.

Smith, A. P. (1981). Growth and population dynamics of Espeletia (Compositae) in the Venezuelan Andes. *Smithsonian Contrib. Bot.* **48.**

Smith, A. P. and Young, T. P. (1987). Tropical alpine plant ecology. *Ann. Rev. Ecol. Syst.* **18,** 137–158.

Smith, W. K. and Carter, G. A. (1988). Shoot structural effects on needle temperature and photosynthesis in conifers. *Am. J. Bot.* **75,** 496–500.

Stewart, W. S. and Bannister, P. (1974). Dark respiration rates in *Vaccinium* spp. in relation to altitude. *Flora* **163,** S415–S421.

Terashima, I., Wong, S.-C., Osmond, C. B. and Farquhar, G. D. (1988). Characterisation of non-uniform photosynthesis induced by abscisic acid in leaves having different mesophyll anatomies. *Plant Cell. Physiol.* **29,** 385–394.

Tieszen, L. L. (1970). Comparisons of chlorophyll content and leaf structure in arctic and alpine grasses. *Am. Midl. Nat.* **83,** 228–235.

Tieszen, L. L. (1978). Photosynthesis in the principle Barrow, Alaskan species: A summary of field and laboratory responses. In: *Vegetation and Production Ecology of an Alaskan Arctic Tundra* (Ed. by L. L. Tieszen), pp. 241–268. Springer-Verlag, New York.

Tieszen, L. L. and Helgager, J. A. (1968). Genetic and physiological adaptation in the Hill Reaction of *Deschampsia caespitosa. Nature* **219,** 1066–1067.

Tranquillini, W. (1964). The physiology of plants at high altitudes. *Ann. Rev. Plant Physiol.* **15,** 345–362.

Tranquillini, W. (1979). *Physiological Ecology of the Alpine Timberline.* Springer-Verlag, Berlin.

Troughton, J. H., Card, K. A. and Hendy, C. H. (1974). Photosynthetic pathways and carbon isotope discrimination by plants. *Carnegie Inst. Wash. Yearbook* **73,** 768–780.

Turesson, G. (1922). The genotypical response of the plant species to the habitat. *Hereditas* **3,** 211–350.

Turesson, G. (1923). The scope and import of genecology. *Hereditas* **4,** 171–176.

Turesson, G. (1925). The plant species in relation to habitat and climate (contributions to the knowledge of genecological units). *Hereditas* **6,** 147–236.

Turesson, G. (1927). Contribution to the genecology of glacial relicts. *Hereditas* **9,** 81–101.

Turesson, G. (1930–31). The selective effect of climate upon the plant species. *Hereditas* **14,** 99–152.

Turesson, G. (1931). The geographical distribution of the alpine ecotype of some Eurasiatic plants. *Hereditas* **15,** 329–346.

Turrill, W. B. (1940). Experimental and synthetic plant taxonomy. In: *The New Synthesis* (Ed. by J. Huxley), pp. 47–71. Clarendon Press, Oxford.

Urbanska, K. M. and Schütz, M. (1986). Reproduction by seed in alpine plants and revegetation research above timberline. *Botanica Helvetica* **96,** 43–60.

Vareschi, V. (1970). *Flora de los Paramos de Venezuela.* Universidad de los Andes, Merida, Venezuela.

Verduin, J. (1953). A table of photosynthetic rates under optimal, near-natural conditions. *Am. J. Bot.* **40,** 675–679.

Vitousek, P. M., Matson, P. A. and Turner, D. R. (1988). Elevational and age gradients in hawaiian montane rainforest: Foliar and soil nutrients. *Oecologia (Berl.)* **77,** 565–570.

Wagner, A. (1892). Zur Kenntniss des Blattbaues der Alpenflanzen und dessen biologischer Bedeutung. *Sitzungsberichte der Kaiserlichen Akademie der Wissenschaften, in Wien, Mathematisch-naturwissenschaftliche Klasse* **100,** 487–547.

Wardlaw, I. F., Begg, J. E., Bagnall, D. and Dunstone, R. L. (1983). Jojoba; Temperature adaptation as expressed in growth and leaf function. *Austr. J. Plant Physiol.* **10,** 299–312.

Wardle, P. (1971). An explanation for alpine timberline. *NZ J. Bot.* **9,** 371–402.

Wardle, P. (1985). New Zealand timberlines 3. A synthesis. *NZ J. Bot.* **23,** 263–271.

Wilson, C., Grace, J., Allen, S. and Slack, F. (1987). Temperature and stature: a study of temperatures in montane vegetation. *Functional Ecol.* **1,** 405–413.

Wilson, J. W. (1966). An analysis of plant growth and its control in arctic environments. *Ann. Bot.* **30,** 383–402.

Wong, S. C., Cowan, I. R. and Farquhar, G. H. (1979). Stomatal conductance correlates with photosynthetic capacity. *Nature* **282,** 424–426.

Woodward, F. I. (1975). The climatic control of the altitudinal distribution of *Sedum rosea* (L.) Scop. and *S. telephium* L. II. The analysis of plant growth in controlled environments. *New Phytol.* **74,** 335–348.

Woodward, F. I. (1979a). The differential temperature responses of the growth of certain plant species from different altitudes. I. Growth analysis of *Phleum alpinum* L., *P. bertolonii* D.C., *Sesleria albicans* Kit. and *Dactylis glomerata* L. *New Phytol.* **82,** 385–395.

Woodward, F. I. (1979b). The differential temperature responses of the growth of certain plant species from different altitudes. II. Analysis of the control and morphology of leaf extension and specific leaf area of *Phleum bertolonii* D.C. and *P. alpinum* L. *New Phytol.* **82,** 397–405.

Woodward, F. I. (1986). Ecophysiological studies on the shrub *Vaccinium myrtillus* L. taken from a wide altitudinal range. *Oecologia (Berl.)* **70,** 580–586.

Woodward, F. I. (1987a). *Climate and Plant Distribution*. Cambridge University Press, Cambridge.

Woodward, F. I. (1987b). Stomatal numbers are sensitive to increases in CO_2 from pre-industrial levels. *Nature* **327,** 617–618.

Woodward, F. I. and Bazzaz, F. A. (1988). The response of stomatal density to CO_2 partial pressure. *J. Exp. Bot.* **39,** 1771–1781.

Woodward, F. I. and Friend, A. D. (1988). Controlled environment studies on the temperature responses of leaf extension in species of *Poa* with diverse altitudinal ranges. *J. Exp. Bot.* **39,** 411–420.

Woodward, F. I. and Jones, N. (1984). Growth studies of selected plant species with well-defined European distributions. I. Field observations and computer simulations on plant life cycles at two altitudes. *J. Ecol.* **72,** 1019–1030.

Woodward, F. I. and Pigott, C. D. (1975). The climatic control of the altitudinal distribution of *Sedum rosea* (L.) Scop. and *S. telephium* L. I. Field observations. *New Phytol.* **74,** 323–334.

Woodward, F. I. and Sheehy, J. E. (1983). *Principals and Methods in Environmental Biology*. Butterworths, London.

Woodward, F. I., Körner, Ch. and Crabtree, R. C. (1986). The dynamics of leaf extension in plants with diverse altitudinal ranges. I. Field observations on temperature responses at one altitude. *Oecologia (Berl.)* **70,** 222–226.

Young, T. P. (1984). The comparative demography of semelparous *Lobelia telekii* and iteroparous *Lobelia keniensis* on Mount Kenya. *J. Ecol.* **72,** 637–650.

Zachhuber, K. and Larcher, W. (1978). Energy content of different alpine species of *Saxifraga* and *Primula* depending on their altitudinal distribution. *Photosynthetica* **12,** 436–439.

Zalensky, O. V. (1954). Photosynthése dans les conditions naturelles. In: *Essais de Botanique I: Moscow-Leningrad.* Acad. des Science de l'URSS, pp. 74–87 (cited by Mooney *et al.*, 1964).

Zelitch, I. and Day, P. R. (1973). The effects of net photosynthesis of pedigree selection for low and high rates of photorespiration in tobacco. *Plant Physiol.* **52,** 33–37.

Mutualistic Interactions in Freshwater Modular Systems with Molluscan Components

J. D. THOMAS

ADVANCES IN ECOLOGICAL RESEARCH VOL. 20
ISBN 0-12-013920-0

I. INTRODUCTION

A. The Concept of the Module

If we were to examine a freshwater ecosystem, we might expect to see pulmonate snails living either on inorganic sediments (e.g. sand, gravel, stones) in association with epilithon, or on the surface of a macrophyte such as *Ceratophyllum demersum* and its associated epiphyton. The biological communities of which the molluscs form a part may be considered as subsystems (May, 1973), co-evolved food webs (Gilbert, 1977) or as modules (Paine, 1980; Thomas *et al.*, 1985; Thomas, 1987, 1989). There is one common theme in all the conceptual frameworks on which the above terms are based, namely that the biological components interact with each other. However, all these authors differ in the methods they used to analyse the systems and in their interpretation of the nature of the mechanisms involved. In the earlier studies (May, 1973; Gilbert, 1977; Paine, 1980), the systems were looked upon essentially as food webs with the emphasis on +/− or −/− interactions, e.g. predation (+/−), herbivory (+/−), parasitism (+/−) and competition (−/−). Thus Paine (1980) states that "a module includes a resource set and their specific consumers which under conditions of maximal coevolution will behave as a functional unit". However, Paine (1980) was also aware of the importance of positive interactions within the module, and states that individuals comprising the module should exhibit sophisticated mutualisms. Thomas *et al.* (1985; Thomas, 1987, 1989) considered two types of modules, both of which may be regarded as components of the whole ecosystem. The first (Type 1) has three components (epilithic bacteria, algae and pulmonate snails) and three subsets, whereas the second (Type 2) has four components (the macrophytes, epiphytic algae, bacteria and the snails) and six subsets (Fig. 1). The concept of the module as developed by Thomas *et al.* (1985) and Thomas (1989) differs from that of Paine (1980) in three important respects. First, it includes the bacteria involved in various processes, including the decomposition of dead organisms. Secondly, it places more emphasis on positive interactions or mutualisms, which occur in each of the subsets. Thirdly, the strongly interacting component, namely the snail, is included as a component of the modular system. In contrast, the marine chiton *Katharina*, which is considered obligatory for the maintenance of the *Corallina* alga, is excluded from the module containing the algal species by Paine (1980). However, despite these differences, the modular concepts of Paine (1980) and Thomas (1989) have much in common, and it is of interest that it is grazing molluscs which are the key components in both cases.

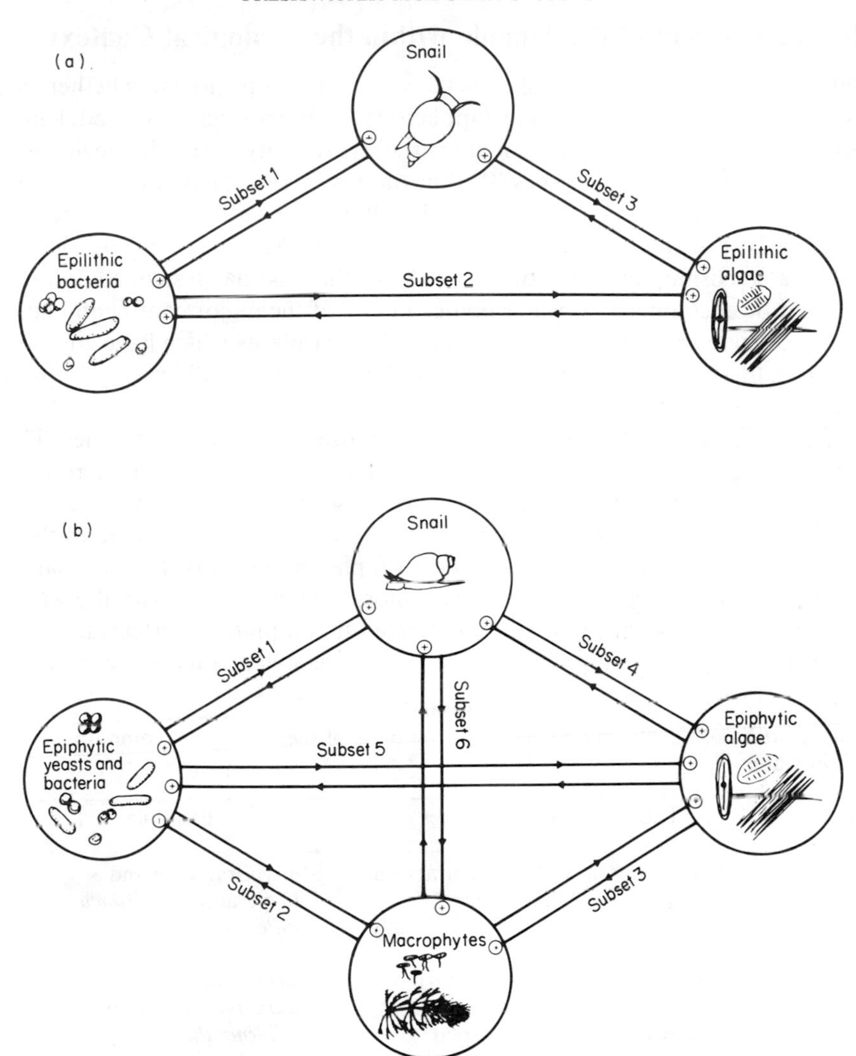

Fig. 1. Snails as components of modular systems. The evolutionarily primitive module (a) consists of epilithic algae, epilithic microorganisms, such as bacteria and fungi, and the snails. It has three subsets and is associated with inorganic sediments such as sand. In contrast, the evolutionarily more advanced module (b) has four major components—aquatic macrophytes, epiphytic bacteria and fungi, epiphytic algae and snails—and has 6 subsets. For convenience, the terms periphyton, periphytic algae or bacteria are used in the text to include both the epiphytic and epilithic forms.

B. The Concept of the Module within the Ecological Context

When considering such modular systems, it is necessary to ask whether the concept is useful and generally applicable in freshwater lotic and lentic ecosystems. Both of these are noted for their variability in size, longevity and physicochemical characteristics. Within these systems, the pulmonate snail–sediment–epilithon system (Type 1) is ubiquitous, although the specific composition of the biotic components may vary. As *Ceratophyllum demersum* is also cosmopolitan (Cook *et al.*, 1974), the modular system containing it could be encountered in water bodies in any of the major continental land masses. There are, however, many kinds of water plants which have adopted different life-history strategies. These are classified into six different groups in Table 1, and illustrated in Fig. 2.

The extent to which these plants are committed to life in water varies. The non-rooting macrophytes, *C. demersum* and *Utricularia* spp. in Group 6 show the highest level of commitment, because they are entirely dependent on the water column for obtaining their resources and for getting rid of their waste products. The other categories of plants (Groups 1–5) exchange chemicals with the sediments and the atmosphere, as well as with the water column. In view of this, it might be expected that aquatic invertebrates such as pulmonate snails, committed to a similar life-style, would interact more

Table 1
Classification of aquatic macrophytes on the basis of their degree of commitment to life in water

Group	Ecological characteristics	Examples
1	Plants living in wetland habitats but not adapted to live under submerged conditions indefinitely	Many Graminae and Cyperacea, e.g. *Juncus inflexus, Acroceras zizanoides*
2	Emergent plants living on margins of more permanent water bodies, usually with well-developed aerenchymatous tissues to convey air to root rhizosphere system	*Sparganium erectum, Glyceria maxima, Equisitum* spp., *Alternanthera*
3	Aquatic plants with roots in sediments, submerged leaves but also leaves floating on surface	*Callitriche* sp., *Nymphaea* sp., *Potomogeton natans*
4	Submerged aquatic plants with roots in sediments, leaves in water column but none on surface	*Potomogetan crispus, Groenlandia densa, Elodea*
5	Aquatic plants restricted to the air–water interface but with roots	*Lemna, Salvinia Azolla, Eichornia*
6	Aquatic plants in the water column but with no roots	*Ceratophyllum demersum, Utricularia* spp.

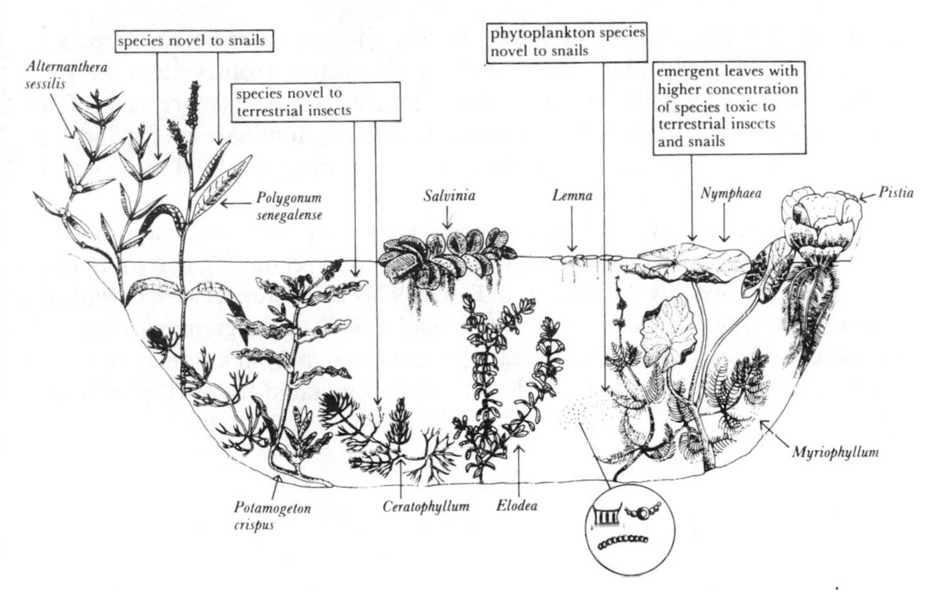

Fig. 2. Diagrammatic representation of some of the macrophytes mentioned in the text. As aquatic pulmonates do not exploit phytoplankton and do not encounter emergent portions of sub-aquatic plants, they may be regarded as being novel to snails. Consequently, such plants are likely to contain secondary plant compounds (SPCs) which are toxic to the snails. Likewise, submerged aquatic macrophytes are novel to terrestrial insects. They are therefore likely to contain SPCs which are toxic to the insects.

strongly with Group 6 plants and their associated epiphyton than plants in other groups. As a result, it can be postulated that the linkages between the subsets in this type of modular system would be stronger than would be the case with plants in other groups. This is one of the reasons why the module with *Ceratophyllum* was selected as a model for study in our laboratory. The other reason is that it is epidemiologically important, as snail hosts of schistosomiasis are associated with this plant, both in Africa and South America (Thomas and Tait, 1984; Thomas, 1987). It is hoped that an understanding of how these systems function will make it possible to develop methods for controlling target snail species.

Much work has been done in an attempt to elucidate the factors influencing the distribution and abundance of the major components of the modular system under discussion, namely the macrophyte and pulmonate snails. Most of this has been directed at higher levels of organization and at the role of negative or excluding factors. In the case of the aquatic macrophytes, physico-chemical factors clearly act as powerful density legislative factors. Group 1 plants occur on land subject to intermittent inundation, whereas Groups 2–6 plants predominate in small or medium-sized, shallow water bodies. Group 5 plants are relatively more important in the warm tropics than in the temperate zone, possibly because wind-generated currents are less of a hazard there. In contrast, phytoplankton tend to dominate in the larger, deeper water bodies, as well as in small temporary pools. These broad concepts are illustrated in Fig. 3.

The various negative biological mechanisms, such as competition, herbivory and pathogenicity, which might be involved in regulating macrophytes, have been reviewed by Thomas and Tait, (1984) and Thomas (1987), within the context of biological control of the snail hosts. Although many of these biological agencies could, theoretically, cause a switch from modular systems containing macrophytes and snails, to those dominated by phytoplankton,

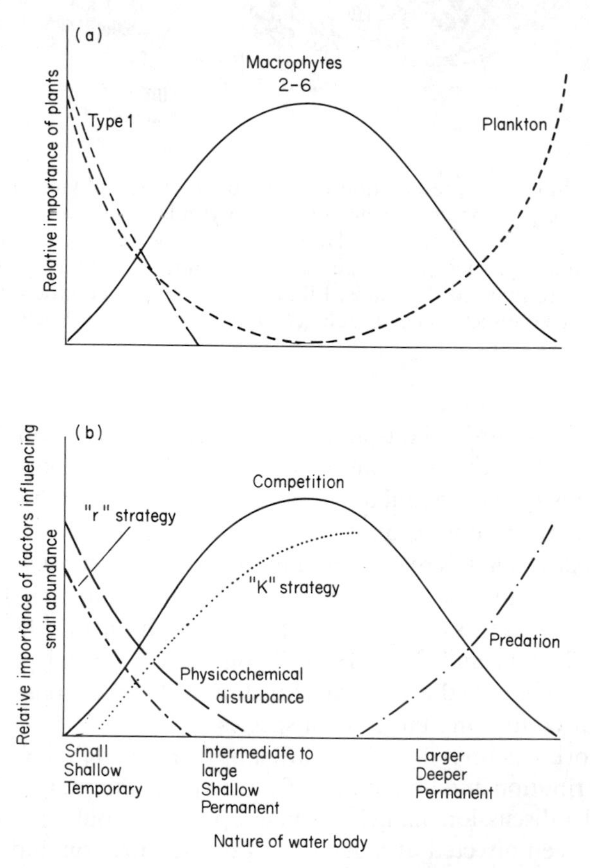

Fig. 3. Diagrammatic representation of the relative importance of the different categories of plants in contrasting aquatic habitats (a) and the various factors involved in determining the presence and abundance of the plant categories and snails in contrasting habitats (b).

this rarely seems to happen in natural systems which are not unduly influenced by man. In other words, the strength of the interactions between herbivores and aquatic macrophytes are generally weak. This is in marked contrast to the terrestrial and marine ecosystems described by Strong *et al.* (1984) and Paine (1980), where the strengths of the interactions between predators and their prey or herbivores and plants are much stronger. As a result, the removal of these key elements can alter a community structure drastically.

Lodge and Kelly (1985), Lodge *et al.* (1987) and Brönmark (1989) have also proposed an interesting model, which again invokes negative mechanisms to explain the distribution and abundance of freshwater snails. They postulate that the population densities of snails in harsh, temporary aquatic habitats will be determined by climatic stress. However, as one proceeds along the gradient towards greater climatic stability, the regulating mechanisms would tend to become dominated, first by competition for resources ("bottom up" regulation) and then by predation ("top down" regulation) (Fig. 3). Eisenberg (1966, 1970) and Brönmark (1989) have produced quantitative evidence in favour of "bottom up" and "top down" concepts, respectively. This unifying model can be linked to the concept of life-history strategies. Thus, *Lymnaea truncatula*, which exhibits many characteristics of an "r" species, provides a good example of a species adapted to life in temporary ecotone habitats. *L. peregra* and *Planorbis vortex*, on the other hand, are more committed to life in water. According to Lodge (1985, 1986), *P. vortex* tends to be associated with fairly long-lived stands of emergent plants, such as *Glyceria maxima* (Group 2). In contrast, *L. peregra* tends to live in submerged beds of *Elodea canadensis* (Group 4), which are more short-lived than *Glyceria*. It might be expected therefore that *P. vortex* or *P. planorbis* and *L. peregra* would exhibit "K" and "r" strategies, respectively. Evidence in support of this hypothesis is given by Lodge (1985, 1986) and Lodge and Kelly (1985).

Two general criticisms can be levelled at these attempts to explain the distribution and abundance of species. First, there are always exceptions to the general rule. For example, Moens (1982) demonstrated that *L. truncatula*, although living in temporary habitats, is regulated by the predatory snail *Zonitoides nitidus*. Furthermore, snails living in temporary, seasonally drying habitats may have evolved sophisticated mechanisms, allowing them to overcome the problems of living in such habitats. This is the case with a polymorphic strain of *Biomphalaria glabrata* in north-east Brazil, which can respond, by diapausing, to signals, such as lowering temperature, which predict the onset of the dry season (Pieri and Thomas, 1986, in press). Although these snails live in temporary habitats, their ability to respond predictably to unfavourable conditions makes it unnecessary for them to resort to "r" strategies.

The main criticism of the above approaches to explain the distribution and abundance of species, is that they only invoke negative processes inherent in the food web concept. The purpose of the present chapter is to show that it is necessary to adopt an approach which takes into account both positive (mutualistic) and negative interactions, before we can hope to understand the mechanisms responsible for determining the distribution, abundance and genetic composition of organisms in the short (ecological) and long (evolutionary) time-scale. In the next section, the various epochs when the components of the modular systems appeared on the evolutionary time-scale will be considered.

C. The Components of the Module on the Evolutionary Time-scale

The approximate times in the evolutionary scale when the various components of the modular system appeared are given in Fig. 4 (Thomas, 1987). The first to appear on the surface of sediments (about 3 billion years ago) were the heterotrophic bacteria which utilized abiotically and biotically synthesized organic molecules. Today, their descendants probably still use essentially the same metabolic strategies to take advantage of the organic bonanzas that follow the death of organisms. The early prokaryotes gave rise to the blue-green prokaryotic algae about 2 billion years ago. These are still prominent on the surfaces of sediment particles and plants, and also in the water column as members of the phytoplankton. Eventually, much later (1.5 billion years ago) the prokaryotes gave rise to the eukaryotes. Among the earliest eukaryotes were the green algae found among periphyton and phytoplankton today. As selective pressures favoured multicellularity and large size, macrophytes eventually appeared, perhaps in the Silurian (400–440 million years ago; Fig. 5). These gave rise to the Characae, liverworts, mosses, club mosses and horsetails, which are still represented in existing freshwater bodies. In the early Cretaceous, the angiosperms or the flowering plants began to radiate and dominate freshwaters. Some of the families, and perhaps genera, which appeared then are still extant (Fig. 5).

The gastropods which appeared in the late Cambrian or Ordovician, about 500–600 million years ago, possessed mouthparts with radulae which were well adapted to exploit the food resources present on the surface of sediments (Fig. 6). When the macrophytes appeared later, the snails were preadapted to utilize the bacteria, algae and dissolved organic matter (DOM) present on the vastly increased surface area presented by them. It is interesting that the radiation of the freshwater pulmonates occurred much later in the Cretaceous, and was coincident with that of the freshwater angiosperms. This also coincided with the co-evolution of the angiosperms and pollinating insects in the terrestrial environment. It is generally agreed that the mutual benefits

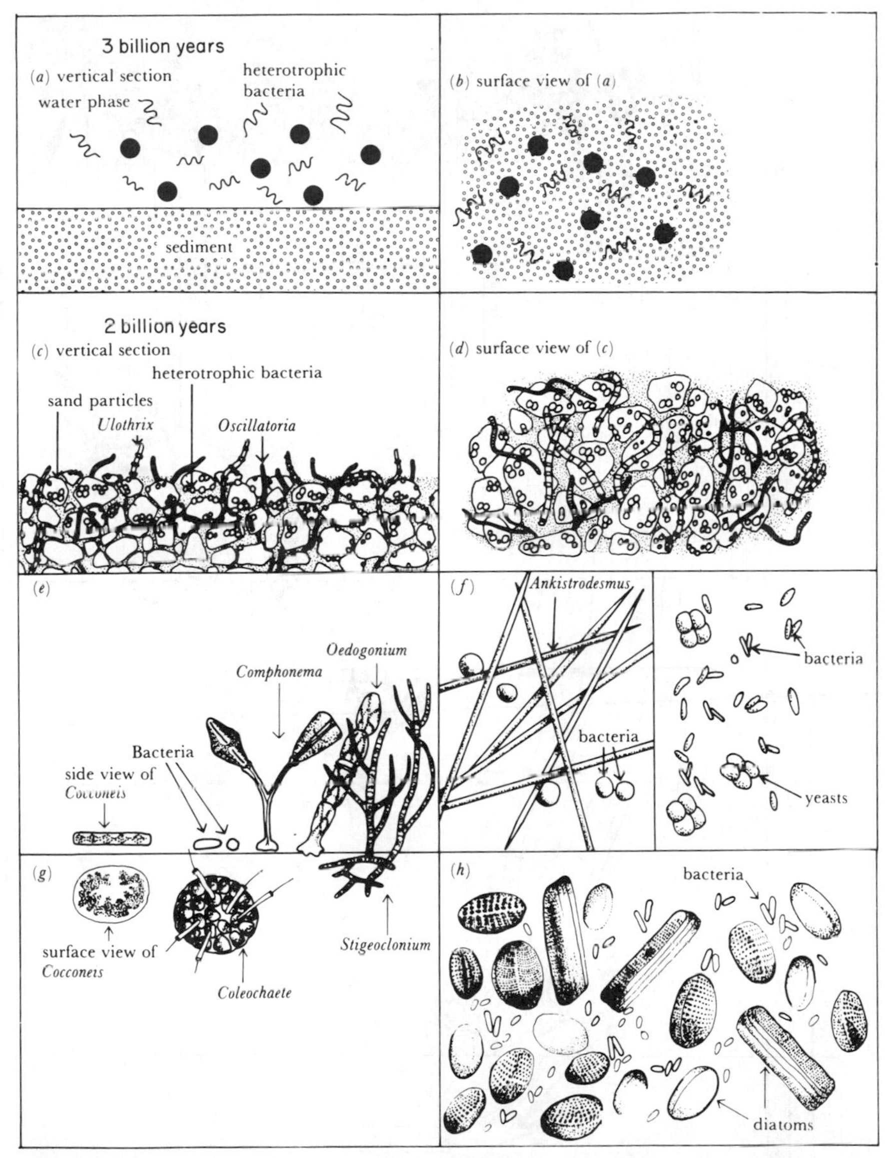

Fig. 4. Evolution of surface-living bacteria and algae. (a,b) Heterotrophic bacteria which first appeared about 3 billion years ago. (c,d) Heterotrophic bacteria and blue-green algae which first appeared as epilithon about 2 billion years ago. (e,f) Various assemblages of microorganisms living on surfaces of macrophytes. (g) Surface view of microorganisms on surface of macrophyte. (h) Diatoms and heterotrophic bacteria on surface of macrophyte.

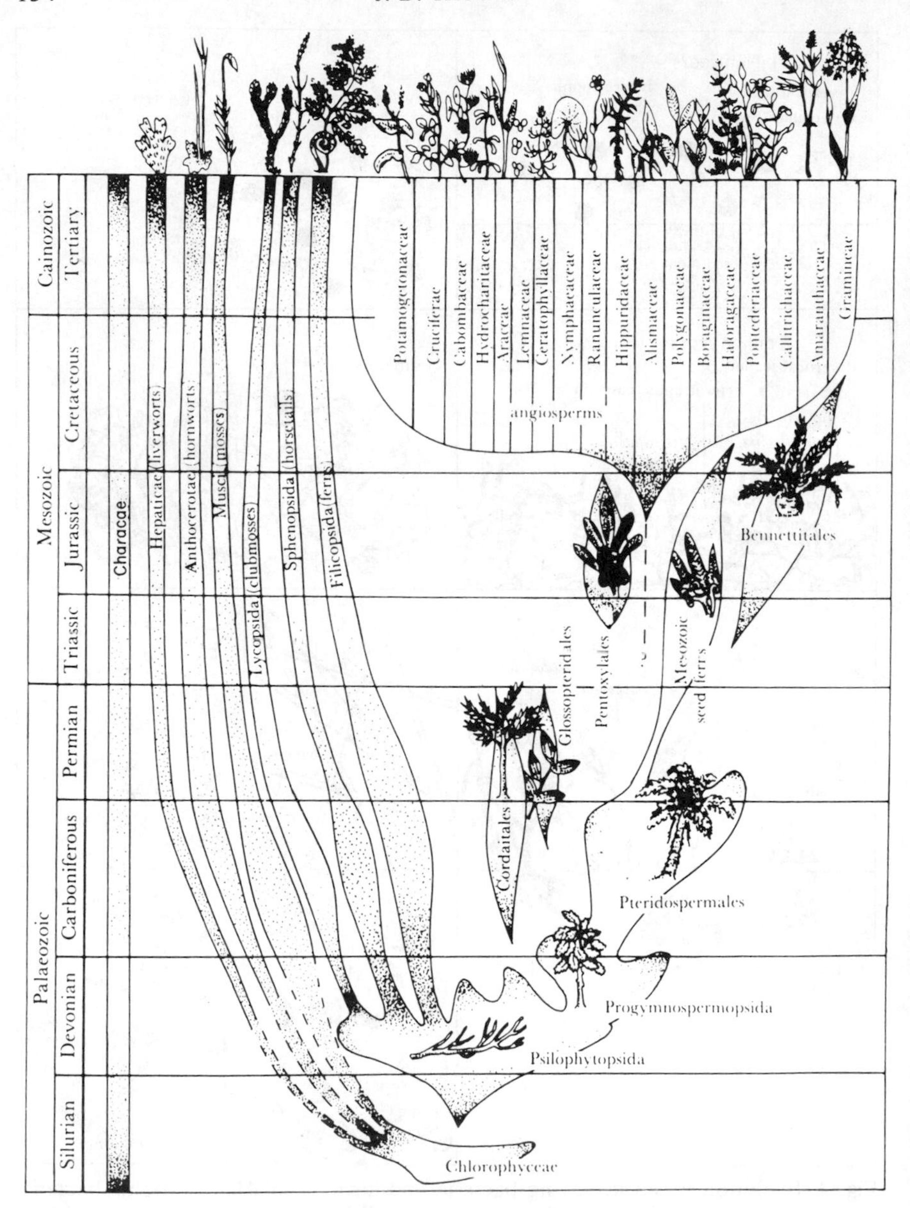

Fig. 5. The evolution of aquatic macrophytes.

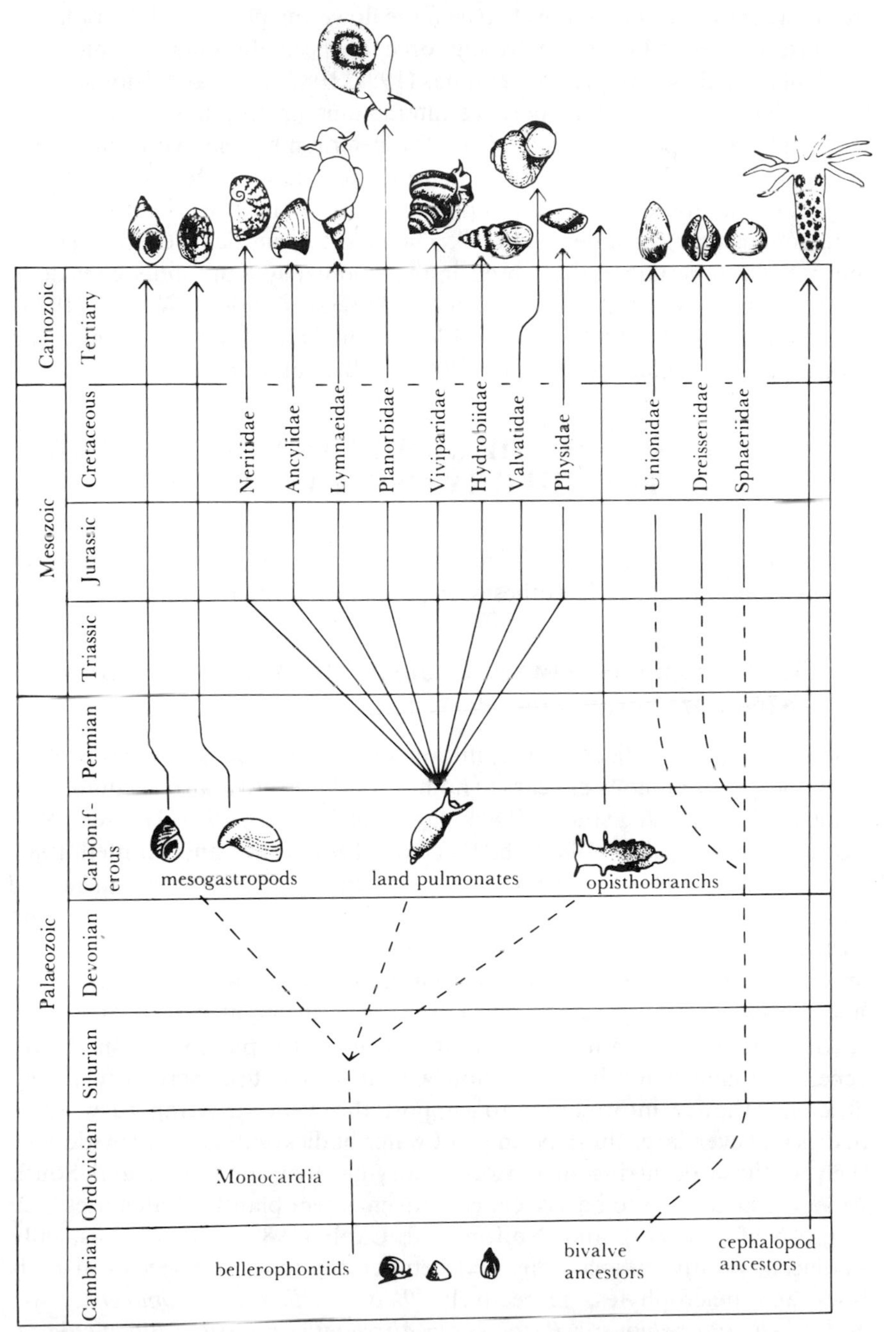

Fig. 6. The evolution of gastropod molluscs.

resulting from the interactions between the flowering plants and the pollinating insects provided the main driving force for the evolutionary changes that took place in these two groups. Thomas (1982, 1987, 1989) and Thomas *et al.* (1985) have suggested that positive interactions or mutualisms have been, and still are, important mechanisms for determining the evolution of the components of the modular system described above. Moens (1989) has suggested the use of the word "apparent mutualism" for "highly indirect interactions that give a net + effect". The need to consider these mutualistic interactions in certain of the subsets in the modular systems under consideration has also been stressed by Allen (1971), Harlin, (1973), Wetzel (1983), Carignan and Kalf (1982), Orth and Van Montfrans (1984), Carpenter and Lodge (1986), Odum and Biever (1984) and Brönmark (1989).

II. ANALYSES OF THE MODULAR SYSTEMS AND THE THEORY OF MUTUALISM

When analysing these systems, various questions need to be asked and the following propositions, hypotheses or predictions tested.

A. The Associations between Components of the Modular Systems are Strong and Persistent

The fossil records indicate that epilithon and aquatic snails have co-evolved for hundreds of millions of years (Thomas, 1987). There is also accumulating evidence from grazing studies (Dickman, 1968; Nicotri, 1977; Hunter, 1980; Gregory, 1980; Cuker, 1983a, b; Peterson, 1987; Lowe and Hunter, 1988; Underwood and Thomas, in press) that particular taxa of epilithic algae tend to be characteristic of grazed substrates. It is, however, necessary to obtain further quantitative information regarding the strengths and degree of specificity and persistence of the associations between particular snail species and the species of algae and bacteria present in the epilithon.

Although there is much anecdotal evidence that pulmonate snails are generally found living in association with macrophytes, there is relatively little quantitative information to support this view. In temporary water bodies and even large more permanent water bodies with large draw-downs, such as those occurring in seasonally drying areas of Africa and South America, snails tend to be associated with emergent plants in categories 1–2. Thus, Ndifon (1979) and Ndifon and Ukoli (1989) found statistically significant positive associations ($\leftrightarrows$) between the following species of molluscs and macrophytes, respectively: *Bulinus globosus*$\leftrightarrows$*Commelina* sp.; *B. forskali*$\leftrightarrows$*Paspalum* sp.; *B. forskali*$\leftrightarrows$*Alternanthera sessilis; Biomphalaria. pfeifferi*$\leftrightarrows$*Commelina* sp., *Acroceras zizanoides* and *Paspalum* sp.; and

B. rohlfsi⇆*A. sessilis*. Only one significant association was found with macrophytes in Groups 3–6, namely *B. forskali*⇆*Lemna paucicostata*. However, this result is largely attributable to the fact that such macrophyte species only occurred in 0·2–2·5% of the 404 habitats investigated in this seasonally drying area. Pieri (1985) and Pieri and Thomas (1987) also found significant positive associations between *Bio. glabrata* and macrophytes in Groups 1 and 2, such as *Commelina* spp. and *Cyperus* spp. in north-east Brazil. In this habitat, *Pistia stratiotes*, which was only found in one habitat, was the only representative of macrophytes encountered in Groups 3–6. The snail *Lymnaea truncatula* is also associated with plants in Groups 1 and 2, such as *Glyceria fluitans, Juncus inflexus* and *Rananculus repens* (Moens, 1981). These plants therefore serve as good biotic indicators for the presence of the snails. However, as the snails in such habitats interact mainly with the sediment and the epilithon, they are best regarded as forming part of the Module Type 1 system. As a result, the strength of interactions between snails and macrophytes in such systems must be regarded as being either very low or zero.

In contrast, pulmonate and other freshwater snails tend to be significantly positively associated with plants in Groups 3–6 in more permanent habitats. As a result of man's activities, such water bodies are becoming ever increasing features of the landscape in the developing nations. In irrigation systems in Egypt, Dazo *et al.* (1966) found *B. truncatus* and *Biomphalaria alexandrina* to be mainly associated with beds of *Potomogetan crispus* and *Eichornia crassipes*, respectively. Statistically significant positive associations between the following plants in Groups 3–6 and snails were also found by Thomas and Tait (1984) in a man-made lake in Nigeria: *Nymphaea lotus* ⇆*Anisus coretus, B. forskali* and *Bio. pfeifferi*⇆*N. lotus*; *Bio. pfeifferi* and *B. forskali*⇆*Ceratophyllum demersum*; and *B. rohlfsi*⇆*Ceratophyllum* and *Commelina* spp. (see Fig. 7). Some of these relations were particularly strong. For example, *A. coretus* seemed to be restricted to *N. lotus*. In all these cases, almost all of the snails were found to be adhering to the plant surface. These results are in accord with those of other workers who have studied pulmonate snails in the more permanent water bodies in West Africa (Paperna, 1969; Odei, 1972, 1973; Ndifon, 1979). Other workers (Pip and Stewart, 1976; Pip, 1978; Lodge 1985, 1986) have also provided evidence of significant positive associations between pulmonate snails and macrophytes in the temperate zones. Lodge (1986) also states that macrophytes tend to have specific periphyton assemblages. However, according to Eminson and Moss (1980), such highly specific associations between macrophytes and epiphytes tend only to be specific in oligotrophic waters and become progressively less so as the nutrient status increases to the mesotrophic and eutrophic levels.

Not all reported associations between macrophytes and aquatic snails are of a positive nature. Thus, Thomas and Tait (1984) found significant negative

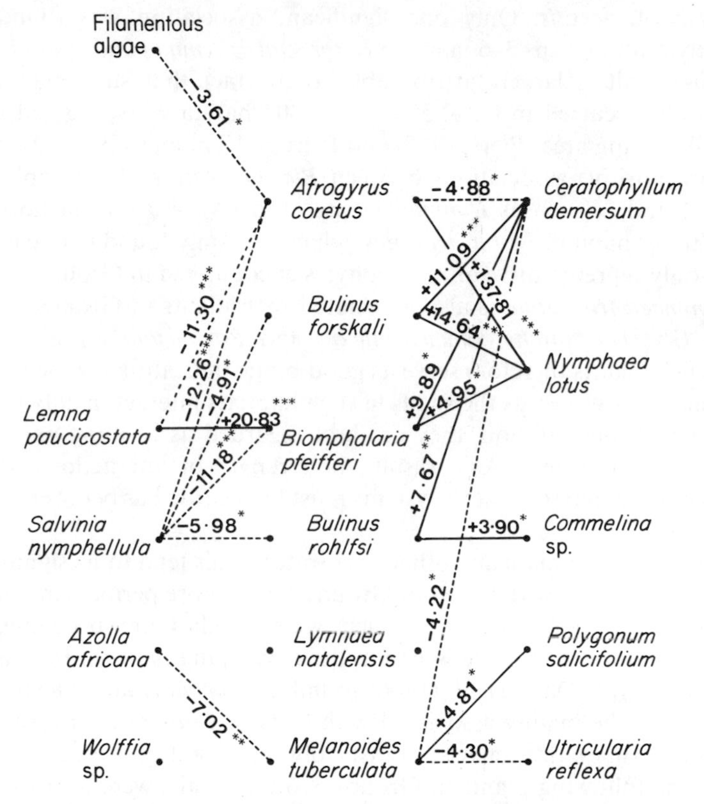

Fig. 7. The nature of the relationships between freshwater molluscs and macrophytes in ITTA Lake, Ibadan, Nigeria. These values are based on 200 samples taken from a cylinder (300 cm surface area) with the aid of a vacuum pump apparatus. The χ^2 values given are based on 2×2 contingency tests. When these are positive, they indicate that the species tend to be found together. Negative values indicate that the two species tend not be be found together*,**,*** signify *P* values less than 0·05, 0·01 and 0·001, respectively.

associations between *B. rohlfsi, Bio. pfeifferi* and *Salvinia nymphellula* (Fig. 7). However, these negative associations could be attributed to the indirect effects resulting from the high plant density. This causes the water column underneath to become anoxic and the tight packaging makes it difficult for the snails to reach the air–water interface. Under low density conditions in the laboratory, it was found that the snails could co-exist with this plant and use the senescent leaves as a resource (Thomas and Tait, 1984).

What are we to conclude from the review of the question regarding the strength and persistence of the associations between pulmonate snails and components of the modular system? Some snail species have remained primarily as components of the Type 1 modular system, and can apparently influence the specific composition of the epilithic algae found in association with them. It would be of interest to ascertain the degree of specificity of such associations. Other snail species tend to be more strongly associated with the surface of macrophytes. However, the strength and persistence of the linkages between the various snail and macrophyte species are likely to vary a great deal both seasonally and on a longer term basis. It can be postulated, however, that such linkages would be at their strongest between plants and snails which are highly committed to life within the water column, such as *Ceratophyllum* and *B. rohlfsi*. However, even in this case, the strength of the associations here are likely to be weaker and less persistent than those reported for coralline algae, their epiphytes and the mollusc *Haliotis*, as reported by Steneck (1977), Barnes and Gonor (1973) and Paine (1980) in the marine environment. These differences can be attributed to the considerable discrepancies between the degree of stability and predictability of the two systems. Nevertheless, the pulmonate snails and the components of the modular systems with which they are associated today have co-evolved together, although perhaps intermittently since the Cretaceous.

B. The More Vulnerable Elements in the Modular System will have Evolved Sophisticated Mechanisms to Reduce Exploitation

Such mechanisms are a prerequisite for the development of mutualistic systems.

1. The Macrophytes in the Snail⇆Macrophyte Subset

The aquatic macrophytes are clearly vulnerable to attack by aquatic snails, including pulmonates. Indeed, many people consider such snails to be simple herbivores. There is now accumulated evidence that when pulmonate snails live in association with macrophytes under natural or simulated natural conditions, they tend to utilize little, if any, of the living plant tissue. Thus, Thomas *et al.* (1985) found that the main components in the diet of *Bio. glabrata*, living in association with *Ceratophyllum*, were dead macrophytic tissue, epiphytic algae and sand grains (Table 2). Only small amounts of what appeared to have been living macrophyte tissue were found, and it is probable that much of this was senescent. Underwood and Thomas (in press) also found that when *Lymnaea peregra* and *Planorbis planorbis* were living in association with this plant, no macrophyte tissue was consumed, the main component in the diet being epiphytic algae. These findings are in agreement with the majority of other workers who have investigated the food of aquatic molluscs, including pulmonate snails (Boycott, 1936; Scheerboom and van Elk, 1978; Reavell, 1980).

Table 2

The relative abundance of various food items in the crop (expressed as a percentage of the total number of cells) of both adult and juvenile *Biomphalaria glabrata* and also the ratios of the number of recognizable cells in the rectum over the number in the crop

	Relative abundance				Ratio of no. cells in rectum/no. in crop	
		Adults		Juveniles	Adults	Juveniles
Macrophytes						
Lemna paucicostata		26·9		17·9	0	0
Ceratophyllum demersum		9·3		0	0	0
Decayed macrophytes		32·2		68·7	0	0
Green algae						
Rhizoclonium		0·02		0·5	1·0	0·6
Scenedesmus		5·6		1·2	0·5	1·0
Diatoms						
Achnanthes	24.5	15·6	7·0	2·4	0·1	0·4
Amphora		0·6		—	0·2	—
Cocconeis		9·1		3·6	0·1	0·3
Cymbella		1·2		—	0·2	—
Fragilaria		0·3		—	0·4	—
Navicula		3·3		0·7	0·2	0·1
Nitzschia		1·3		—	0	—
Synedra		3·1		0·3	0·1	0·5
Sand Grains		1·1		4·6	0·4	0·2

The possible defensive strategies which macrophytes might deploy to counteract possible harmful effects of pulmonate snails and other potential grazers (Thomas, 1987) are as follows:

(1) Shielding of meristem and rapid regeneration.
(2) Nutrient deficiency.
(3) Production of secondary plant compounds (SPC) which may act as repellents, feeding inhibitors and toxicants.
(4) An outer envelope which would be difficult for the snails to penetrate.
(5) Tolerating decoys in the form of epiphytic algae.
(6) Releasing exogenous organic substances.

By acting as an alternative food source to the snails, these could reduce the probability of the plant itself being attacked.

The evidence presented by Thomas (1987) suggests that aquatic macrophytes rely mainly on strategies (1), (4) and (5), although the possibility that feeding inhibitors may be activated when the epidermal cells are penetrated cannot be ruled out. Evidence in support of this possibility is given by Sterry *et al.* (1983), as they found that freshly homogenized plants like *Elodea, Ceratophyllum, Nasturtium, Hippuris* and *Epilobium* spp. were significantly repellent to *Bio. glabrata.* Although submerged aquatic macrophytes are generally well endowed with potentially toxic SPCs (Ostrofsky and Zettler, 1986; Thomas, 1987), there is no evidence that they are toxic to snails (Thomas, 1987). This is because the snails have evolved mechanisms to detoxify them. However, SPCs present in emergent leaves of plants in Group 2 (e.g. *Alternanthera sessilis, Polygonum* sp.) and Group 3 (*Nymphaea*) were toxic to the snails. This effect can be attributed to their novelty, as snails would not normally encounter them. By the time they enter the water column as dead plants, the toxic SPCs will have been degraded by bacteria (Fig. 2).

2. The Periphytic Algae in the Snail⇆Periphytic Algae Subset

The snail's mouthparts, particularly the radula, are very well adapted for taking up surface algae. As a result, the latter are much more vulnerable than the macrophytes. However, the various species of algae present on surfaces are not taken up to the same extent by the grazing action of the snails. Thus it has been shown, by determining electivity indices (Ivlev, 1961), that both positive and negative selection can occur during the ingestion process (Underwood & Thomas, in press).

In addition, it has been shown that many species of algae are resistant to digestion after being ingested by snails (Van Aardt and Wolmarans, 1981; Cuker, 1983a, b; Thomas *et al.*, 1985; Underwood and Thomas, in press). The latter authors have also shown that a high percentage remains viable after egestion.

Some of the defensive strategies which may be deployed by periphytic algae and other plants to increase the probability of them surviving when interacting with pulmonate snails or other grazing invertebrates have been discussed by Porter (1977), Otto and Svenson (1981) and Gregory (1983). A comprehensive list of algal defensive strategies is given in Table 3. Porter (1977) suggested that indigestibility might serve as a preadaptation allowing algae to establish obligatory symbiosis with aquatic invertebrates. Cuker (1983a, b) implies that algae that resist digestion continue to function metabolically as they pass through the gut of the snail and suggests that both the snails and the algae may exchange resources during the process. If indeed the algae are being "milked" of some of their resources during gut passage, this would reduce the need for snails to expend energy in producing enzymes to digest them. It follows, therefore, that the selective pressures on the snails to produce efficient systems to kill and digest the algae would be reduced.

Table 3
Spatio-temporal classification of the defensive strategies available to periphytic algae

1 *Pre-contact*
1.1 Release of chemical repellants or toxicants
1.2 Visual or chemical crypsis
2 *Post-contact*
2.1 Anatomical features which may reduce the probability of being ingested
2.1.1 Adnate growth form
2.1.2 Adhesion to substrate
2.1.3 Hard outer envelope
2.1.4 Small size allowing use of refugia near to surface of the macrophyte
2.1.5 Large size making ingestion difficult, e.g. large cylindrical shape
2.1.6 Anatomical processes, e.g. spines and hairs
2.2 Release anti-feedants, including mucilagenous material
2.3 Patchy distribution, i.e. rare, occurring singly rather than in patches
3 *Post-ingestion*
3.1 Hard, digestion-resistant outer envelope
3.2 Release of toxicants, causing regurgitation or rapid gut passage
3.3 Release of nutrients, thus reducing need for digestion
4 *Life-history strategies*
4.1 High capacity to reproduce either asexually (e.g. budding, etc.) or sexually.
4.2 Good powers of dispersal
4.3 Polymorphisms, allowing escape into other systems, e.g. plankton, epilithon, neuston or symbiosis

This may account for the observations that the ability of many invertebrates to digest cellulose is limited (Calow and Calow, 1975). This brief review indicates that periphytic algae have evolved sophisticated strategies to decrease the probability of being ingested and digested by invertebrate grazers. It is to be expected that these strategies would be better developed in the algal taxa which have co-evolved with the snails.

3. The Periphytic Bacteria in the Snail⇋Periphytic Bacteria Subset

The question of the importance of bacteria found on living or dead plants as a food for pulmonate snails and other grazing invertebrates still remains unanswered. Thomas *et al.* (1985) have shown that the morphology of the radula and its mode of action preclude the snail from exploiting adpressed bacteria on the leaf phylloplane to any marked extent. As a result, their relative abundance was found to increase on grazed substrates. In contrast, the bacteria present in decaying plant material are much more readily ingested. However, the following considerations would appear to suggest that it is unlikely that they could be readily killed and digested by snails.

First, they have evolved to exploit chemical resources in decaying systems under conditions similar to those in the gut. They are, therefore, preadapted for survival in the gut under such conditions. Secondly, they have been passing through the gut of invertebrates for hundreds of millions of years. It is to be expected, therefore, that the selective pressures would favour those which would be resistant to digestion.

However, there are two schools of thought regarding the importance of ingested and digested bacteria as a food source for invertebrates. According to one, the ingested bacteria are readily digested and are more important as a food source for invertebrate detritivores than the organic matter in which they are located (Newell, 1965; Calow, 1974; Barlocher and Kendrick, 1975; Fenchel, 1975; Kofoed, 1975; Fenchel and Jørgensen 1977; Harper *et al.*, 1981a, b). Although Harper *et al.* (1981a, b) have produced evidence that the detritivorous invertebrate *Nais* assimilates DNA of bacterial origin and that bacterial viability appears to decline during passage through the gut, it remains to be shown how much nutrient these worms actually obtain by killing and digesting the bacteria. Other workers take the opposite view. Thus Baker and Bradnam (1976) concluded that bacteria are not as important quantitatively as other components of the detrital food material to species of *Baetis, Ephemeralla, Simulium* and *Chirononomus*. Similar conclusions have also been recorded by Tunnicliffe and Risk (1977), Cammen *et al.* (1978), Jensen and Siegismund (1980), Lopez and Levinton (1978), Bianchi and Levinton (1981, 1984) and Levinton and Bianchi (1981) regarding the nutrition of deposit feeding macrofauna. More recently, Willoughby and Earnshaw (1982) found that fungal spores and unicellular bacteria in fine detritus are not digested by *Gammarus pulex*. Although Willoughby and Marcus (1979) found that *Asellus* could digest actinomycetes, the growth rates achieved by animals fed on pure actinomycete cultures was less than those of animals fed on macroscopic food. It must be concluded, therefore, that the general consensus among freshwater biologists is that other components of the detritus are more important as food than the bacteria themselves (Tenore, 1977; Tenore *et al.*, 1979). This view is put succinctly by Ladle (1982) who states that "The actual gain of material from ingestion of bacteria is insufficient for the energetic requirements of most of the feeding organisms in streams". It is evident from these conflicting views that further detailed quantitative work is needed to resolve these questions.

However, on balance, it can be concluded that there is good evidence that the macrophytes, epiphytic algae and bacteria are well endowed with efficient mechanisms which protect them, to varying degrees, from being exploited by consumer species like snails. This opens up possibilities for mutualistic interactions involving the exchange of nutrients between organisms which have complementary requirements.

C. If Components of the Modular System Derive Mutual Benefits, then it would be to the Advantage of Sedentary Forms to Produce Kairomonal Attractants

As little work has been done to test this hypothesis, the evidence for or against is sparse, and only seems to be available for the macrophyte⇆snail subset. In the case of freshwater modular systems, Brönmark (1989) has shown that *Ceratophyllum demersum* does release kairomones which attract species of freshwater pulmonate snails, whereas the epiphytic algae associated with this plant do not. This kind of interaction resembles that described between the marine coralline algae *Corallina* and molluscs such as *Haliotis* and *Tonicella*. In this case, the kairomone has more far-reaching effects, as it causes the competent larvae of these molluscs to settle and undergo metamorphosis on the plant (Steneck 1977; Barnes and Gonor, 1973; Paine, 1980). This kairomone has been characterized and shown to be γ-aminobutyric acid, which also acts as a neurotransmitter in the brain of vertebrates (Morse *et al.*, 1979). This finding reflects the tight nature of the linkages in this marine modular system. Paine (1980) makes no mention of the response shown by the larvae of the chiton *Katharina* to the *Corallina* kairomonc. This is surprising because he has shown that it is the presence of *Katherina* as a cleaning symbiont on *Corallina* that allows the *Corallina* algae and other molluscs in the module to exist. *Tonicella*, one of the other molluscs present, grazes directly on *Corallina* itself. It can be hypothesized, therefore, on evolutionary grounds, that the γ-aminobutyric acid kairomonal system might have evolved in the first place to attract *Katharina*, and that the other chiton *Tonicella* subsequently latched on to the system. Further work is required to characterize kairomones released by freshwater macrophytes and to elucidate their effects on the snails. It is likely that the picture that will eventually emerge from such studies could be quite complex, as it is known that some plants may produce repellents under certain conditions. Thus, de Coster and Persoone (1970) found that some gastropod species leave *Ceratophyllum demersum* beds in the spring. They suggested that this might be due to the production of a repellent or antifeedant, called myriophillin, by the growing buds to prevent over-exploitation at this time.

D. The Growth, Reproduction and Longevity of One of the Components in the Subset should be Enhanced by the Presence of the Others

Only some of the components in the modular system will be considered in detail in this section. These will include the macrophytes and the snails in the snail⇆macrophyte subset and the periphytic algae and bacteria in the snail⇆periphytic algae and the snail⇆periphytic bacteria subsets.

1. *The Plant* (Ceratophyllum) *in the Macrophyte⇆Snail Subset*

This hypothesis receives support from the work of Brönmark (1985, 1989), as he showed that the growth of *Ceratophyllum*, measured in terms of stem length and number of nodes, was significantly enhanced when it was grown in the presence of two snail species, *Planorbis planorbis* and *Lymnaea peregra*. These results have recently been corroborated by detailed studies carried out by Underwood (1989). Underwood measured the growth of *Ceratophyllum* in terms of stem length, number of leaf nodes and growing tips and was able to show that each one of these parameters was significantly enhanced when the plant was cultured separately in the presence of either *L. peregra* or *P. planorbis* under both laboratory (Fig. 8) and field conditions. As the growing tips readily separate from the main stem, they serve as a means of asexual reproduction. Underwood (1989) has also demonstrated that the longevity of *Ceratophyllum* is also enhanced by the presence of *P. planorbis* (Fig. 9).

The mechanisms which result in the growth of the macrophyte being enhanced by the presence of snails have yet to be fully elucidated. Brönmark (1985, 1989) suggested that the effects were due to the snails acting as cleaning symbionts. However, recent work by Underwood (see Fig. 10) has shown that even when the snails are prevented from making direct contact with the plant, the medium conditioned by them can cause the growth of the plant to be enhanced to the same extent as when the snails are in direct contact. It seems likely, therefore, that the enhancement of plant growth in the presence of snails is caused, in part at least, by excretory and egested factors released by the snails. These include ammonia, urea, phosphate and various cations. Snails could also benefit the macrophytes by removing epiphytic algae capable of releasing toxic allelopathic compounds and of competing with them for resources. Snail grazing may also increase the longevity of the plants by removing potentially pathogenic bacteria or fungi. Both Rogers and Breen (1983) and Underwood (1989) have produced evidence based on their studies of *Potomogeton crispus* and *C. demersum*, respectively, in support of this hypothesis. The benefits which macrophytes may derive from co-existence with snails are summarized in Fig. 11.

Many other workers (Howard, 1982; van Montfrans *et al.*, 1982a, b, 1984; Hootsmans and Vermaat, 1985; Howard and Short, 1986) have also produced evidence, based on field studies, that aquatic macrophytes may benefit from the presence of molluscs. There is, however, another school of thought which takes the opposite view; namely, that many molluscs including pulmonate snails, utilize green, photosynthesizing macrophyte tissue as a major resource (Morton, 1958; Frömming, 1953, 1956; Gaevskaya, 1969; Pip and Stewart, 1976; Sheldon, 1987). The latter author even suggests that freshwater pulmononate snails, such as *Physa gyrina* (Say), may be responsible for major reductions in biomass of some macrophyte species and even in

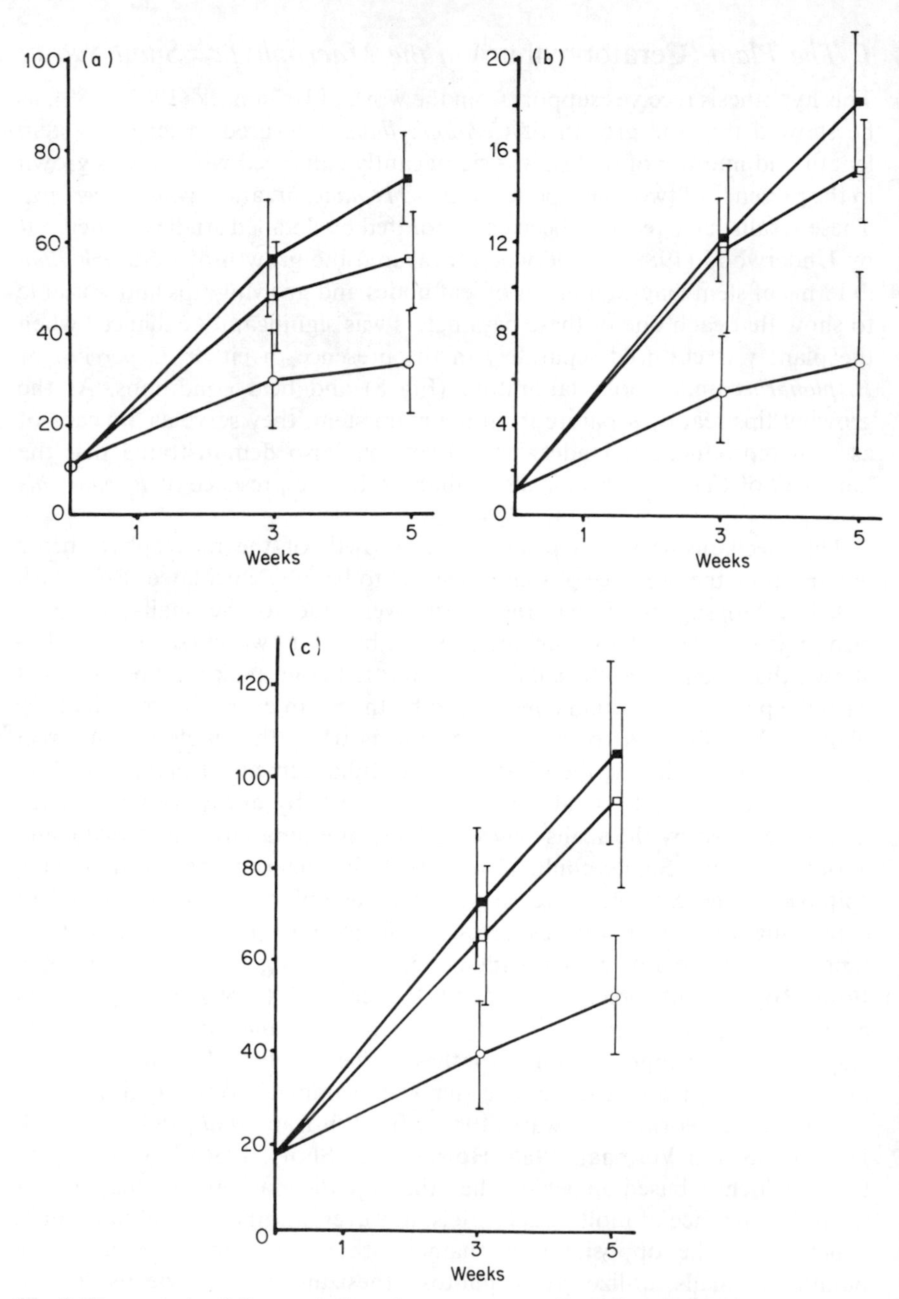

Fig. 8. The growth of *Ceratophyllum demersum* over a 5-week period under laboratory conditions, when incubated in the absence of snails (○) and in the presence of *Lymnaea peregra* (□) or *Planorbis planorbis* (■). Growth was measured in terms of (a) stem length (cm), (b) numbers of growing tips and (c) number of leaf nodes. The vertical bars represent the 95% confidence limits.

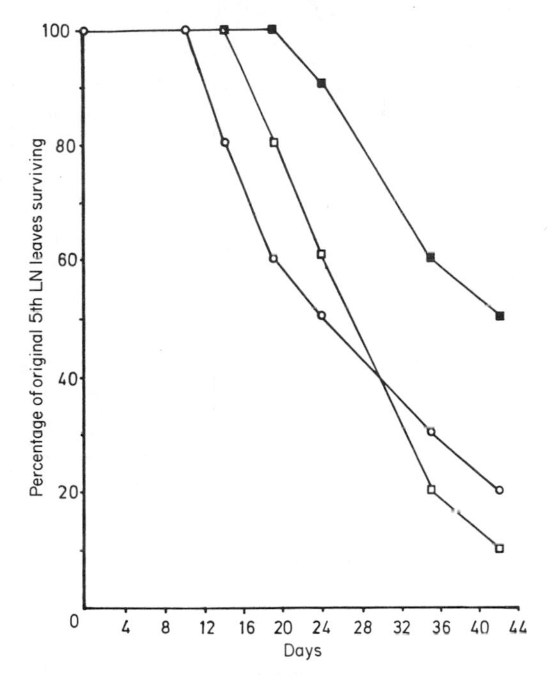

Fig. 9. The percentage of the original fifth leaf nodes (LN) surviving in three treatments over a period of up to 44 days under laboratory conditions. −G/−C (○), control, no snails; +G/+C (■), in the presence of grazing (G) and conditioning (C) *Planorbis planorbis*; −G/+C (□), in the presence of media conditioned by *Planorbis planorbis*.

their disappearance from some water bodies. In other words, freshwater pulmonate snails are cast in the same role as phytophagous insects in the terrestrial environment (Strong *et al.*, 1984).

In view of this conflict of opinions, a re-examination of the evidence on which the above claims are based is necessary. A closer scrutiny of Gaevskaya (1969) reveals that snail species living under natural conditions were observed to consume large amounts of decomposing macrophytes. As senescent plant material may still have remains of chlorophyll, it is possible that it might have been wrongly identified as living tissue. Such an error would result in an over-estimation of the amount of living plant tissues present in the stomach of the snails. The direct evidence cited by Sheldon (1987) in support of her claim that *Physa gyrina* harms some aquatic macrophytes is based on laboratory bioassays involving very small portions of plants, with up to one leaf, and relatively high densities of snails (2, 4 and 8 per plant) in an unspecified volume. Under these conditions, she found that the plants suffered a net loss in weight when the snails were present. The field

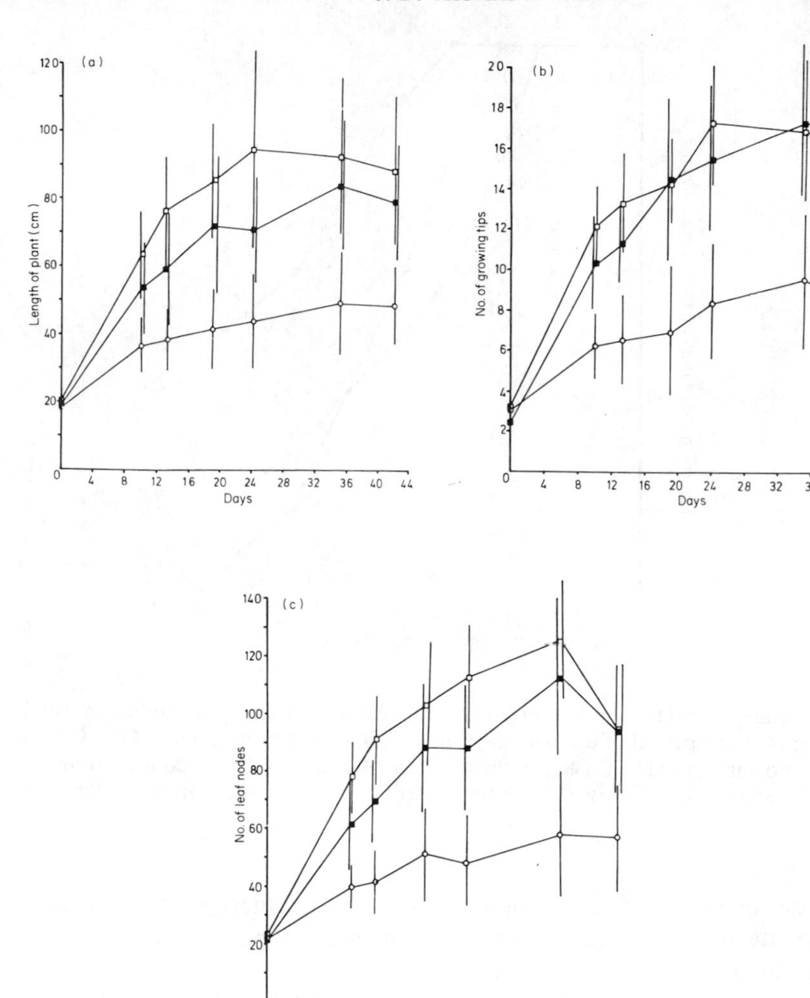

Fig. 10. Growth as measured in terms of (a) stem length, (b) number of growing tips, and (c) number of nodes achieved by *Ceratophyllum demersum* over a period of 44 days in three treatments. −G/−C (○), control without snails; +G/+C (■), in the presence of grazing (G) and conditioning (C) *Planorbis planorbis*; −G/+C (□), in the presence of media conditioned by *Planorbis planorbis*. The vertical bars represent the 95% confidence limits.

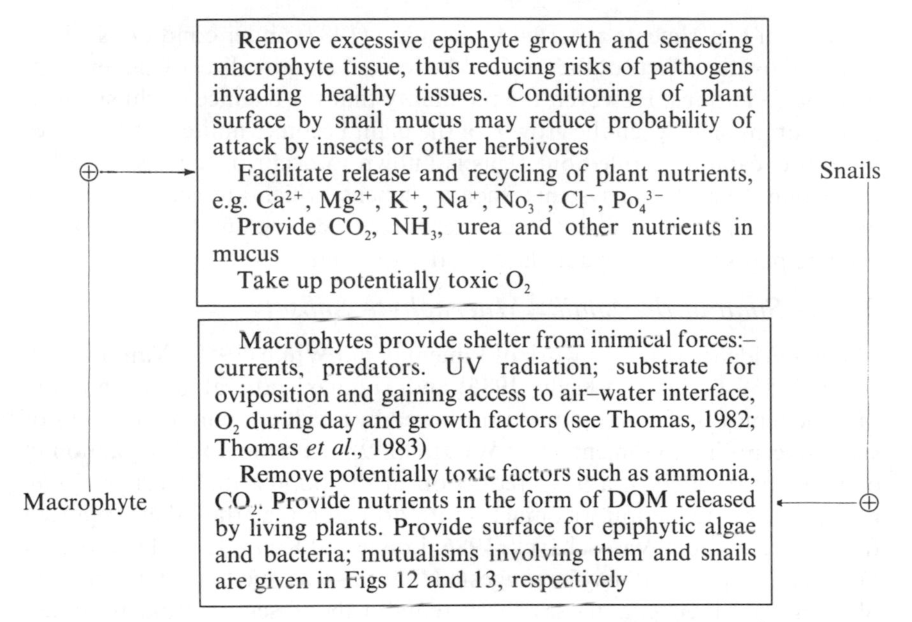

Fig. 11. An analysis of mutualistic interactions in the snail ⇆ macrophyte subset of module type 2.

evidence cited by Sheldon (1987) that snails harm plants is of a circumstantial nature and rests on the observation that plant species in cylinders from which fish were excluded performed less well than plants kept in cylinders with slits to allow fish to enter. It was argued that the plants in the open cylinders performed better than those in the closed cylinders because they allowed fish predators to remove herbivorous snails. These results can, however, be criticized for the following reasons.

First, many aquatic macrophytes do not grow well under laboratory conditions and tend to degenerate, particularly when cut into portions. Secondly, there are considerable errors in determining the wet weight of aquatic plants. Thirdly, the snail densities per unit of plant biomass used in the laboratory experiments are likely to be higher than those found in nature. Fourthly, the plants in the open cylinders might have grown better because they allowed nutrient and gas exchange to occur more freely. Fifthly, if it is assumed that the fish in the field experiments did act as predators, it is likely that they would also prey on other invertebrates as well as molluscs. This possibility receives support from observations made by Soszka (1975 a, b), as he found that damage to *Potomogeton* leaves in nature was caused mainly by trichopteran and lepidopteran larvae. An examination of snails' stomach contents by Soszka (1975a,b) revealed that they consumed only very small amounts of living macrophytes. It would seem desirable therefore to repeat the kind of experiments reported by Sheldon (1987) under conditions which would allow the effects of the snails to be looked at more directly under more realistic conditions. However, the possibility that the balance might swing in favour of the snail when the growth of the plant becomes limited by resources in nature, cannot be ruled out. This is known to occur in the case of other autotroph–heterotroph systems, such as lichens, which are generally recognized as being mutualistic. When resources become limiting, the fungi become parasitic and devour their algal symbionts.

2. The Snail in the Snail⇆Macrophyte Subset

There is evidence from the work of Pimental and White (1959), Van Schayck (1985, 1986), Lodge and Kelly (1985) and Thomas and Daldorph (in press) that the removal of macrophytes is followed by a reduction in populations of snails. According to Pimentel and White (1959), the removal of vegetation in one stream resulted in a 90% reduction in the population density of *Bio. glabrata*. This reduction was much more dramatic that the others reported from lentic systems (Van Schayck, 1986; Lodge and Kelly, 1985; Thomas and Daldorph, in press). It may be suggested that this was due to the vulnerability of snails and their egg masses to currents in the absence of macrophytes, particularly under spate conditions. However, the results obtained in both lentic and lotic systems are not unexpected when the benefits which the snails may derive by co-existing with macrophytes are considered (see Fig. 11). It is noteworthy that the results obtained in lentic systems (Van Schayck, 1986; Thomas and Daldorph, in press) were based on relatively short-term observations. It is possible that the effects on the snail populations would have been more marked had the experiments been carried out over a longer period. Further quantitative work to test this hypothesis should be undertaken in the developing nations in water bodies where transmission of schistosomiasis is taking place. The information given here provides the rationale for including macrophyte control as part of an environmental management package, aimed at reducing the transmission of schistosomiasis. Field studies involving the removal of macrophytes by various means, as well as their additions, would provide information of the cost-effectiveness of particular measures.

3. The Periphytic Algae and Bacteria in the Periphytic Algae⇆Snail and Periphytic Bacteria⇆Snail Subsets

There is now accumulating evidence that the relative abundance of certain algae, typically those with adnate growth forms, small size and the ability to resist digestion, increases when the substrate is grazed by snails or other organisms (Hunter, 1980; Patrick, 1970; Cuker, 1983a, b; Dickman, 1968; Lamberti and Resh, 1983; Jacoby, 1987; Gregory, 1980; Nicotri, 1977; Peterson, 1987; Underwood and Thomas, in press). It has also been shown that when grazer densities increase to a critical threshold, the rate of production of such algae (Hunter, 1980; Cuker, 1983a; Jacoby, 1985) or their primary production (Hunter, 1980; Lamberti and Resh, 1983) may be enhanced, as predicted by Lamberti and Moore's (1984) model. This may also be accompanied by an increase in algal species richness (Eichenberger and Schlatter, 1978; Sumner and McIntire, 1982). Such results are not unexpected when the benefits which particular species of periphytic algae may derive from co-existing with such grazers are considered (Fig. 12).

There appears to be no information on the effects of grazers like pulmonate snails on the species composition of the periphytic bacterial community. This is not unexpected in view of the taxonomic difficulties. However, Levinton and Bianchi (1981) and Bianchi and Levinton (1981, 1984) found that the steady-state bacterial standing stock in sediment grazed by the mud snail *Hydrobia* was not disturbed even when the snail density was greatly increased. There is also evidence that the introduction of grazing invertebrates, such as amphipods in marine sediments, resulted in a doubling of the microbial population (Fenchel and Jørgensen, 1977). Such results are not unexpected when the benefits which the periphytic bacteria may derive from co-existing with detritivorous invertebrates, such as snails and amphipods or isopods, are considered (Fig. 13). Further studies along these lines are needed to ascertain the effects of adding or excluding particular detritivorous invertebrates, such as snails, to the distribution and abundance of particular bacterial taxa in the periphyton or sediments.

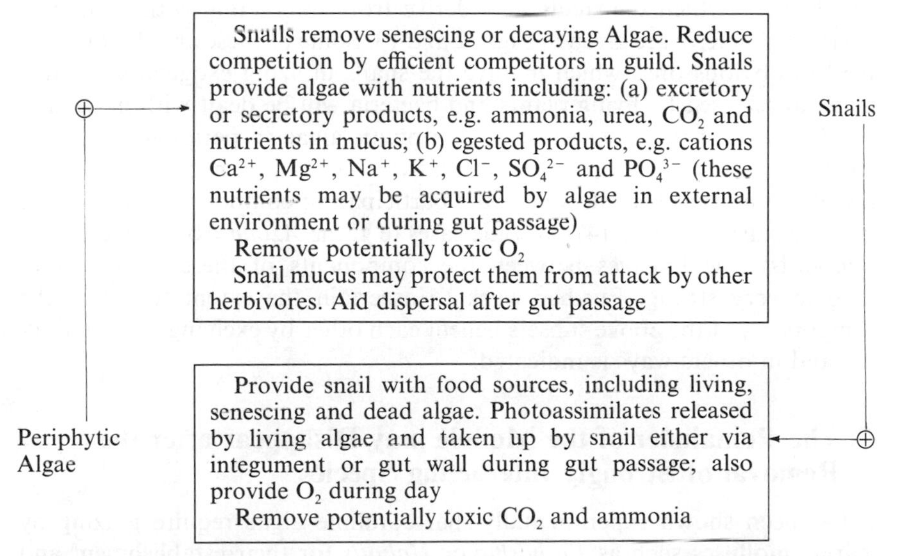

Fig. 12. An analysis of mutualistic interactions in the snail ⇆ periphytic algae subset (these include epilithic or epiphytic algae in module types 1 and 2, respectively).

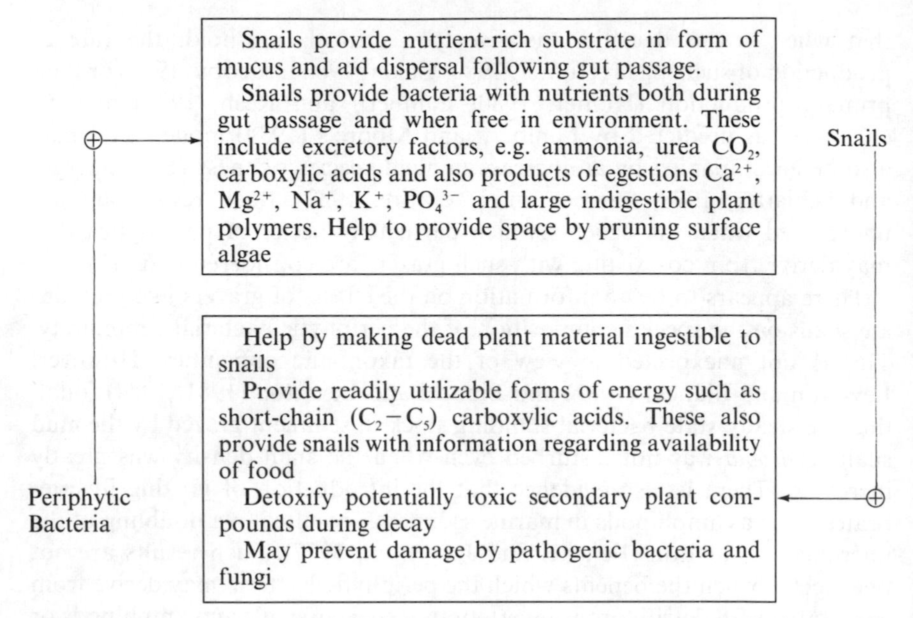

Fig. 13. An analysis of mutualistic interactions in the snail ⇆ periphytic bacteria subset (these include epilithic or epiphytic bacteria in module types 1 and 2, respectively).

4. Other Components in the Subsets of the Modular System

The benefits which the snails may derive from interacting with periphytic algae and bacteria are given in Figs 12 and 13. Some of these are obvious, but the less obvious ones which involve the snails utilizing exogenous organic material released by living plants and bacteria will be dealt with in Section III. Analyses of the other subsets—periphytic algae ⇆ periphytic bacteria, epiphytic algae ⇆ macrophytes and epiphytic bacteria ⇆ macrophytes—also reveal the mutual benefits which the participating components may derive from co-existence (Fig. 14). In some cases (e.g. the *Azolla*⇆blue-green algae symbiosis), the linkages between the components of these subsets have become very strong. Further work to ascertain the extent to which the components of the above subsets benefit each other by exchange of metabolites and in others ways is indicated.

E. The Remainder of the Module may Disappear after the Removal of Strongly Interacting Species

It has been shown experimentally that coralline algae require grazing by marine molluscs such as *Katharina* or *Haliotis* for their establishment and persistence (Paine and Vadas, 1969; Vine, 1974). In the absence of such

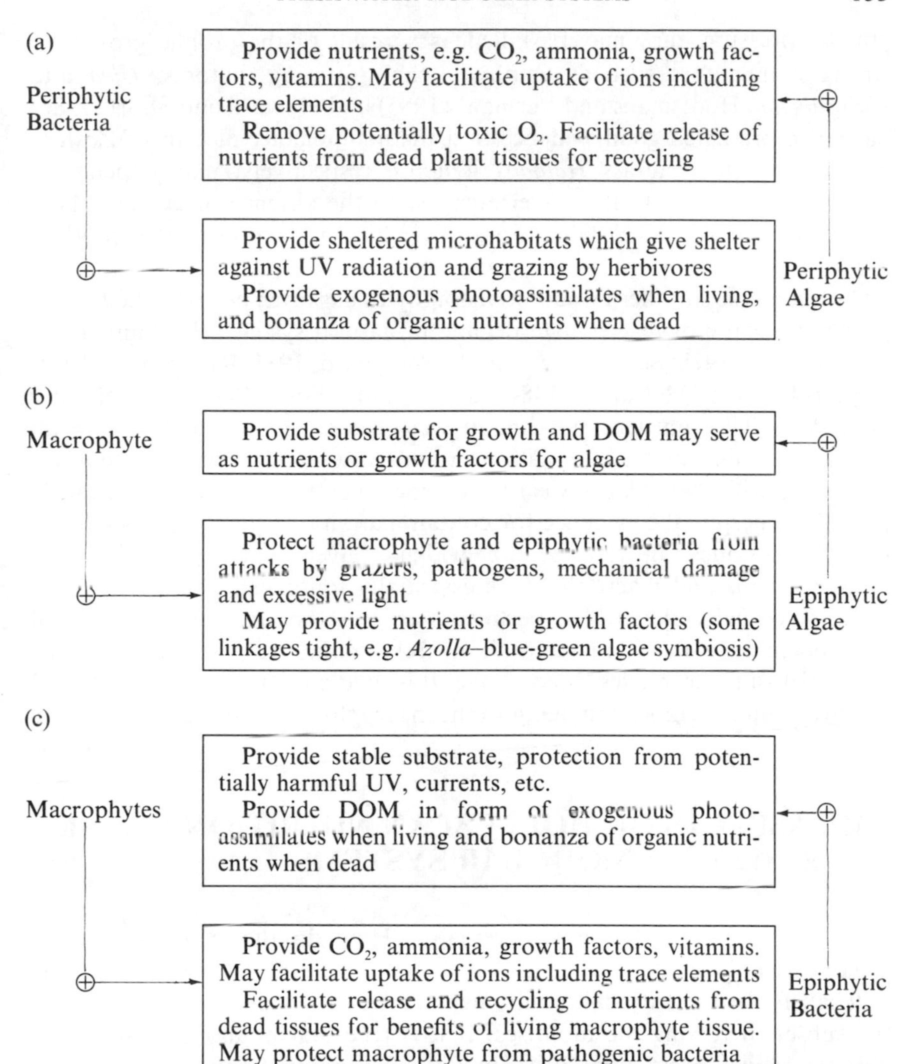

Fig. 14. An analysis of mutualistic interactions in other subsets of the modules. (a) The periphytic algae ⇆ periphytic bacteria subset (includes the epilithic and epiphytic forms); (b) the epiphytic algae ⇆ macrophyte subset; (c) the macrophyte ⇆ epiphytic bacteria subset.

grazing pressure, they may be killed as a result of the prolific growth of strong competitors such as epiphytic chain-forming diatoms, *Ulva* and *Halosaccion*. Hootsmans and Vermaat (1985) and Howard and Short (1986) have also presented good evidence that the maintenance of beds of *Zostera marina* and the seagrass *Halodus wrightii*, respectively, may depend on grazing by snails and other invertebrates. In the absence of grazers, these plants suffered increased defoliation, lower biomass and decreased productivity.

Although the smothering action of epiphytes has often been invoked as the proximate cause of macrophyte decline in natural systems (Larkum, 1976; Phillips *et al.*, 1978; Sand-Jensen and Søndergaard, 1981; Kemp *et al.*, 1983; Cambridge and McComb, 1984; Short and Short, 1984; Graneli and Solander, 1988), nutrient enrichment was seen as the causal factor in all cases. A decline in the populations of molluscs and other invertebrates associated with these plants was not considered as a possible contribution to this effect, despite the evidence for contamination of aquatic ecosystems by pesticide residues. In view of the evidence given that the relationships between snails and macrophytes is essentially mutualistic, it would be of interest to undertake field experiments to test the strength of the snail component as a factor in maintaining representatives of each of the six categories of macrophytes listed above. It is predicted that the effect of snail removal would be more damaging to the macrophytes in Group 6 than to the others.

III. ASPECTS OF MOLLUSCAN NUTRITION IN THE MODULAR SYSTEMS

One vitally important question remains. How do the snails and other organisms in the modular systems, where mutualistic interactions are important, satisfy their nutritional requirements? Attention here will be focused on the subsets involving the molluscs. If it is true that pulmonate and other aquatic snails ingest very little living macrophyte tissue, and that an appreciable proportion of the algae and bacteria ingested remain viable during gut passage, is it possible that they may supplement their nutritional requirements in some other way? One possibility is that they obtain some nutrients from chemicals leached out of dead organisms. However, bacteria and fungi have an advantage here and there is now accumulating evidence which suggests that snails may also obtain an appreciable amount of their metabolic requirements from living organisms. These include the metabolic end-products of bacteria as well as photoassimilates released by periphytic algae and macrophytes. These possibilities are discussed below.

A. Contributions to the Snail's Metabolism by Dissolved Organic Matter (DOM) of Bacterial Origin

When macrophytes, periphytic algae and animals die, they are immediately invaded by periphytic bacteria. As these grow exponentially, they are likely to obtain the lion's share of the free metabolites and also the more readily available organic polymers. The rapid growth of the bacteria results in the environment within the decaying plant envelope becoming anoxic. As a result, the bacteria are obliged to use glycolytic fermentation as a metabolic strategy. This results in the production of appreciable amounts of short-chain (C_2–C_5) carboxylic acids as end-products of their metabolism (Patience *et al.*, 1983; Sterry *et al.*, 1985; Hoffman *et al.*, 1986; Hipp *et al.* 1986 a, b); Daldorph, 1988; Daldorph and Thomas, in press a, b) (Fig. 15). In the case of decaying *Lemna*, for example, it has been shown that C_2–C_5 acids may constitute 24·2% of the original dry weight (Patience *et al.*, 1983; Table 4). Some species of pulmonate snails such as *Bio. glabrata* are attracted to these acids at low concentrations of 10^{-6}–10^{-7} M (Fig. 16) over the range of pH

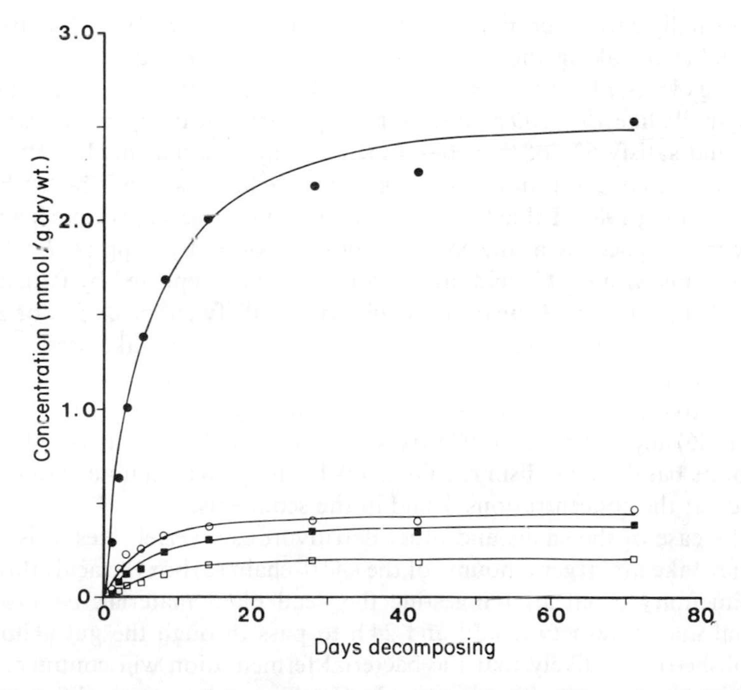

Fig. 15. Concentrations of short-chain carboxylic acids (C_2–C_5) produced during the decomposition of *Lemna paucicostata* (see Patience *et al.*, 1983, for details). ●, Acetate; ○, propanoate; ■, butanoate; □, pentanoate.

Table 4
The chemical composition of *Lemna* supernatant after decomposition over 71 days (see Patience *et al.*, 1983)

	% g/100 g (dry wt)	mmol/g (dry wt)
Acetate	15·2	2·53
Propanoate	3·4	0·47
Butanoate	3·5	0·40
Pentanoate	2·1	0·21
Ammonia	5·0	2·94
Amino compounds	6·7[a]	0·45
Protein	0·4	
	36·3	
"Solids"	*c.* 45 ± (5)	
Total	*c.* 80% of original dry wt	

[a]Assuming mean molecular weight of 150.

they normally encounter in nature (Fig. 17; Thomas *et al.*, 1980, 1983b), and are capable of taking them up through the integument and metabolizing them (Figs 18 and 19; Thomas *et al.*, 1984; Daldorph, 1988). It was calculated that if snails like *Bio. galbrata* were relying solely on integumental uptake, they could satisfy 6% of their basal metabolism requirements by taking up acetate at the concentrations found in stream sediments in north-east Brazil (Thomas *et al.*, 1984; Table 5). On the other hand, if the snails were living on sediments with acetate at the concentrations reported by Hipp *et al.* (1986b) or in detritus with C_2–C_4 acids at the concentrations reported by Patience *et al.* (1983) in decaying *Lemna*, the snails could satisfy either 28·5% or all of their basal metabolic requirements by integumental uptake, respectively (Table 5). Short-chain carboxylic acids are also important sources of nutrients to other benthic invertebrates such as oligochaetes. Thus, Hoffman *et al.* (1986) and Hipp *et al.* (1986b) calculated that a *Tubifex* sp. could satisfy 40% of its basal metabolism requirements by integumental uptake of C_2 and C_3 acids at the concentrations found in the sediments.

In the case of the snails and other detritivorous invertebrates, it is likely that they take far larger amounts of the short-chain carboxylic acids through the alimentary canal after ingesting the dead plant material. As ingested material may take between 10 and 24 h to pass through the gut (Thomas, unpublished), it is likely that the bacterial fermentation will continue, thus increasing the amount of food available. Further work is required to quantity the contributions made by these acids present in the alimentary canal to the snail's metabolism. In addition, it is to be expected that the snails will also

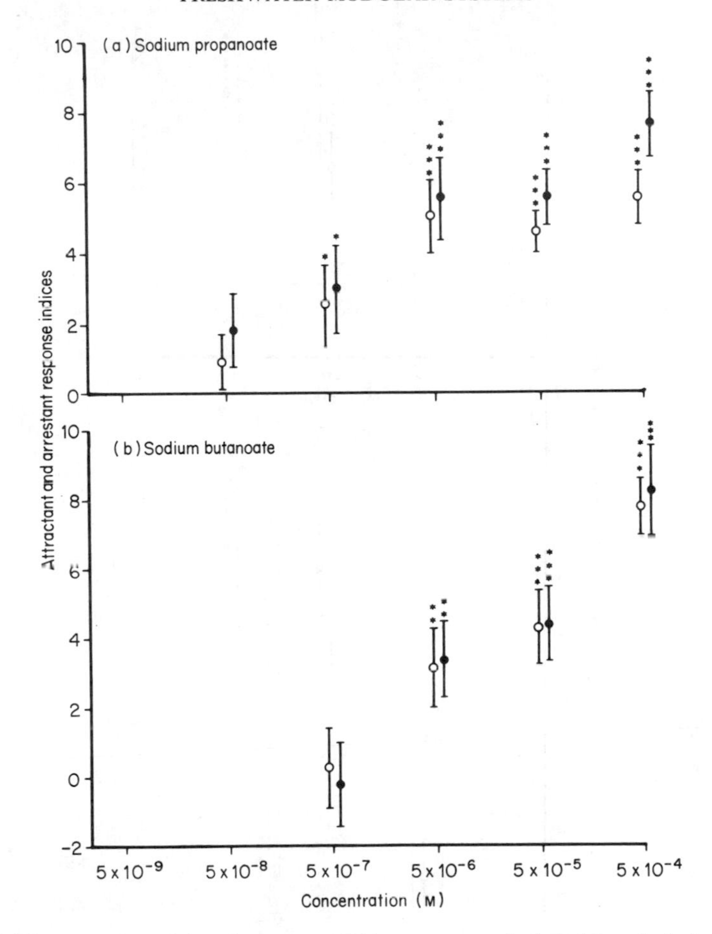

Fig. 16. The attractant (●) and arrestant (○) responses of adult *Biomphalaria glabrata* to varying concentrations of (a) propanoic and (b) butanoic acids when tested in diffusion olfactometers (see Thomas *et al.*, 1983b). *,**,***: *P* values, $<0{\cdot}05$, $<0{\cdot}01$ and $<0{\cdot}001$ respectively.

derive some nutrients from the dead plants and bacteria. The system as envisaged, therefore, bears some resemblance to that described for ruminants. These obtain most of their energy requirements in the form of C_2–C_5 acids produced by bacteria fermenting ingested plant material in the rumen (Latham, 1979). Unlike ruminants, which ingest living plant tissue, the snails rely heavily on identifying microcosms of decay containing bacteria and their metabolic end-products, including short-chain carboxylic acids. In view of the importance of the latter to snails such as *Bio. glabrata*, it is not

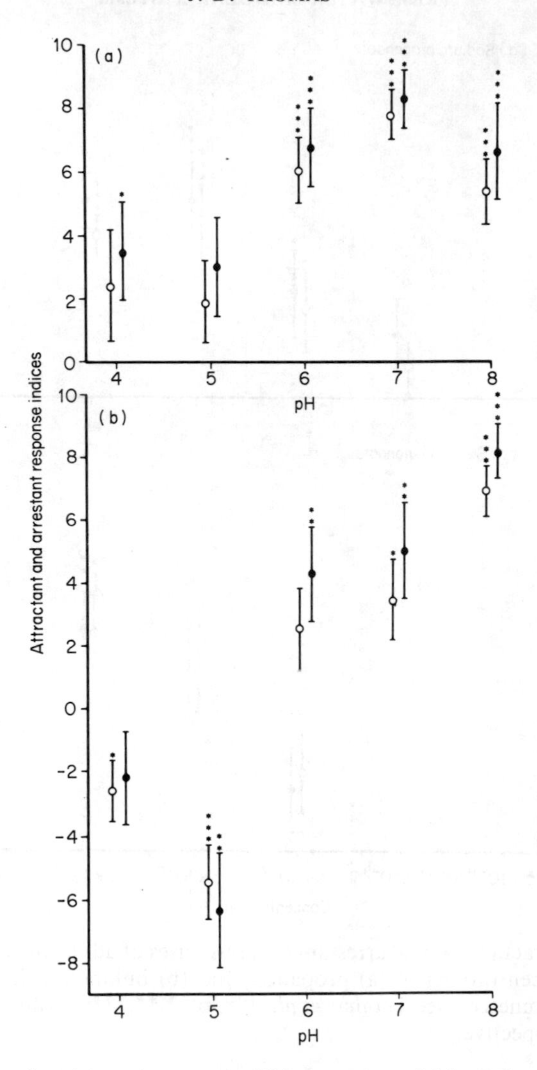

Fig. 17. The attractant (●) and arrestant (○) responses of (a) adult and (b) juvenile *Bio. glabrata* to pulmonate snails tested at various pH values in diffusion olfactometers (see Thomas *et al.*, 1983b). *,**,***: *P* values, $<0{\cdot}05$, $<0{\cdot}01$ and $<0{\cdot}001$ respectively.

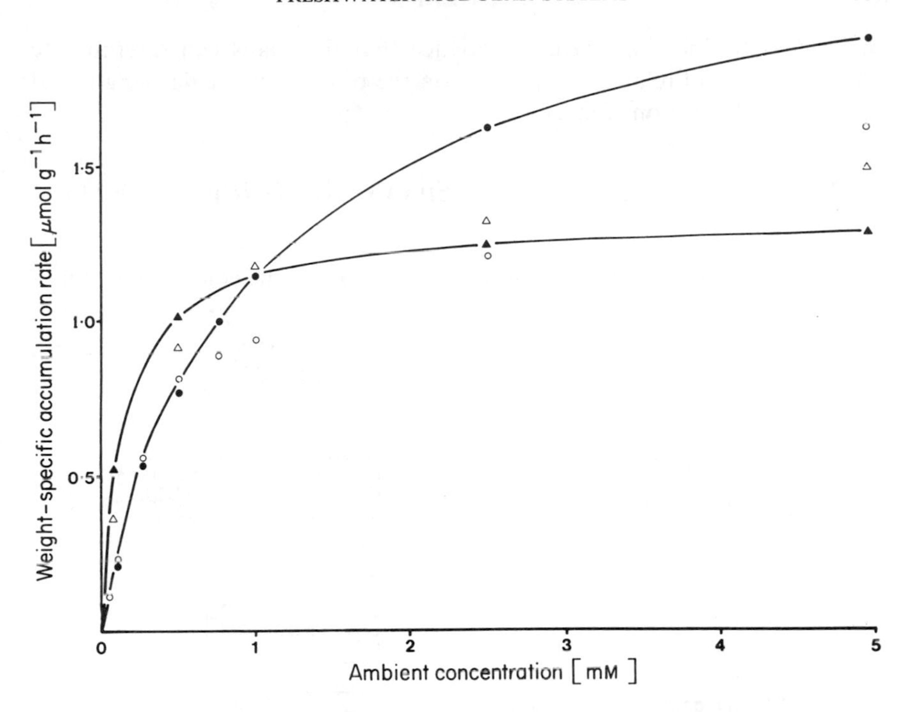

Fig. 18. The mass-specific accumulation rate (in μmol equivalents per g wet mass per h) by 35–400 mg of *Bio. glabrata* incubated in acetate and butanoate at various concentrations. The calculated values are derived from the best fit regression lines in Fig. 19 (see Thomas *et al.*, 1984). Acetate: ○, actual values; ●, calculated values; butanoate: △, actual values: ▲, calculated values.

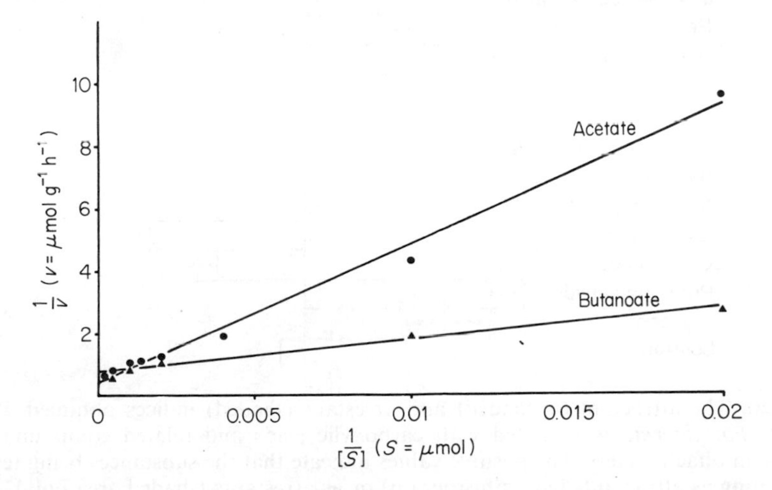

Fig. 19. Double reciprocal plot of $1/v$ against $1/[S]$ for acetate (●) and butanoate (▲). v, Accumulation rate in μmol equivalents per g wet mass per h; S, concentrations in μmol l^{-1}. The regression equations are $1/v = 439{\cdot}2/[S] + 0{\cdot}44$ for acetate and $1/v = 100{\cdot}4/[S] + 0{\cdot}77$ for butanoate. J_{max} and K_s values are 2·29 and 10·05, respectively, for acetate and 1·29 and 129·5 for butanoate. (see Thomas *et al.*, 1984).

unexpected to find that there is evidence that the snails can discriminate between them and related compounds on the basis of molecular length and chemical configuration (Fig. 20).

B. Possible Contributions to the Snail's Metabolism by DOM of Algal Origin

As pointed out by Jüttner and Matuschek (1978), Mague *et al.* (1980), Jones

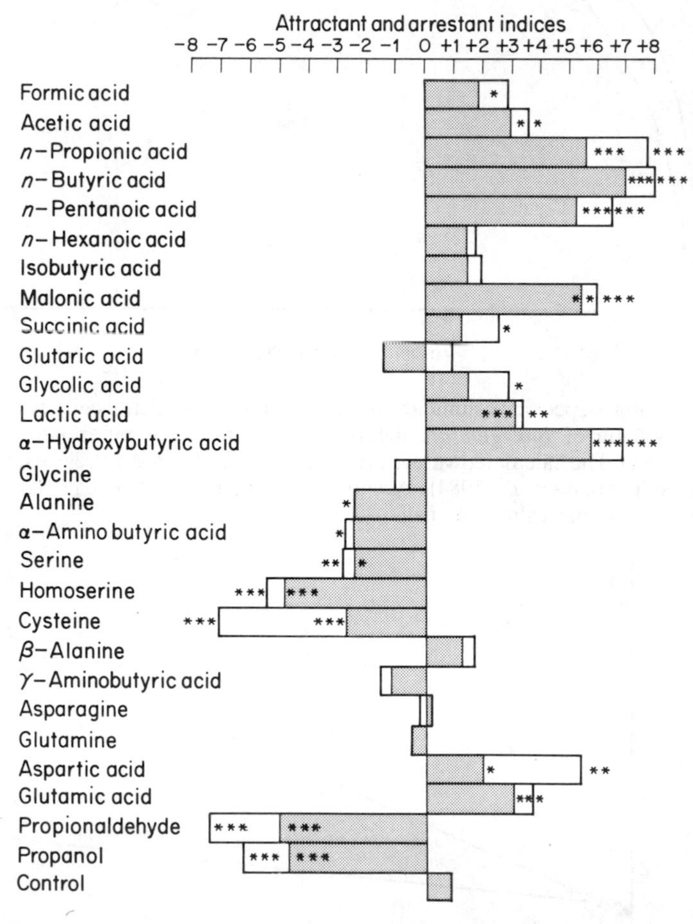

Fig. 20. The attractant (unshaded) and arrestant (shaded) indices obtained when adult *Bio. glabrata* were tested with carboxylic acids and related compounds in diffusion olfactometers. The positive values indicate that the substances being tested are acting as attractants (entire histogram) or as arrestants (shaded area only). The negative scores indicate that the substances are acting as repellents. *,**,***, indicate P values of $<0{\cdot}5$, $<0{\cdot}01$ and $<0{\cdot}001$, respectively (see Thomas *et al.*, 1980).

Table 5

Possible contributions of dissolved carboxylic acid to the energy budget of *Biomphalaria glabrata* (250–300 mg wet mass) (see Thomas *et al.*, 1984)

Parameters	Sources substrate	Decaying *Lemna* Acetate	Decaying *Lemna* Butanoate	Sediment acetate
J_{max} μm equiv. g^{-1} dry mass h^{-1}		22·9	12·9	22·9
K_s (mM)		1·0	0·129	1·0
Estimated max conc.		18·0	2·9	0·1
R(influx) in μmol g^{-1} h^{-1}		21·69	12·35	2·08
O_2 required for metabolism of substrate in μmol		43·38	49·4	4·16
Percentage of O_2 requirements satisfied by available substrate concentration		64·0	92·0	6·0

et al. (1983) and Søndergaard *et al.* (1985) there is now ample evidence, from carefully controlled experiments, that aquatic algae secrete significant quantities of DOM. The chemical species involved include glycolic acid (Fogg, 1971; Weinemann, 1970) pyrrolidone carboxylic acids (Jüttner and Fritz, 1974) free amino acids (Pomeroy, 1974, 1984; Ducklow, 1983; Stephens, 1982) and sugars such as maltose (Douglas and Smith, 1984). It has been shown that both the ciliate *Euplotes* and *Hydra* are capable of "milking" species of symbiotic *Chlorella* within its tissues by adjusting the pH (Finlay and Fenchel, 1989; Mews and Smith, 1982). It is possible that pulmonate snails may be able to use this strategy with ingested periphytic algae, many of which remain viable while passing through their alimentary canal (Underwood and Thomas, in press). Cuker (1983b) has also suggested that nutrients such as phosphates, nitrates and various cations may be taken up by the algae during this process. Experimental work is needed to quantify the nutrient exchange likely to occur between the algae and the snail during the transit of the former through the gut of the latter. In addition to this, there is evidence that snails such as *Bio. glabrata* are capable of taking up exogenous sugars such as glucose and maltose that might be released by periphytic algae or macrophytes, either by drinking (Fig. 21) or through the integument (Figs 22 and 23; Thomas *et al.*, 1990). The results given in Figs 22 and 23, in conjunction with published values for the concentrations of glucose found in natural habitats (probably released by epilithic algae), have made it possible

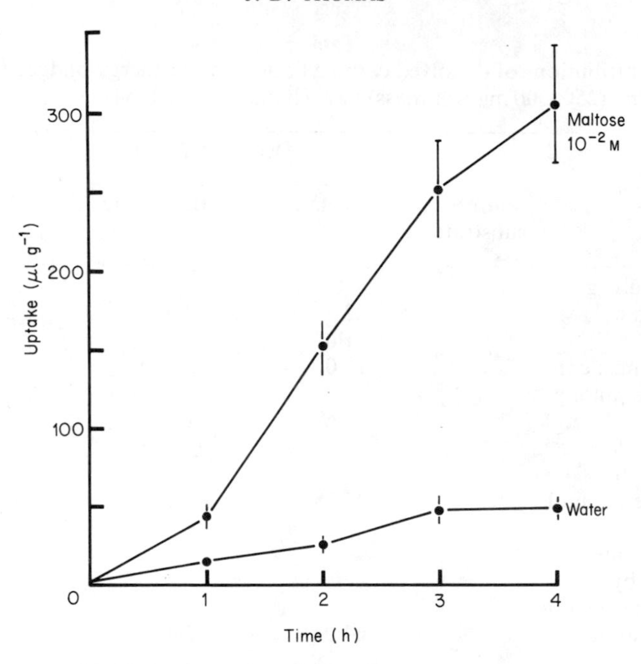

Fig. 21. The mean drinking rate in μl g^{-1} (wet weight of snail) over various time intervals for up to 4 h, by adult *Bio. glabrata* placed in a standard medium (SSW2) without maltose and with maltose at concentrations of 10 mM. The bars indicate ± standard errors.

to calculate that the glucose present in the water column, or in the interstitial water in sediments, could provide the snails with 0·48–13·7 and 10·3–14·8% of their basal metabolism, respectively (Table 6). In view of the importance of sugars both as nutrients and as sources of information for the presence of important targets in the snail's environment, it is not unexpected to find evidence that the snails can discriminate between different kinds of sugars. Thus some sugars, such as maltose, are much more potent as attractants, arrestants and phagostimulants than others (Figs 24–27). It has been shown that the snails can respond to maltose at concentrations as low as 5×10^{-7} to 5×10^{-6} M (Fig. 25). As has already been suggested, the release by the algae of a proportion of their photoassimilates for the benefit of the snails may reduce the selective pressures on the latter to evolve more efficient mechanisms to ingest and digest the algae.

C. Possible Contributions to the Snail's Metabolism by DOM of Macrophyte Origin

Axenic freshwater macrophytes are known to release between 4 and 5% of their photoassimilates as DOM (Wetzel and Manny, 1972; Baker and Farr,

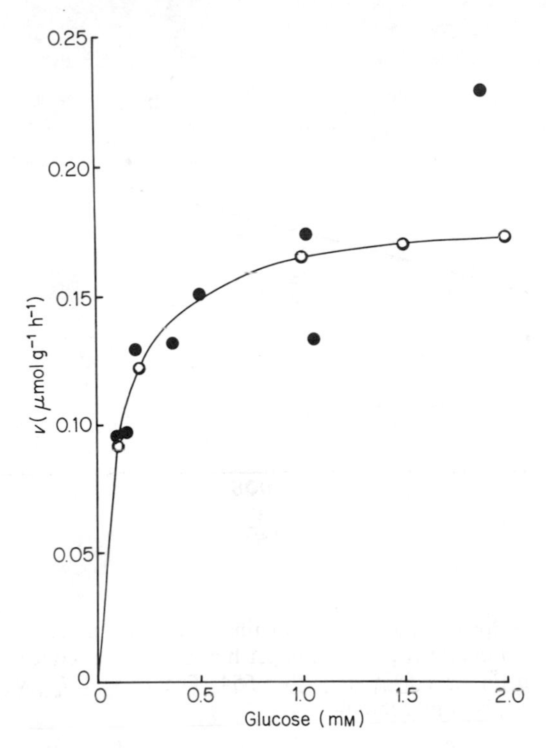

Fig. 22. The empirically determined mass-specific accumulation rates (ν) in μmol equivalents per g wet mass per h (●) by adult *Bio. glabrata* incubated in glucose at concentrations below those which induce drinking. The calculated (○) values are derived from the best fit regression line given in Fig. 23.

1987). Among the chemical species identified in the DOM by Wetzel and Manny were glucose, sucrose, fructose and xylose. Erokin (1972) and Morris and Culkin (1975) also found sugars to be present in DOM released by marine algae. It has been shown that the pulmonate snail *Bio. glabrata* is capable of taking up a part of this DOM (Table 7). It is likely that when the snails glide over the surface of the macrophyte within the modular systems, they will encounter locally high concentrations of DOM, of which some may also be taken up by epiphytic algae and bacteria. It would be of interest to measure how much of this DOM contributes to the snail's metabolism, but this can only be done when the chemical species involved have been identified and quantified. Within the evolutionary context, it can be suggested that it might be cost-effective for the plants to release DOM for the benefit of surface-living organisms like snails, which can enhance their fitness. This strategy would be analogous to the release of sugars from extrafloral nectaries by terrestrial plants for the benefit of symbiotic ants (Strong *et al.*,

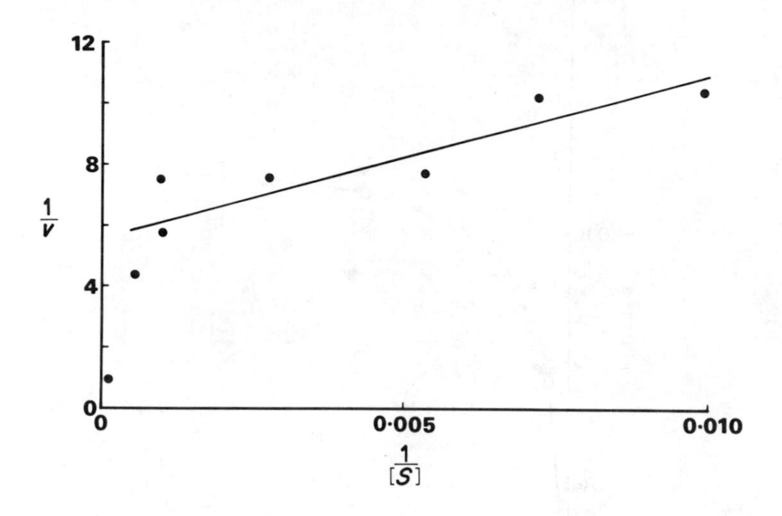

Fig. 23. Double reciprocal plot to $1/v$ against $1/[S]$ for glucose, where v is the accumulation rate in μmol per g wet mass per h and $[S]$ is the concentration in µmol l^{-1}. The regression line is $1/v = 534{\cdot}7[S] + 5{\cdot}554$. The K_s and J_{max} values calculated from these data are given in Table 6.

Table 6

Possible contribution of glucose present as dissolved organic matter in the water column to the energy budget of *Biomphalaria glabrata* (250–300 mg wet weight) (see Thomas *et al.*, 1990)

Sources	Water column[a]		Sediments[b]	
	min.	max.	min.	max.
Estimated conc. (mM)	0·003	0·55	0·17	1·05
Influc (μmol g^{-1} h^{-1})	0·56	1·53	1·15	1·65
O_2 required for metabolism of substrate (μmol g^{-1} h^{-1})	0·32	9·19	6·89	9·89
Percentage of O_2 requirement for basal metabolism satisfied by available substrate concentrations	0·48	13·7	10·28	14·8

Calculations based on $J_{max} = 1{\cdot}8$ μmol g^{-1} dry mass h^{-1} and $K_t = 96/3$ μM; basal respiration rate of 67 μmol g^{-1} dry mass body h^{-1} (Thomas *et al.*, 1984).

[a] Based on values given by Riemann *et al.* (1982).

[b] Based on values given by Vallentyne and Bidwell (1956).

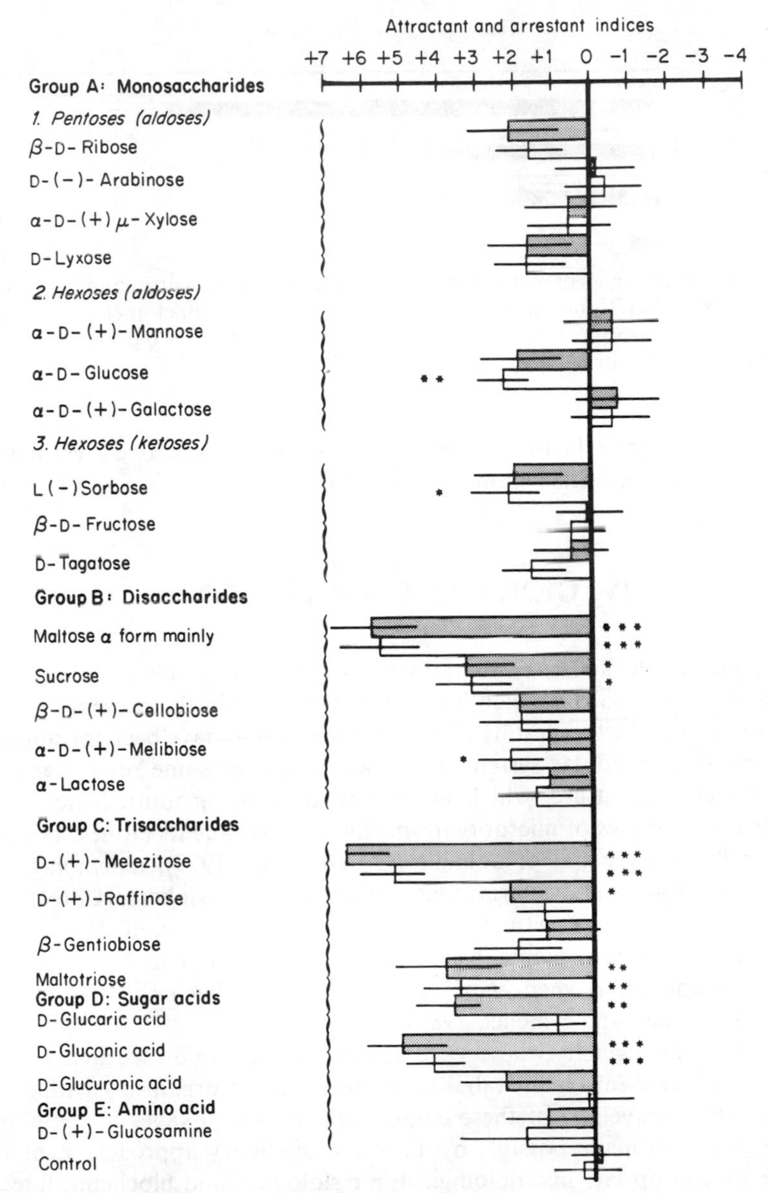

Fig. 24. The effectiveness of various sugars as attractants (shaded), arrestants (unshaded) (positive values) or as repellents (negative values) to adult *Bio. glabrata* in diffusion olfactometers. All the sugars were tested at effective concentrations of 5×10^{-4} M. Standard errors of means are represented by bars. *,**,***, $P<0{\cdot}05$, $<0{\cdot}01$, $<0{\cdot}001$, respectively. None of the sugars tested was repellent (see Thomas, 1986).

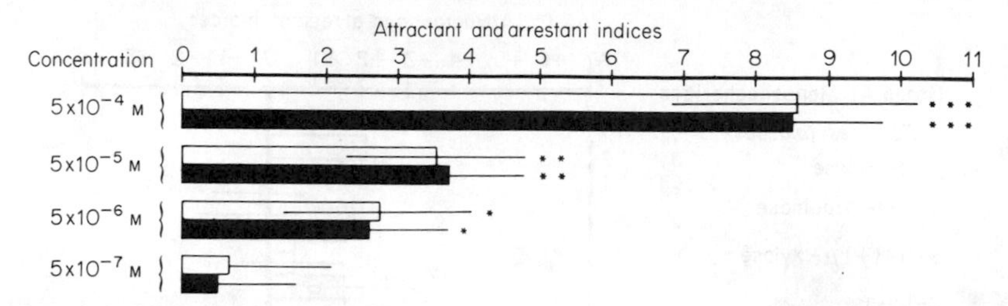

Fig. 25. The effectiveness of maltose at various concentrations as attractants (□) or as arrestants (■) to adult *Bio. glabrata* as measured in diffusion olfactometers. Standard errors of means are represented by bars. *,**,***, $P<0{\cdot}05$, $<0{\cdot}01$ and $<0{\cdot}001$, respectively (see Thomas, 1986).

1984). It would have the further benefit of reducing the selective pressures on the snails to produce efficient mechanisms for their destruction and exploitation.

IV. GENERAL CONCLUSIONS

A great deal of work remains to be done to evaluate the modular concept and the theory of mutualism which have been developed in the present chapter. This will involve further testing of the hypotheses that have been formulated. The freshwater modular systems described here bear some resemblance to other systems in nature, which are regarded as being mutualistic. These include communities of microorganisms living in sediments (Jones, 1985) or in the alimentary canal of animals (Lynch *et al.*, 1979), assemblages of planktonic algae and bacteria constituting the phycosphere (Chrost and Faust, 1983), *Rhizobium* and higher plants (Postgate and Hill, 1979), mycorrhiza and higher plants, bacteria in the alimentary canal of ruminants and other animals (Lynch *et al.*, 1979), protozoa, bacteria and algae in sediments (Finlay and Fenchel, 1989).

Despite their diversity, these systems have one common feature, i.e. the exchange of nutrients or metabolites between living organisms within the system. The unravelling of these complex metabolic linkages in the above system has been made possible by an interdisciplinary approach involving the use of appropriate microbiological, physiological and biochemical techniques. A similar approach is now needed to further our understanding of the modular systems containing higher invertebrates which have been described in this chapter. Enough evidence has now been presented to show that living organisms within these systems do release dissolved organic matter, including carboxylic acids and sugars, which can then be utilized by others.

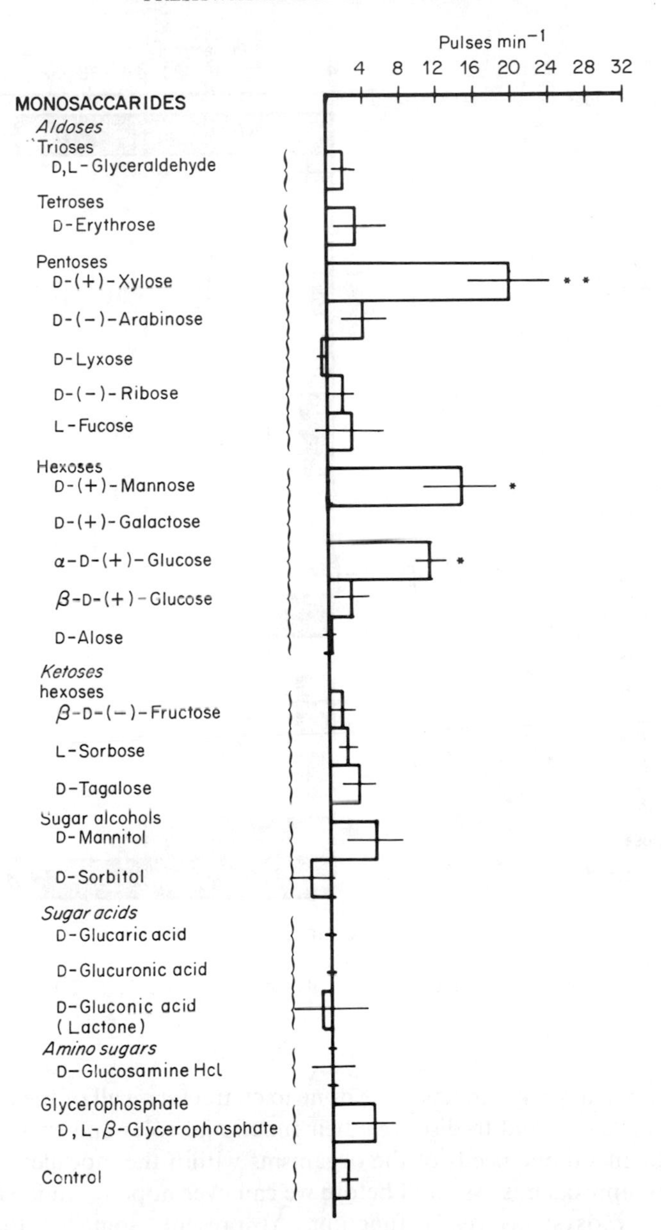

Fig. 26. Influence of various monosaccharides on the buccal mass response of adult *Bio. glabrata* (250 mg approximately) measured in pulses min^{-1} ($\pm$ S.E.). *,**,***, $P<0{\cdot}05$, $<0{\cdot}01$, $<0{\cdot}001$, respectively. (See Thomas *et al.*, 1986)

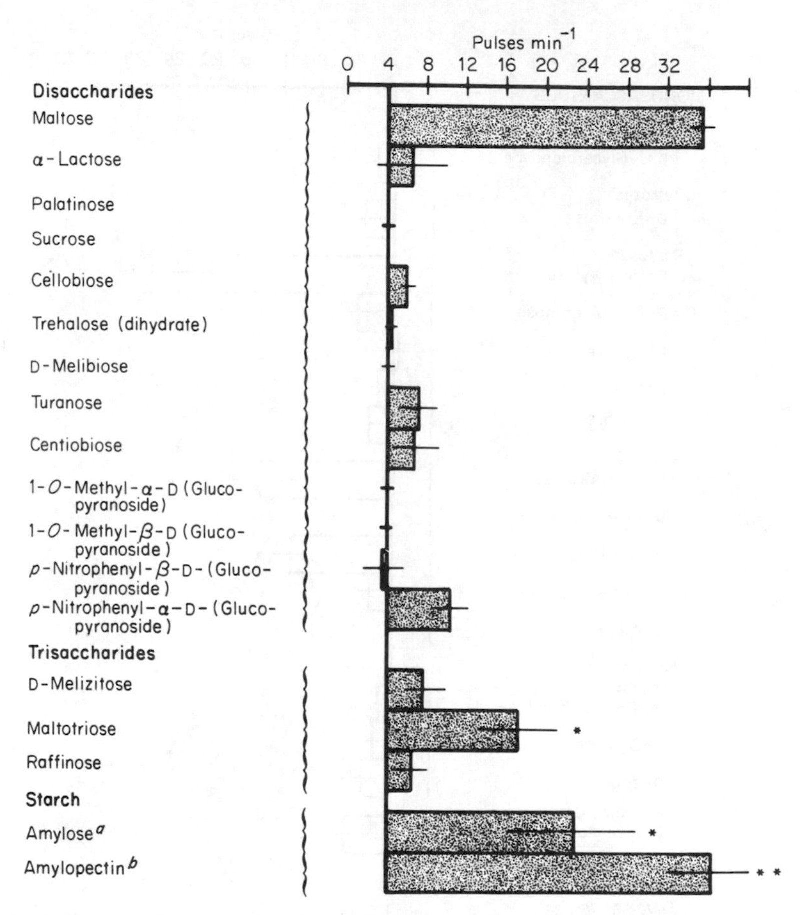

Fig. 27. Influence of various disaccharide sugars on the buccal mass response of adult *Bio. glabrata* (250 mg approximately) measured in pulses min^{-1} ($\pm$S.E.). *,**,***, $P<0{\cdot}05$, 0·01, 0·001, respectively. [a] Hydrolysed starch 0·2%; [b]0·2 g 100 ml^{-1}. (See Thomas *et al.*, 1986)

However, much work remains to be done to characterize all of the exogenous chemicals released and to discover their functional roles and their contribution to the metabolic needs of the organisms within the modules.

Such an approach is essential before we can ever hope to understand how freshwater ecosystems really function. At present, some of the central dogmas of ecological and evolutionary biology place much emphasis on interactions with a negative component, such as competition (−/−), predation (+/−), herbivory (+/−) and parasitism (+/−). This has led to the dominance of the concept of food web and energy flow, measured in terms of calories. Thus, Pimm (1982) regards the food web as "the centrepiece of the

Table 7

The mean ^{14}C activity expressed as a percentage of the total in various compartments after the snails *Bio. glabrata* had been incubated in media containing ^{14}C-labelled dissolved organic matter released by non-axenic *Lemna* for 38 h, (Thomas and Tait, 1984)

	Treatment		
	Snails with alcohol-swabbed shell	Snails with normal shell	Killed snails (control)
Compartments	$\bar{x}$ S.D.	$\bar{x}$ S.D.	$\bar{x}$ S.D.
Body	9·5±5·3	10·0±4·7	0·0±0·1
Haemolymph	6·3±2·6	5·7±2·7	0·3±0·2
Shell	0·4±0·3	0·4±0·2	0·0±0·0
Faeces	3·1±2·2	3·4±3·3	1·1±0·3
Mucus	0·2±0·2	0·2±0·1	0·0±0·0
Total organic carbon in water column	46·5±12·7	45·5±5·6	30·2±4·5
Dissolved inorganic carbon	9·8±1·8	10·0±5·5	9·0±2·0
CO_2	24·9±16·0	24·8±5·7	59·3±6·5

Note: The swabbing of snail shells with 50% ethanol prior to the experiments did not influence the uptake characteristics of the dissolved organic carbon by the snails to any extent. The values for ^{14}C activity in the dead snail compartments were either zero or very low. The activity in the faeces may be due to colonization by microorganisms.

biotic community part of ecosystem models", while Fretwell (1987) considers food chain dynamics to be the central theory of ecology. Implicit in this concept is the belief that all higher eukaryotic organisms obtain their energy entirely by eating and digesting other organisms. It follows from this that all the information required to explain how these systems function can be obtained by examining the stomach contents of higher heterotrophs and determining the calorific values of the component parts. This chapter (see also Odum and Biever, 1984) has shown that this approach is over-simplistic and may give misleading information. Thus many small organisms remain viable after being ingested, and dissolved organic matter from living organisms (e.g. DOM in aquatic medium, nectar and plant vascular material) is an important component in the diet of many organisms (Odum and Biever, 1984). The modular systems which have been described provide a realistic structure to examine the role of DOM in freshwater ecosystems. However, this can only be achieved by an interdisciplinary approach involving the use of appropriate microbiological and biochemical techniques within this modular framework.

ACKNOWLEDGMENTS

The author wishes to thank the UNDP/World Bank/WHO Special Programme for Research in Training in Tropical Diseases, the MRC and NERC for financial support, which made some of the work reviewed here possible. Thanks are also due to Dr M. Wallis for providing facilities, M. J. Stenning and C. Kowalczyk for technical assistance, G. Underwood for allowing the use of some of his figures, and Mrs S. T. Laurence for typing the manuscript.

REFERENCES

Allen, H. L. (1971). Primary production, chemo-organotrophy, and nutritional interactions of epiphytic algae and bacteria on macrophytes in the littoral of a lake. *Ecol. Monogr.* **1,** 97–126.

Baker, J. H. and Bradnam, L. A. (1976). The role of bacteria in the nutrition of aquatic detritivores. *Oecologia (Berl.)* **24,** 95–104.

Baker, J. H. and Farr, I. S. (1987). Importance of dissolved organic matter produced by duckweed (*Lemna minor*) in a southern English river. *Freshwater Biology* **17,** 325–330.

Barlocher, F. and Kendrick, B. (1975). Assimilation efficiency of *Gammarus pseudolimnaeus* (Amphipoda) feeding on fungal mycelium or autumn shed leaves. *Oikos* **26,** 55–59.

Barnes, J. R. and Gonor, J. J. (1973). The larval settling response of the lined chiton, *Tonicella lineata. Mar. Biol.* **20,** 259–264.

Begon, M. and Mortimer, M. (1981). *Population Ecology*. Blackwell, Oxford.

Bianchi, T. S. and Levinton, J. S. (1981). Nutrition and food limitation of deposit feeders. II. Differential effects of *Hydrobia totteni* and *Ilyanassa obsoleta* on the microbial community. *J. Mar. Res.* **39,** 547–556.

Bianchi, T. S. and Levinton, J. S. (1984). The importance of microalgae, bacteria and particulate organic matter in the somatic growth of *Hydrobia totteni. J. Mar. Res.* **42,** 431–443.

Boycott, A. E. (1936). The habitats of freshwater Mollusca in Britain. *J. Anim. Ecol.* **5,** 116–186.

Brønmark, C. (1985). Interactions between macrophytes, epiphytes and herbivores: An experimental approach. *Oikos* **45,** 26–30.

Brønmark, C. (1989). Interactions between epiphytes, macrophytes and freshwater snails: A review. *J. Moll. Stud.* **55,** 299–322.

Calow, P. (1974). Evidence for bacterial feeding in *Planorbis contortus* Linn. Gastropoda: Pulmonata). *Proc. Malac. Soc. Lond.* **41,** 145–156.

Calow, P. and Calow, L. J. (1975). Cellulase activity and niche separation in freshwater gastropods. *Nature (Lond.)* **255,** 478–480.

Cambridge, M. L. and McComb, A. J. (1984). The loss of seagrass in Cockburn Sound, Western Australia. I. The time course and magnitude of seagrass decline in relation to industrial development. *Aquat. Bot.* **20,** 229–243.

Cammen, L., Rublee, P. and Hobbie, J. (1978). The significance of microbial carbon in the nutrition of the polychaete *Nereis succinea* and other aquatic deposit feeders. *Sea Grant Publication* No. UNC-SG-12, pp. 1–84, University of North Carolina.

Carignan, R. and Kalf, J. (1982). Phosphorus release of submerged macrophytes: Significance to epiphyton and phytoplankton. *Limnol. Oceanogr.* **27,** 419–427.

Carpenter, S. R. and Lodge, D. M. (1986). Effects of submersed macrophytes on ecosystem processes. *Aquat. Bot.* **26,** 341–370.

Chrost, R. H. and Faust, M. A. C. (1983). Organic carbon release by phytoplankton: Its composition and utilisation by bacterioplankton. *J. Plankton. Res.* **5,** 477–493.

Cook, D. K. C., Gut, B. J., Rix, E. M., Schneller, J. and Seitz, M. (1974). *Water Plants of the World.* W. Junk, The Hague.

Cuker, B. E. (1983a). Grazing and nutrient interactions in controlling activity and composition of the epilithic algal community in an arctic lake. *Limnol. Oceanogr.* **28,** 133–141.

Cuker, B. E. (1983b). Competition and coexistence among the grazing snail *Lymnaea*, Chironomidae, and microcrustacea in an arctic epilithic lacustrine community. *Ecology* **64,** 10–15.

Daldorph, P. W. G. (1988). The role of short-chain carboxylic acids in the ecology of freshwater snails. Ph.D. Thesis, University of Sussex.

Daldorph, P. W. G. and Thomas, J. D. (in press a). The distribution of short-chain carboxylic acids entrophic drainage channels. *Hydrobiologia.*

Daldorph, P. W. G. and Thomas, J. D. (in press b). Snail cadavers as sources of short-chain carboxylic acid to scavenging freshwater invertebrates. *Hydrobiologia.*

Dazo, B. C., Hairston, N. G. and Dawood, I. K. (1966). The ecology of *Bulinus truncatus* and *Biomphalaria alexandrina* and its implications for the control of Bilharziasis in the Egypt 49 Project Area. *Bull. WHO* **35,** 339–356.

de Coster, W. and Persoone, G. C. (1970). Ecological study of Gastropoda in a swamp in the neighbourhood of Ghent (Belgium). *Hydrobiologia* **36,** 65–80.

Dickman, M. (1968). The effect of grazing by tadpoles on the structure of a periphyton community. *Ecology* **49,** 1188–1193.

Douglas, A. and Smith, D. C. (1984). The green hydra symbiosis VIII. Mechanisms in symbiont regulation. *Proc. R. Soc. Lond.* **B221,** 291–319.

Ducklow, H. W. (1983). Production and fate of bacteria in oceans. *Bioscience* **33,** 494–501.

Eichenberger, E. and Schlatter, A. (1978). Effect of herbivorous insects on the production of benthic algal vegetation in outdoor channels. *Verh. Internat. Ver. Theor. Ange. Limnol.* **20,** 1806–1810.

Eisenberg, R. M. (1966). The regulation of density in a natural population of the pond snail *Lymnaea elodes. Ecology* **47,** 889–906.

Eisenberg, R. M. (1970). The role of food in the regulation of the pond snail *Lymnaea elodes. Ecology* **51,** 680–684.

Eminson, D. and Moss, B. (1980). The composition and ecology of periphyton communities in freshwaters. I. The influence of host type and external environment on community composition. *Br. Phytocol. J.* **15,** 429–446.

Erokin, V. Y. (1972). Dissolved carbohydrates in some marine coastal biotopes. *Oceanology* **12,** 246–253.

Fenchel, T. (1975). Character displacement and coexistence in mud snails (*Hydrobiidae). Oecologia* **20,** 19–32.

Fenchel, T. M. and Jørgensen, B. B. (1977). Detritus food chains of aquatic ecosystems: The role of bacteria. *Adv. Microb. Ecol.* **1,** 1–58.

Finlay, B. and Fenchel, T. (1989). Everlasting picnic for protozoa. *New Sci.* July, No. 1676, pp. 66–69.

Fogg. G. E. (1971). Extracellular products of algae in freshwater. *Arch. Hydrobiol. Beih. Ergeb* **5,** 1–25.

Fretwell, S. D. (1987). Food chain dynamics: The central theory of ecology. *Oikos* **50,** 291–301.

Frömming, E. (1953). Quantitative untersuchungen über die Nahrungsaufnahme der Susswasserlungeschnecke *Lymnaea stagnalis* L. *Z. Fisch.* **2,** 451–456.
Frömming, E. (1956). *Biologie der mitteleuropaischen Susswasserschnecken.* Duncker and Humbolt, Berlin.
Gaevskaya, N. S. (1969). *The Role of Higher Aquatic Plants in the Nutrition of the Animals of Freshwater Basins* (Ed. by K. H. Mann), Vol. 1. Chs 1 and 2 (translated from Russian by D. G. Maitland Müller). Boston Spa National Lending library, Boston.
Gilbert, L. E. (1977). The role of insect–plant coevolution in the organisation of ecosystems. *Coll. Internat. CNRS,* **265,** 399–413.
Graneli, W. and Solander, D. (1988). Influence of aquatic macrophytes on phosphorus cycling in lakes. *Hydrobiologia* **170,** 245–266.
Gregory, S. V. (1980). Effect of light, nutrients, and grazing on periphyton communities in streams. Ph.D. Thesis, Oregon State University, Corvalis, Oregon.
Gregory, S. V. (1983). Plant–herbivore interactions in stream systems. In: *Stream Ecology, Application and Testing of General Ecological Theory* (Ed. by J. R. Barnes and G. W. Minshall), pp. 157–189. Plenum Press, New York.
Harlin, M. M. (1973). Epiphyte–host relations in seagrass communities. *Aquat. Bot.* **1,** 125–131.
Harper, R. M., Fry, J. C. and Learner, M. A. (1981a). Digestion of bacteria by *Nais virabilis* (Oligochaeta) as established by autoradiography. *Oikos* **36,** 211–218.
Harper, R. M., Fry, J. C. and Learner, M. A. (1981b). A bacteriological investigation to elucidate the feeding biology of *Nais variabilis* (Oligochaeta Naididae). *Freshw. Biol.* **11,** 227–236.
Hipp, E., Mustapha, T., Bickel, U. and Hoffman, K. H. (1986a). Integumentary uptake of acetate and propionate (VFA) by *Tubifex* sp., a freshwater oligochaete. I. Uptake rates and transport kinetics. *J. Exp. Zool.* **240,** 289–297.
Hipp, E., Bickel, U., Mustapha, T. and Hoffman, K. H. (1986b). Integumentary uptake of acetate and propionate (VFA) by *Tubifex* sp., a freshwater oligochaete. II. Role of VFA as nutritional resources and effects of anaerobiosis. *J. Exp. Zool.* **240,** 299–308.
Hoffman, K. L., Seuss, J., Hipp, E. and Sedlmeier, U. A. (1986). Aerobic and anaerobic metabolism in *Tubifex*, a freshwater oligochaete. *Zool. Beitr. N.F.* **30,** 153–170.
Hootsmans, M. J. M. and Vermaat, J. E. (1985). The effects of periphyton-grazing by three epifaunal species on the growth of *Zostera marina.* L. under experimental conditions. *Aquat. Bot.* **22,** 83–88.
Howard, R. K. (1982). Impact of feeding activities of epibenthic amphipods on surface-fouling of eelgrass leaves. *Aquat. Bot.* **14,** 91–97.
Howard, R. K. and Short, F. T. (1986). Seagrass growth and survivorship under the influence of epiphyte grazers. *Aquat. Bot.* **24,** 287–302.
Hunter, R. D. (1980). Effects of grazing on the quantity and quality of freshwater aufwuchs. *Hydrobiologia* **69,** 251–259.
Ivlev, V. S. (1961). *Experimental Ecology of the Feeding of Fishes.* Yale University Press, Newhaven, Conn., 302 pp.
Jacoby, J. M. (1985). Grazing effects on the periphyton by *Theodoxus fluviatilis* (Gastropoda) in a lowland stream. *J. Freshw. Ecol.* **3,** 265–274.
Jacoby, J. M. (1987). Alterations in periphyton characteristics due to grazing in a cascade foothill stream. *Freshw. Biol.* **18,** 495–508.
Jensen, K. T. and Siegismund, H. R. (1980). The importance of diatoms and bacteria in the diet of *Hydrobia* species. *Ophelia* 17 (suppl.), 193–199.

Jones, J. G. (1985). Microbes and microbial processes in sediments. *Phil. Trans. R. Soc. Lond.* **29,** 283–288.

Jones, J. G., Simon, B. M. and Cunningham, C. R. (1983). Bacterial uptake of algal extracellular products: An experimental approach. *J. Appl. Bacteriol.* **54,** 355–365.

Jüttner, F. and Fritz, F. (1974). Excretion products of *Ochromonas* with special reference to pyrroliodone carboxylic acids. *Arch. Microbiol.* **96,** 223–232.

Jüttner, F. and Matuschek, T. (1978). The release of low molecular weight compounds by the phytoplankton in an eutrophic lake. *Wat. Res.* **12,** 251–255.

Kemp, W. M., Twilley, R. R., Stevenson, J. C., Boynton, W. R. and Means, J. C. (1983). The decline of submerged vascular plants in upper Chesapeake Bay: Summary of results concerning possible causes. *Mar. Tech. Soc. J.* **17,** 78–89.

Kofoed, L. H. (1975). The feeding biology of *Hydrobia ventrosa* (Montague) I. The assimilation of differing components of the food. *J. Exp. Mar. Biol. Ecol.* **19,** 233–241.

Ladle, M. (1982). Organic detritus and its role as a food resource in chalk streams. *50th Annual Report of the Freshwater Biological Association*, pp. 30–37. Titus Wilson, Kendall.

Lamberti, G. A. and Moore, J. W. (1984). Aquatic insects as primary consumers. In: *The Ecology of Aquatic Insects* (Ed. by V. H. Resh and D. M. Rosenberg), pp. 164–195. Praeger, New York.

Lamberti, G. A. and Resh, V. H. (1983). Stream periphyton and insect herbivores: An experimental study of grazing by a caddisfly population. *Ecology* **64,** 1124–1153.

Larkum, A. W. D. (1976). Ecology of Botany Bay, I. Growth of *Posidonia australis* (Brown) Hook. f. in Botany Bay and other bays in the Sydney Basin. *Austr. J. Mar. Freshw. Res.* **27,** 117–127.

Latham, M. J. (1979). The animal as an environment. In: *Microbial Ecology, a conceptual approach* (Ed. by J. M. Lynch and N. J. Poole), pp. 115–137. Blackwell, Oxford.

Levinton, J. S. and Bianchi, T. S. (1981). Nutrition and food limitation of deposit feeders. I. The role of microbes in the growth of the mud snails (*Hydrobiidae*). *J. Mar. Res.* **39,** 531–545.

Lodge, D. M. (1985). Macrophyte–gastropod associations: Observations and experiments on macrophyte choice by gastropods. *Freshw. Biol.* **15,** 695–708.

Lodge, D. M. (1986). Selective grazing on periphyton: A determinant of freshwater gastropod microdistributions. *Freshw. Biol.* **16,** 831–841.

Lodge, D. M. and Kelly, P. (1985). Habitat disturbance and the stability of freshwater gastropod populations. *Oecologia* **68,** 111–117.

Lodge, D. M., Brown, K. M., Klosiewski, S. P., Stein, R. A., Covich, A. P., Leathers, B. K. and Brönmark, C. (1987). Distribution of freshwater snails: Spatial scale and the relative importance of physicochemical and biotic factors. *Ann. Malac. Bull.* **5,** 73–84.

Lopez, G. R. and Levinton, J. S. (1978). The availability of micro-organisms attached to sediment particles as food for *Hydrobia ventrosa Montague* (Gastropoda: Prosobranchia). *Oecologia* **32,** 263–275.

Lowe, R. L. and Hunter, R. D. (1988). Effects of grazing by *Physa integra* on periphyton community structure. *J. Nth. Am. Benth. Soc.* **7,** 29–36.

Lynch, J. M., Fletcher, M. and Latham, M. J. (1979). Biological interactions. In: *Microbial Ecology, A Conceptual Approach* (Ed. by J. M. Lynch and N. J. Poole), pp. 171–187. Blackwell, Oxford.

Mague, T. H., Friberg, E., Hughes, D. J. and Morris, I. (1980). Extracellular release of carbon by marine phytoplankton: A physiological approach. *Limnol. Oceanogr.* **20,** 183–190.

May, R. M. (1973). *Stability and Complexity in Model Ecosystems.* Princeton University Press, Princeton, N.J.

Mews, L. and Smith, D. C. (1982). The green hydra symbiosis. VI. What is the role of maltose transfer from alga to animal? *Proc. R. Soc. Lond.* **B216,** 397–413.

Moens, R. (1981). Les habitats de *Lymnaea truncculata* hote intermediaire de *Fasciola hepatica. Rev. Agric.* **34,** 1563–1580.

Moens, R. (1982). Observations au sujet de la predation de *Zonitoides nitidus* Müller sure *Lymnaea truncculata. Bull. Ecol.* **13**(3), 273–282.

Moens, J. (1989). Diffuse competition–a diffuse concept. *Oikos* **54,** 260–263.

Morris, R. J. and Culkin, F. (1975). Environmental organic chemistry of oceans, fjiords and anoxic basins. In: *Environmental Chemistry* (Reporter G. Eglinton), Vol. 1, pp. 81–108. Chemical Society, London.

Morse, D. E., Hooker, N., Duncan, H. and Jensen, L. (1979). γ-Aminobutyric acid, a neuro-transmitter, induces planktonic abalone larvae to settle and begin metamorphosis. *Science* **204,** 407–410.

Morton, J. E. (1958). *Molluscs.* Hutchinson, London.

Ndifon, G. T. (1979). Studies on the feeding biology, anatomical variations and ecology of the vectors of Schistosomiasis and other freshwater snails in South Western Nigeria. Ph.D. Thesis, University of Ibadan.

Ndifon, G. T. and Ukoli, F. M. A. (1989). Ecology of freshwater snails in South Western Nigeria. I. Distribution and habitat preferences. *Hydrobiologia* **171,** 231–253.

Newell, R. C. (1965). The role of detritus in the nutrition of two marine deposit feeders, the prosobranch *Hydrobia ulvae* and the bivalve *Macoma balthica. Proc. Zool. Soc. Lond.* **144,** 25–45.

Nicotri, M. E. (1977). Grazing effects of four marine intertidal herbivores on the microflora. *Ecologia* **58,** 1020–1032.

Odei, M. A. (1972). Some preliminary observations on the distribution of Bilharzia host snails in the Volta Lake. *Bull. Inst. Français Afr. Noire* **34A,** 534–543.

Odei, M. A. (1973). Observations on some weeds of malacological importance in the Volta Lake. *Bull. Inst. Français. Afr. Noire* **35A,** 57–66.

Odum, E. P. and Biever, L. J. (1984). Resource quality, mutualism and energy partitioning in food chains. *Am. Nat.* **124,** 360–376.

Orth, R. J. and Van Montfrans, J. (1984). Epiphyte–seagrass relationships with an emphasis on the role of micrograzing. *Aquat. Bot.* **18,** 43–69.

Ostrofsky, M. L. and Zettler, E. R. (1986). Chemical defences in aquatic plants. *J. Ecol.* **74,** 279–287.

Otto, C. and Svenson, B. S. (1981). How do macrophytes growing in or close to water reduce their consumption by aquatic herbivores? *Hydrobiologia* **78,** 107–112.

Paine, R. T. (1980). Food webs: Linkage, interaction, strength and community infrastructure. *J. Anim. Ecol.* **49,** 667–685.

Paine, R. T. and Vadas, R. L. (1969). The effects of grazing by sea urchins *Strongylocentrotus* spp. on faunal communities of Barnegat Inlet, New Jersey. *Oecologia* **39,** 1–24.

Paperna, I. (1969). Aquatic weeds, snails and transmission of Bilharzia in the new man-made Volta Lake in Ghana. *Bull. Inst. Français Afr. Noire* **31A,** 487–499.

Patience, R. L., Sterry, P. R. and Thomas, J. D. (1983). Changes in the chemical composition of a decomposing aquatic macrophyte *Lemna paucicostata. J. Chem. Ecol.* **9,** 889–991.

Patrick, R. (1970). Benthic stream communities. *Am. Sci.* **58,** 546–549.

Peterson, C. G. (1987). Gut passage and insect grazer selectivity of lotic diatoms. *Freshw. Biol.* **18,** 455–460.
Phillips, G. L., Eminson, D. and Moss, B. (1978). A mechanism to account for macrophyte decline in progressively eutrophicated freshwaters. *Aquat. Bot.* **4,** 103–126.
Pieri, O. S. (1985). Studies on the host snails of schistosomiasis from North East Brazil, with special reference to diapause in *Biomphalaria glabrata* (Say). Ph.D. Thesis, University of Sussex.
Pieri, O. S. and Thomas, J. D. (1986). Polymorphism in a laboratory population of *Biomphalaria glabrata* from a seasonally drying habitat in N. E. Brazil. *Malacologia* **27,** 313–321.
Pieri, O. S. and Thomas, J. D. (1987). Snail–host control in the eastern coastal areas of North East Brazil. *Mem. Inst. Oswaldo Cruz.* **82,** 197–201.
Pieri, O. S. and Thomas, J. D. (in press). Induction of morphological behavioural and physiological changes in polymorphic populations of *Biomphalaria glabrata* (Say) by an environmental factor of predictive value. Paper presented to the *9th International Malacol. Congress*, Edinburgh, August (1986).
Pimentel, D. and White, P. C. (1959). Biological environments and habits of *Australorbis glabratus. Ecology* **40,** 541–550.
Pimm, S. L. (1982). *Foodwebs.* Chapman and Hall, London.
Pip, E. (1978). A survey of the ecology and composition of submerged aquatic snail–plant communities. *Can. J. Zool.* **56,** 2263–2279.
Pip, E. and Stewart, J. M. (1976). The dynamics of two aquatic plant–snail associations. *Can. J. Zool.* **54,** 1192–1209.
Pomeroy, L. R. (1974). The ocean web, a changing paradigm. *BioSci.* **24,** 499–504.
Pomeroy, L. R., Hanson, R. B., McGillin, P. A., Sherr, B. F., Kirchman, U. and Deibel, D. (1984). Microbiology and chemistry of faecal products of pelagic tunicates–rates and fates. *Bulletin of Marine Science* **35,** 426–439.
Porter, K. G. (1977). The plant–animal interface in freshwater ecology *Am. Sci.* **65,** 159–170.
Postgate, J. R. and Hill, S. (1979). Nitrogen fixation. In: *Microbial Ecology, A Conceptual Approach* (Ed. by J. M. Lynch, and N. J. Poole), pp. 191–213. Blackwell, Oxford.
Reavell, P. E. (1980). A study of the diets of some British freshwater gastropods. *J. Conchol.* **30,** 253–271.
Riemann, B., Søndergaard, M., Schierup, H. H., Bosselmann, S., Christiansen, G., Hanson, J. and Nielsen, B. (1982). Carbon metabolism during a spring diatom bloom in the eutrophic lake Massø. *Rev. Hydrobiol.* **67,** 145–185.
Rogers, K. H. and Breen, C. M. (1983). An investigation of macrophyte, epiphyte and grazer interactions. In: *Periphyton of Freshwater Ecosystems* (Ed. by R. G. Wetzel). *Developments in Hydrobiology* No. 17. Dr W. Junk, The Hague.
Sand-Jensen, K. and Søndergaard, M. (1981). Phytoplankton and epiphytic development and their shading effect on submerged macrophytes in lakes of different nutrient status. *Int. Rev. Ges. Hydrobiol.* **66,** 529–552.
Scheerboom, J. E. M. and van Elk, R. (1978). Field observations on the seasonal variation in the natural diet of the haemolymph-glucose concentration of the pond snail *Lymnaea stagnalis* L. *Proc. Ned. Akad. Wet.* **81,** 365–376.
Sheldon, S. P. (1987). The effects of herbivorous snails on submerged communities in Minnesota lakes. *Ecology* **68,** 1920–1931.
Short, F. T. and Short, C. A. (1984). The seagrass filter: Purification of estuarine and coastal waters. In: *The Estuary as a Filter* (Ed. by V. S. Kennedy), pp. 395–413. Academic Press, London.

Søndergaard, M., Riemann, B. and Jørgensen, N. O. G. (1985). Extracellular organic carbon (EOC) released by plankton and bacterial production. *Oikos* **45,** 323–332.

Soszka, G. J. (1975a). The invertebrates on submerged macrophytes in three Masurian lakes. *Ekologia Polska* **23,** 271–391.

Soszka, G. J. (1975b). Ecological relations between invertebrates and submerged macrophytes in the lake littoral. *Ekologia Polska* **24,** 393–415.

Steneck, R. (1977). Crustose coralline–limpet interactions in the Gulf of Maine. *J. Phycol.* **13,** 65.

Stephens, G. C. (1982). Recent progress in the study of Die Ernahrung der wassertiere und der stoffhausholt der Gewasser. *Am. Zool.* **22,** 611–620.

Sterry, P. R., Thomas, J. D. and Patience, R. L. (1983). Behavioural responses of *Biomphalaria glabrata* (Say) to chemical factors from aquatic macrophytes including decaying *Lemna paucicostata* (Hegelm ex Engelm). *Freshw. Biol.* **13,** 465–476.

Sterry, P. R., Thomas, J. D. and Patience, R. L. (1985). Changes in the concentrations of short-chain carboxylic acids and gases during decomposition of the aquatic macrophytes *Lemna paucicostata* and *Ceratophyllum demersum*. *Freshw. Biol.* **15,** 139–153.

Strong, P. R., Lawton, J. H. and Southwood, R. (1984). *Insects and Plants: Community Patterns and Mechanisms*. Blackwell, Oxford.

Sumner, W. T. and McIntire, C. D. (1982). Grazer–periphyton interactions in laboratory streams. *Arch. Hydrobiol.* **93,** 135–157.

Tenore, K. R. (1977). Growth of the polychaete *Capitella capitata*, cultured on different levels of detritus. *Limnol. Oceanogr.* **24,** 350–355.

Tenore, K. R., Hanson, R. B., Dornseif, B. E. and Weiderhold, C. N. (1979). The effect of organic nitrogen supplement on the utilisation of different sources of detritus. *Limnol. Oceanogr.* **24,** 350–355.

Thomas, J. D. (1982). Chemical ecology of the snail hosts of schistosomiasis: snail–snail and snail–plant interactions. *Malacologia* **22,** 81–91.

Thomas, J. D. (1986). The chemical ecology of *Biomphalaria glabrata* (Say): sugars as attractants and arrestants. *Comp. Biochem. Physiol.* **83A,** 457–460.

Thomas, J. D. (1987). An evaluation of the interactions between freshwater pulmonate snail hosts of human schistosomes and macrophytes. *Phil. Trans. R. Soc. Lond.* **B315,** 75–125.

Thomas, J. D. (1989). The comparative ecological biochemistry of sugar chemoreception and transport in freshwater snails and other aquatic organisms. *Comp. Biochem. Physiol.* **93A,** 353–374.

Thomas, J. D. and Daldorph, P. W G. (in press). Evaluation of some bioengineering approaches to the control of pulmonate snails: A preliminary account of the effects of light extinction and mechanical removal of macrophytes on populations of freshwater gatropods. *10th International Malacol. Congress, August 1989*, Tubingen.

Thomas, J. D. and Tait, A. I. (1984). Control of the snail hosts of schistosomiasis by environmental manipulation: A field and laboratory appraisal in the Ibadan area of Nigeria. *Phil. Trans. R. Soc. Lond.* **B305,** 201–253.

Thomas, J. D., Assefa, B., Cowley, C. and Ofosu-Barko, J. (1980). Behavioural responses to amino acids and related compounds, including propionic acid by *Biomphalaria glabrata* (Say), a snail host of *Schistosoma mansoni*. *Comp. Biochem. Physiol.* **66C,** 17–27.

Thomas, J. D., Grealy, B. and Fennell, C. (1983a). Plant–snail interactions: The effects of varying quantity and quality of macrophytes on the feeding and growth of *Biomphalaria glabrata*. *Oikos* **41,** 77–90.

Thomas, J. D., Ofosu-Barko, J. and Patience, R. L. (1983b). Behavioural responses to carboxylic acids and amino acids, by *Biomphalaria glabrata* (Say), the snail host of *Schistosoma mansoni* (Sambon) and other freshwater molluscs. *Comp. Biochem. Physiol.* **75C,** 57–76.

Thomas, J. D., Patience, R. L. and Sterry, P. R. (1984). Uptake and assimilation of short-chain carboxylic acids by the freshwater snail host of *Schistosoma mansoni. Phil. Trans. R. Soc. Lond.* **B222,** 447–476.

Thomas, J. D., Nwanko, D. I. and Sterry, P. R. (1985). The feeding strategies of juvenile and adult *Biomphalaria glabrata* (Say) under simulated natural conditions, and their ecological significancc. *Phil. Trans. R. Soc. Lond.* **B226,** 177–209.

Thomas, J. D., Sterry, P. R., Jones, H., Gubala, M. and Grealy, B. M. (1986). The chemical ecology of *Biomphalaria glabrata*: sugars on phagostimulants. *Comp. Biochem. Physiol.* **83A,** 461–475.

Thomas, J. D., Kowalczyk, C. and Somasundaram, B. (1990). The biochemical ecology of *Biomphalaria glabrata*, a freshwater pulmonate mollusc: the uptake and assimilation of exogenous glucose and maltose. *Comp. Biochem. Physiol.* **95A,** 511–528.

Tunnicliffe, V. and Risk, M. J. (1977). Relationships between the bivalve *Macoma balthica* and bacteria in intertidal sediments: Minas Basin, Bay of Fundy. *J. Mar. Res.* **35,** 499–507.

Underwood, G. J. C. (1989). Interactions between freshwater pulmonate snails, macrophytes and epiphytes. D.Phil Thesis, University of Sussex.

Underwood, G. J. C. and Thomas, J. D. (in press). Grazing interactions between pulmonate snails and epiphytic algae and bacteria. *Freshw. Biol.*

Vallentyne, J. R. and Bidwell, R. G. S. (1956). The relation between sedimentary chlorophyll and free sugars in lake muds. *Ecology* **37,** 495–500.

Van Aardt, W. J. and Wolmarans, C. T. (1981). Evidence for non-assimilation of *Chlorella* by the African snail *Bulinus (Physopsis) globusus. S. Afr. J. Sci.* **77,** 319–320.

Van Montfrans, J., Orth, R. J. and Ryer, C. H. (1982a). The role of grazing on eelgrass periphyton: Implication for plant vigor. In: *Submerged Aquatic Vegetation: Distribution and Abundance in the Lower Chesapeake Bay, and the Interactive Effects of Light, Epiphytes and Grazers*, pp. 108–136. Final report to the US Environmental Protection Agency, Contract No. X003246, 136 pp.

Van Montfrans, J., Orth, R. J. and Vay, S. A. (1982b). Preliminary studies of grazing by *Bittium varium* on eelgrass periphyton. *Aquat. Bot.* **14,** 75–89.

Van Montfrans, J., Wetzel, R. L. and Orth, R. J. (1984). Epiphyte–grazer relationships in seagrass meadows: Consequences for seagrass growth and production. *Estuaries* **7,** 289–309.

Van Schayck, C. P. (1985). Laboratory studies on the relation between aquatic vegetation and the presence of two *Bilharzia*-bearing snail species. *J. Aquat. Pl. Mgmt* **23,** 87–91.

Van Schayck, C. P. (1986). The effect of several methods of aquatic plant control on two *Bilharzia*-bearing snail species. *Aquat. Bot.* **24,** 303–309.

Vine, P. J. (1974). Effect of algal grazing and aggressive behaviour of the fishes *Pomacentrus lividus* and *Acanthurus sohal* on coral reef ecology. *Mar. Biol.* **24,** 131–136.

Weinemann, G. (1970). Geloste kophlenhydrate und ander organische stoffe in naturlichen geswassern und kulturen von *Scenedesmus quadricauda. Arch. Hydrobiol.* **37** (Suppl.), 164–242.

Wetzel, R. G. (1983). Attached algal–substrata interactions: Fact or myth, and when and how? In: *Periphyton of Freshwater Ecosystems* (Ed. by R. G. Wetzel). W. Junk, The Hague.

Wetzel, R. G. and Manny, B. A. (1972). Secretion of dissolved organic carbon and nitrogen by aquatic macrophytes. *Vehr. Internat. Limnol.* **18,** 162–170.

Willoughby, L. G. and Earnshaw, R. (1982). Gut passage time in *Gammarus pulex* (Crustacea: Amphipoda): An aspect of summer feeding in a stony stream. *Hydrobiologia* **97,** 105–117.

Willoughby, L. G. and Marcus, J. H. (1979). Feeding and growth of the isopod *Asellus aquaticus* on actinomycetes considered as a model filamentous bacteria. *Freshw. Biol.* **9,** 441–449.

Structure and Function of Microphytic Soil Crusts in Wildland Ecosystems of Arid to Semi-arid Regions

NEIL E. WEST

ADVANCES IN ECOLOGICAL RESEARCH VOL. 20
ISBN 0-12-013920-0

I. INTRODUCTION AND DEFINITIONS

Most observers perceive desert and semi-desert landscapes as having only scarce plant life with sparseness of vegetation, a criterion commonly used to define deserts and semi-deserts. Many observers consider only vascular plants, however. The inclusion of non-vascular plants provides quite a different view of the plant cover of deserts and semi-deserts. Most of the soil surfaces of deserts and semi-deserts that are not frequently flooded, tilled or burned, excessively stony, mobile sand or intensively traversed by animals or vehicles are covered by communities of small, non-vascular plants. Although there are no accurate estimates of the area involved, reports on such soil surface crusts from all continents indicate that non-vascular plant cover is common in deserts and semi-deserts, though they have not been studied as extensively as the vascular vegetation.

The complex of mosses, lichens, liverworts, algae, fungi and bacteria at the soil surface has been variously called organogenic (Evenari, 1985), biologic, cryptogamic (Skujins, 1984), microfloral (Loope and Gifford, 1972) or microphytic (Cameron, 1978) crust. Cameron (1978) points out that ferns and club mosses are also cryptogams, but are not components of these crusts. I prefer the term microphyte, which separates the total terrestrial vegetation into macro- and microphytic components, and thus draws more attention to the diminutive and less conspicuous microphytes and is consistent with the term employed by aquatic ecologists.

This chapter reviews the structure, function, environmental responses, interactions with other ecosystem components and indicator values of microphytic crust on non-tilled, extensively managed (wildlands) land in arid to semi-arid regions. A greater awareness of the characteristics and roles of these crusts in such ecosystems will help ensure that they are observed and accounted for in inventories, monitoring of change, basic field ecology research and land management decisions.

"Structure" involves the study of kinds and numbers of organisms and their spatial distribution, whereas "function" deals with processes of capture and transfer of energy and uptake and circulation of nutrients and water (Richards, 1974). I will also consider how these organisms react to environmental factors, and interact with other organisms and the physical components of wildland ecosystems in arid to semi-arid environments. These relationships will be useful in determining the significance of the various kinds of microphytic crusts as indicators for monitoring the condition (health) of wildlands in arid and semi-arid regions.

Only the soil surface covered with microphytes (terricolous organisms in the epiedaphic habitat: Friedmann and Galun, 1974) will be considered. It is important to remember, however, that microphytes are often associated with other kinds of soil surface and near-surface features, variously called crusts, caps, carapaces (Cooke and Warren, 1973), films, veils, pans (Houerou, 1984), mats, skins or scaly micro-horizons (Kovda *et al.*, 1979) that may interact with microphytes. For instance, microphytic crusts often overlay or intermingle with vesicular horizons (foam soils) (Eckert *et al.*, 1978), patterned ground (Hugie and Passey, 1964), structural (desiccation or rain) crusts (Epstein and Grant, 1973; Hillel, 1982; Tsarchitzchy *et al.*, 1984), depositional or laminar crusts (Shainberg and Singer, 1985; Levy *et al.*, 1986, Imeson and Verstraten, 1989), chloride (Nickling, 1984), silica (Gifford and Thran, 1971; Lishers and Jones, 1974), carbonate (Campbell, 1979) and gypsum crusts (Watson, 1979). Biologists have usually ignored or minimized the importance of physical and chemical crusts. Soil scientists have tended to ignore or minimize the role of organisms.

Some microphytic crust organisms, particularly the cyanobacteria (blue-green algae), exude mucilaginous materials that "glue" the organisms, organic matter and soil physical particles in place (Campbell *et al.*, 1989) but destroy alumino-silicates (Kovda, 1980). The mix of these features may greatly influence infiltration, percolation, retention and evaporation of water, soil erodibility, overland flow, seedling emergence and other types of resistance to mechanical deformation. Microphytic crusts may aid or inhibit development of these physiochemical crusts and near-surface horizons and thus may play a role in desertification. Much remains to be learned about these potential interactions, however. This review concerns only microphytic soil crusts.

Literature published after about 1972 is emphasized. Friedmann and Galun (1974) thoroughly summarized the older literature, which is largely descriptive and taxonomic, floristic or physiological. Only recently have community structure and dynamics and functions at the ecosystem level attracted researchers' attention.

II. STRUCTURE

The kinds of organisms that can be involved in microphytic crusts are very diverse. There are two broad groups:

(1) Those visible with the unaided human eye (mosses, lichens and liverworts), which could be termed "thallophytic" crusts (Box, 1981).

(2) Those parts of crusts whose individuals are visible only microscopically (algae, fungi and bacteria), which could be termed "microscopic" crusts.

The presence of the latter can sometimes be detected by soil surface color, consistency and tensile strength.

Because of the taxonomic complexity of the organisms in the total microphyte community, the most common approaches are for specialists to identify only those groups of organisms with which they are familiar, e.g. mosses alone, lichens alone or algae alone (see, e.g. Ashley *et al.*, 1985). Thus, it is uncommon for an inventory of a complete microphytic community to have been established (Margulis *et al.*, 1986).

Many groups are poorly known taxonomically. As Friedmann and Galun (1974) emphasize: "A great number of inadequately researched taxa have received recognition, causing chaos in the literature." Different authorities have used different species concepts, "lumping" or "splitting" taxa back and forth. This situation causes considerable difficulty when one tries to compare studies.

The microscopic components have been traditionally identified via enriched media and unispecies cultures (Metting, 1981). Direct counts are difficult, yet prolonged culture to obtain more identifiable stages are not always successful. In addition, laboratory cultures employ conditions far different from those found in the field. All of the above mentioned difficulties have inhibited traditional ecological research. A general lack of autecological and ecophysiological studies of these organisms under realistic conditions has hindered synecologic inferences as well.

Because it is difficult to identify discrete individuals, especially those in the short-lived microscopic portion, densities are rarely recorded. Abundance is usually indexed by cover. The amount of ground covered can be nearly complete (Rauh, 1985, fig. 7.19; Walter, 1985, figs 7.9 and 7.10; West, 1983, fig. 16.6), intermediate (Loope and Gifford, 1972; Kleiner, 1983), or more commonly below 10% (Nash *et al.*, 1977). Frequency (occurrence in a set of small plots) has also been estimated in a number of studies. This variable is, however, a lagging indicator of change in cover and beset with numerous statistical pitfalls (West, 1985). Plot size must be carefully matched to plant size or otherwise microscale patterns are incomprehensibly mixed together and masked.

Biomass estimates require destructive sampling, because the material must be ashed to subtract the intermingled mineral soil. Even then, part of the recovered organic matter may be derived from other sources, e.g. from vascular plants and animal defecation.

By correlating chlorophyll densities to the total phytomass of algae and lichens in field-derived samples, Lynn and Cameron (1973) were able to estimate standing crops of 10–24 g m^{-2}, even during the summer dry period in the northern Great Basin of the USA. Greater standing crops are expected during the cooler, wetter parts of the year. Nash *et al.* (1977) found microphytes contributed 2–38 g m^{-2} of phytomass to sites scattered across the southern or "hot" deserts of North America. Orshan (1985) gives standing crops of lichens as 6 to 12 and algae as 0·2 g m^{-2} in his model of a Negev highlands ecosystem. Kappen (quoted in Walter, 1985) estimates that standing crops of up to 275 g m^{-2} of lichens exist in the fog-influenced areas of the Namib Desert in south-west Africa. It is unclear what portion of this astounding non-vascular phytomass is from microphytes on vascular plants and rocks. Novichkova-Ivanova (1972) reports that 50–140 g m^{-2} dry organic matter may be formed by algae on the takyrs (salt flats = playas) of south-western Turkmenian SSR.

Spatial distributions can be examined from a variety of scales. One of the most interesting observations is the ubiquity of some taxa, even on a worldwide scale (Starks *et al.*, 1981; Rogers, 1977). If the same taxonomic rules are employed, then one usually finds that the majority of species (especially the microscopic ones) are the same when homologous environments are compared between continents. This is probably due to the ease of diaspore transport in the atmosphere. Pye (1987), Pewe (1981) and Gillette (1981) point out that particles less than 20 μm in diameter are easily dispersed over thousands of kilometers via dust storms. "Dust devils" and clear thermal vortices that frequent desert landscapes can carry particles up to an altitude of 5000 m (Sinclair, 1976). From there, the jet streams could carry them between continents. Thus, unlike vascular plants and animals that may differ greatly in a phylogenetic sense between homologous environments on different continents, microphytes are much more taxonomically similar. There is, however, some endemism, especially among the larger thallophytes (Bailey, 1976).

The portion of stone-free ground covered by microphytic crusts seems to be inversely related to the percentages of ground covered by vascular plants and their litter, if blowing sand and other disturbances are minor or absent. In the most extremely arid deserts of the world, Atacama (Rauh, 1985) and Namib (Walter, 1985), crusts dominated by Cyanophyceae may be the only plant growth present. Rigorous soil surface environments within semi-arid climatic zones are also noted for their microphytes (North America: West, 1983; MacMahon and Wagner, 1985; Eurasia: Petrov, 1979; Walter and Box, 1983; Breckle, 1983; Gupta, 1985; South Africa: Kaiser, 1983; Walter, 1985; Werger, 1986; North Africa: Houerou, 1985; Aayad and Ghabbour, 1985; the Middle East: Orshan, 1985; Ahb El Rahman, 1985; South America: Thompson and Iltis, 1968; Soriano, 1983; Mares *et al.*, 1985; Australia: Rogers, 1989).

Complex patterns of microphyte distribution around individual trees, shrubs and large rocks are apparently related to microclimatic patterns. Careful attention has to be paid to these microscales if realistic characterization of the patchiness of microphyte distribution is to be conveyed (Schofield, 1985). Some researchers (Bell, 1983) believe that microphytes are largely responsible for maintaining the mounding so common around trees and shrubs in drier environments (West, 1989). At all of these spatial scales, the mosses and liverworts seem to favor the comparatively more mesic sites. Cyanobacteria favor the harshest sites and lichens dominate in intermediate sites.

Microphytes also sort out on even smaller spatial scales. Cameron and Blank (1966) distinguished between three types of growth of epiedaphic microphytes in the deserts of the south-western USA:

(1) Raincrusts appearing in slight depressions (even ungulate hoof-prints) in the soil micro-relief where water stands for a short period after rains. This is a smooth, thin stratum of algae intermixed with clay which characteristically warps upon drying, curving upward and breaking into polygonal fragments.

(2) Algal soil crusts of reddish, blackish or brownish color when dry (some are fuzzy because of the incorporation of filamentous cyanobacteria). These more conspicuous crusts typically occur on a slightly elevated micro-relief.

(3) On the apparently most stable and relatively elevated micro-relief are found the parasitized algae or crustose lichens which become folded and reticulate as they extend above the soil surface (up to 10 cm) and separate from the substrate. Cameron (1972) later added mosses to this progression and pointed out that succession generally follows phylogeny.

These categories are similar to the micro-profiles shown by Eckert *et al.* (1978, 1987) for the Great Basin of the western USA. To this list, we must add the wandering (*vagrans*), foliose lichens found in some shrub steppes and grasslands (MacCracken *et al.*, 1983).

III. FUNCTIONS

A. Primary Production

Microphytic crusts *in situ* usually occur in a difficult to separate mix of autotrophs (mosses, liverworts, lichens and algae) and heterotrophs (fungi and most bacteria). Algae, however, can also obtain carbon from organic sources when light is unavailable (Starks *et al.*, 1981). All microphytes lack cuticles and stomates and are poikilohydrous. Microphytes are apparently capable of existing on resources that are not utilized by vascular plants (Rickard and Vaughn, 1988).

Microphytes contribute primary production and other functions to that portion of a desert landscape that is spatially large, temporally longlasting, but low in production and energetic reward (Shmida *et al.*, 1986). Man has traditionally been more concerned with other desert subsystems, such as oases or those dominated by ephemeral vascular plants, the latter yielding high energetic rewards, but are difficult to predict in space and time.

The contribution of soil surface inhabiting microphytes to net primary production in desert ecosystems has rarely been estimated in the field (Levi *et al.*, 1981; Jefferies *et al.*, 1989). Other groups investigating such phenomena (Smith and Nobel, 1986) have concentrated on bark and rock-occupying lichens. This is apparently because the influence of vascular plants and other soil microbiota on CO_2 dynamics cannot be easily separated from that due only to the terricolous microphytic autotrophs.

B. Nutrient Cycling

Microphytes have long been connected to this ecosystem function, primarily as fixers of nitrogen. The cyanobacteria, both free-living and as symbionts in lichens and mosses, have been the focus of studies in deserts (Shields, 1957; Shields *et al.*, 1957; Cameron and Fuller, 1960; MacGregor and Johnson, 1971; Marathe and Anantani, 1975; Eskew and Ting, 1978; Skarpe and Henricksson, 1987; Sprent, 1985), shrub steppes (Rogers *et al.*, 1966; Snyder and Wullstein, 1973; Rychert and Skujins, 1974; Rychert *et al.*, 1978; West and Skujins, 1977; Skujins and Klubek, 1978a, b; Skujins, 1981; Terry and Burns, 1987) and arid grasslands (Mayland *et al.*, 1966; Loftis and Kurtz, 1980; Isichei, 1980). The fixation rates that have been quoted have varied from 2 to 41 kg N ha^{-1} $year^{-1}$.

It should also be noted that the carbon inputs of microphytes may enhance heterotrophic microbial fixation, which Klubek and Skujins (1980) term a "phycosphere-like effect". Because the warm wet periods when activity is possible may vary greatly between years, Crawford and Gosz (1982, 1986) argue that reporting of such figures may be deceptive. Frequent wetting and drying seems to enhance nitrogen fixation (Coxson and Kershaw, 1983a, b, c).

Unfortunately, the majority of this work on N-fixation probably reports inflated rates due to the now discounted (Weaver, 1986) acetylene reduction method. Linkages of N fixed by the microphytes showing up in vascular plant tissues have only been made in the greenhouse and growth chamber (Mayland and McIntosh, 1966; Snyder and Wullstein, 1973). High rates of denitrification (MacGregor, 1972; Skujins and Klubek, 1978b; Westerman and Tucker, 1978; Burns, 1983) and volatilization of ammonia in some desert (Klubek *et al.*, 1978) and grassland (Woodmansee *et al.*, 1981) contexts warn us that a high fixation rate does not automatically mean that the productivity of higher plants will be enhanced.

C. Hydrologic Cycle

Although microphytes probably use little moisture in metabolism and transpire less water than vascular plants, they may nevertheless influence the hydrologic cycle in other ways where they contribute substantively to gound cover.

1. Interception and Infiltration

Interception is the catching of falling precipitation by plants that prevents or slows movement of water into the soil. This diminishes the amount of precipitation that becomes soil moisture for vascular plant growth. Although no direct research could be found on this process, it is conceivable that microphytic crusts could intercept some precipitation, especially from light showers. This, along with soil hydrophobicity that some microphytes are thought to produce (Walker, 1979), could lead to less infiltration — the movement of water through the air–soil interface and percolation — and the movement of water once it has entered the soil (Wood, 1988). Greater infiltration of water on microphyte-covered patches of soil compared to nearby bare, physiochemically crusted or "scalded" areas has been observed on several continents (Fletcher and Martin, 1948; Gifford, 1972; Blackburn, 1975; Malcolm, 1972; Hacker, 1986).

Others think that microphytic crusts negatively influence infiltration (Rozanov, 1951; Jackson, 1958; Bolyshev, 1962; Rogers, 1977; Danin, 1978; Stanley, 1983), yet others found no significant difference (Booth, 1941; Faust, 1971; Walker, 1979; Penning deVries and Djiteye, 1982).

Loope and Gifford (1972) reported reduced intrinsic permeability with increased lichen crusting, but greater infiltration with increased microphytic cover. Loope and Gifford (1972) and Rogers (1977) believed this is possible because the rugosity of the crust creates more surface detention storage. Brotherson and Rushforth (1983) found less infiltration under lichen and algae crusts, but more under mosses than in the bare interspaces. Graetz and Tongway (1986) found three-fold greater infiltration and lesser overland flow in the absence of a lichen-dominated crust on a heavily grazed chenopod shrubland in South Australia compared to the nearby ungrazed site with abundant microphytes. They attributed these differences largely to changes in texture and chemistry of the soil surface. Their heavily grazed site had apparent losses of silt, clay and salt. The remaining blow sand allowed water to readily infiltrate. Scalping of the lichens from the lightly grazed area resulted in pronounced increases in infiltration and lessened overland flow.

Because of assumed reductions in infiltration where microphytes prevail, some conclude that treading by livestock may be desirable from a forage plant standpoint (Lee, 1977; Savory, 1988). Hacker (1986) found that infiltration decreases under unrestricted grazing were partly mediated by structural changes in soil microphytic crusts when grazing was relaxed.

2. Moisture Storage

Schwabe (1960) has observed two species of Cyanophyceae in the Atacama

Desert which cement soil particles and form mats which store liquid water. Rogers (1977) also noted "succulence" in some crustose lichens in Australia, which may be a result of holding water, at least temporarily. Galun *et al.* (1982) observed that cyanolichens of desert regions in Israel can absorb 3–13 times their dry weight or volume in water within a few minutes.

3. Evaporation

Brotherson and Rushforth (1983) found that microphytic crusts in northern Arizona allow deeper penetration of applied water and reduced overland flow. They reason that the crusts seal the surface, reducing evaporation. The only direct evidence of microphytic alteration of evaporation of water from soil surfaces comes from Harper and Marble's (1988) studies involving dark-colored cyanobacteria and lichen crusts in Utah. They found that heavily crusted sites lost significantly more water from the upper 7·5 cm of soil than intermixed, scalped plots. They concluded that the dark crusts absorbed more solar radiation and reached higher temperatures than did the light-colored soil of scalped plots. This would lead to higher evaporation.

D. Soil Erosion

1. General

Soil erosion is defined as the detachment and removal of soil material from the ground by either water or wind (Finkel, 1986). About equal amounts of sediment movement worldwide can be attributed to water and wind transportation (Goudie, 1978). Wind erosion predominates in arid regions, whereas water erosion is the main force in semi-arid regions (Marshall, 1973).

Because microphytic crusts variously cover landscapes with generally sparse vascular plant growth and litter, but high rates of natural erosion, it has been intuitively appealing to assume that soil surface microphytes have an important role in slowing down both wind and water erosion. One can find numerous written statements that this is indeed the case (e.g. Fletcher and Martin, 1948; Alexander, 1969; Marshall, 1972; Fuller, 1974; Crawford and Gosz, 1982, 1986; Rushforth and Brotherson, 1982; Mücher *et al.*, 1988; Kinnell *et al.*, 1990; Harper and Marble, 1988; Campbell *et al.*, 1989), but unfortunately none of these contentions is supported by evidence from well-designed experiments. No-one has yet reported results of studies where comparisons have been made after initial inventories and removals or reductions were randomly assigned and the erosion was subsequently compared to that on untreated controls. The best information we have so far is from unintentional experiments where differences are thought to be largely due to the impacts studied. Furthermore, many other unmonitored differences in the study sites could explain the differences in the amount of erosion observed.

We also must be very careful when trying to scale up results from small plots to whole watersheds. Large landscapes respond to much different influences than smaller ones (Bull, 1981; Graf, 1986). I must therefore repeat what Dregne concluded in 1968: the roles of microphytic crusts in erosion in general and particular processes are still unresolved.

2. Water Erosion

Drop splash erosion occurs when a drop of water falls upon the soil and breaks up into a ring of droplets in a crown-like structure (Finkel, 1986). Raindrop impact has from 8 to 25 times more erosive energy than surface flow (Wood *et al.*, 1987). If a slope exists, a net downward movement occurs from rain splash alone. Osborn (1952) considered microphytic crusts on poor grassland range more desirable than bare soil because they were thought to reduce rain splash. Tchoupopnou (1989) has recently shown through direct experiments than rain splash erosion is indeed less on moss-dominated desert soil surfaces than on lichen-dominated plots, and less again than algae-dominated or bare surfaces.

Microphytes contribute to soil surface roughness and probably have a role in causing water to move non-uniformly down a slope. Increasing sinuosity of the pathway of water dissipates erosive energy. It is entirely possible that the microphytes may, on a microscale, form aggrading and degrading surfaces in various proportions such as to influence the net movement of sediment from a hill slope. Stanley (1983) reasons that microphytic crusts, because of the assumed increased surface flow that they produce at the tops of slopes, cause greater water erosion on the lower slopes on calcareous soils of semi-arid rangelands in southern Australia. Bolyshev (1962) uses similar logic for explaining like patterns in the south-eastern USSR. Alexander and Calvo (in press) initially found more detention storage but, with time, increased surface flow, albeit less laden with sediment on crustose lichen-covered slopes than where they simulated rainfall on bare slopes in a Spanish badland. Cooke and Warren (1973) reviewed this literature earlier and concluded that microphytic crusts generally lead to increased water erosion.

The moisture status of crust may determine whether surface flow or infiltration is influenced. For instance, in Arizona, Fletcher (1960) found that a drying period following growth of microphytic crusts results in cracks in the crust and thus a marked increase in initial infiltration. Loope and Gifford (1972) found microphytic crusts in south-eastern Utah reduced potential erosion on small plots, but tell us little about the moisture status of their crusts. If no drying period occurs (and no consequent cracking), such growth may depress infiltration capacity.

The compositional differences in microphytic crusts may seriously influence any of the hydrologic responses. For instance, Faust (1971) found significantly weaker concentrations of suspended sediments with several levels of rainfall similation on cyanobacteria-dominated soils, but detected no differences in surface flow, or infiltration on soils covered with cyanobacteria compared to Arizona soils denuded of that growth. Rozanov (1951) concluded that algal crusts cause considerable surface flow by sealing soil surfaces. Booth (1941) studied algal crusts in the southern Great Plains of the USA, and concluded that the rate of infiltration was not slowed by the presence of an algal crust and that sediment production was minimized. He postulated that algal crusts bind surface soil particles into an erosion-resistant layer. Shields *et al.* (1957) reported similar findings.

Realized water erosion is naturally greatest in the semi-arid regions compared to other environments because of the sparse vascular plant cover and the resultant litter relative to precipitation (Branson *et al.*, 1981). Johnson and Gordon (1986) and Johnson *et al.* (1984, 1985) have observed wide variations in infiltration, overland flow and sediment production obtained from infiltrometer runs on sagebrush steppes in south-western Idaho. The variations could not be adequately explained by soil physical and chemical characteristics, or vascular plant cover or its litter. Because microphytes were abundant, but also variable in composition, they may have contributed to the variable responses (Johnson and Blackburn, 1987).

Satterlund (1972) contends that there are critical points in land deterioration due to soil erosion. For semi-arid lands, these thresholds may be relatively low. The abundance of microphytes may influence these thresholds on some sites.

Although microphytes may be less effective than vascular plants in breaking kinetic energy of both falling and flowing water, their areal abundance in arid to semi-arid regions may slow erosion compared to bare mineral soils. For instance, Harper and Marble (1988) believe that the recovery of microphytic crusts was probably the major reason for the slowing of soil erosion following exclusion of livestock from experimental watersheds on salt shrub desert ranges near Grand Junction, Colorado. The original researcher (Turner, 1971) could not detect a significant improvement in the vascular plant cover and litter (although his measurements were crude and replications few) to correspond with lessened sediment production from small watersheds. He and later investigators (Lusby, 1979) attributed the sediment reduction to the removal of livestock trampling and lowered soil bulk density due to frost heaving. Unfortunately, no-one thought to assess the microphytic cover at the start of that experiment. The inside (no livestock) and outside (continued livestock use) fenced small watershed comparisons now possible cannot be legitimately subjected to statistical tests because of *post-hoc*, non-random assignments of "treatments" and the lack of their spatial interspersion (Hurlbert, 1984). Similar problems occur in most other natural experiments (e.g. Brotherson *et al.*, 1983).

Johansen (1984) demonstrated that extreme surface flow from convectional storms can temporarily destroy algal crusts, but Cyanophyceae were damaged least. I have observed hail to be particularly damaging to soil microphytic crusts. I have also observed that periodic flooding erodes or buries microphytic crust.

3. Wind Erosion

Worldwide increases in wind-borne sediments over the past 150 years (Goudie, 1978, 1983; Kovda, 1980; Tsoar and Pye, 1987) may in some way be related to man-influenced reductions in microphytic crusts on wildlands in arid to semi-arid regions. Brazel and Nickling (1987) and Gillette *et al.* (1980, 1982) acknowledge that some of the unexplained variations in the ability of climatic and meteorologic-based models to predict dust storms may be due to variations in soil surface crusts. This creates opportunities for scientists willing to address this issue.

Dulieu *et al.* (1977) feel that the abundance of sand-binding cyanobacteria is related to wind erosion in Chad. Tsoar and Tyge Moller (1986) acknowledge the role that mosses and algae play in the stabilization of sands in the Halutza sand sea along the Egyptian–Israeli border. They even offer a model of feedbacks between destruction of vegetation (including microphytes) and movement of sand. Trampling by humans, livestock and vehicles breaks down microphytic crusts and allows the smaller particles to be detached and moved by wind (Marshall, 1972).

Andrew and Lange (1986a) have attempted to address experimentally the role of livestock trampling on the alteration of microphytic crusts and dust production. They placed sheep on a previously ungrazed area in the bluebush–saltbush (*Maireana–Atriplex*) steppe region of South Australia and monitored changes in soil and vegetation, radiating out from a new watering point. Sheep trampling was enough to quickly increase soil bulk density by 20% near to the watering point. However, despite considerable reductions in both micro- and macrophytic cover (Andrew and Lange, 1986b), only a weak increase in dust fall was detected. They acknowledged that the stocking rate was light by Australian and world standards. Similar studies at more realistic stocking rates, using other kinds of animals at different sites, are encouraged.

IV. RESPONSES TO ENVIRONMENT

A. Climate

Worldwide patterns of vascular vegetation are most commonly attributed to climate (Walter, 1973). Much less has been said about non-vascular plants. Microphyte dominance on soil surfaces is certainly most apparent in the climatic extremes of deserts and tundras. Microphytes are also found abundantly on soil surfaces in open woodlands, savannas and shrub steppes where litter is sparse and patchy. Where vascular plants are more continuous, and bare soil surfaces less available, microphytes are found on the comparatively drier surfaces of roofs, walls, cliffs, river sand-bars, dunes, new volcanic materials, rocks, stems, branches and sometimes even leaves. A small stature and slow growth rates apparently cause most soil surface-inhabiting thallophytes to lose out whenever flooding, deep shade, heavy litter fall or deposition of mineral soil occurs frequently.

Only a few continent-wide surveys of microphytes have been attempted. Rogers (1972) first studied the relationships of soil surface lichen diversity and abundance in relation to environmental factors in south-eastern Australia. He later extended his analysis to the entire continent (Rogers, 1977, 1982). At both scales, he speculated that the distribution of these organisms prior to occupation of that continent by European man was primarily explained by climatic patterns. The greatest diversity was in wetter regions, but abundance of these organisms is greatest in semi-arid portions along the southern part of the continent. Such areas have a limited cover of vascular plants, but a reasonable frequency of dew to extend the period of adequate soil surface moisture. Anomalies in the current distribution can be explained largely by differences in land-use history, particularly livestock grazing.

Nash *et al.* (1979) and Nash and Moser (1982) found a much greater abundance of all types of lichens in deserts closer to the western coast of North America. This trend is apparently related to more frequent fogs and dews, and more moderate temperatures near the coasts. The relative contribution to total lichen cover and weight from terricolous lichens increases toward the interior of the continent. However, northern slopes in the interior have 10–100 times as many lichens as southern exposures. Kappen *et al.* (1975) noted similar patterns in the central Negev region of Israel. Nash and colleagues also noted increases in all kinds of lichens with increasing elevation on the interior mountains of the south-western USA desert regions that are strongly related to macroclimate. Harper and Marble (1988) noted, however, that soil microphytic cover in the more northerly Great Basin and Colorado Plateau regions was greatest in the comparatively lower elevation desert than in the mountains.

Microphytes have evolved ways of coping with enormous variations in climatic features. Microphytes have rapid physiological responses and can use surface soil moisture that comes in pulses (Noy-Meir, 1973). Rather than depending solely on rain, lichens and mosses commonly survive on dew or even air of high humidity (Evenari, 1985). They can also become active during relatively cool conditions, but are also capable of withstanding very hot and desiccating conditions. Algae are apparently more sensitive than mosses and lichens to annual variations in precipitation and temperature. Scherer *et al.* (1984) have recently shown that *Nostoc flagelliforme* from hot, arid areas of China can survive drought for 2 years, and yet upon rewetting it takes up water very rapidly. Johansen and Rushforth (1985) reported the abundance of soil algae at their Utah site correlated positively with antecedent precipitation but negatively with temperature, thus creating considerable variation in seasonal abundance. Scott (1982) has shown that ephemeral mosses contribute considerable seasonal variability to the microphytic soil cover in semi-arid reaches of southern Australia.

Desiccation usually occurs as temperatures rise. Many microphytes have dark pigmentation or other surface morphology which serves to filter light. With all these adaptations to extreme climatic conditions, it is more likely that they will be more influenced by the indirect rather than the direct effects of climatic change, e.g. dust deposition, vascular plant litter, fire, trampling, etc.

Manuel Molles (pers. comm.) has observed differing plant growth efficiencies related to differences in seasonality of precipitation in southern New Mexico. El Niño climatic events increase winter–spring moisture in this region, apparently reducing the hydrophobicity of the surface soil and leading to greater infiltration of "monsoonal" rains the following summer. During years with little winter–spring precipitation, it appears that hydrophobicity is less mitigated, and therefore lower effective soil moisture input is realized for vascular plant growth. The climatic changes anticipated from the "greenhouse effect" will likely lead to more intense late summer storms and consequent increased erosion in desert regions (Westoby, 1987). The growth and activity of microphytic crusts is greater during the cool season and thus they could be involved in the winter–spring reduction of hydrophobicity. Microphytes may thus interact importantly with possible climatic changes and their influences on desert ecosystems.

B. Soils

It is difficult to separate the interactions between microclimate, vascular plant abundance and various disturbances from purely edaphic influences on microphytes. Nevertheless, correlative studies in arid to semi-arid regions have consistently shown more microphytic cover of the soil when the texture trended to finer particles (Shields and Drouet, 1962; Anderson *et al.*, 1982a; Graetz and Tongway, 1986). Danin (1983) observed that the sandy deserts of Israel and the Sinai have to have at least 4–5% clay and silt to support measurable microphytic cover.

Microphytes occur on soils with a wide variety of bulk densities. I have observed usually higher bulk densities and flatter microphytic cover in the subtropics. Generally, lower bulk densities with more rugose microtopography is associated with microphytic surfaces in the temperate zone. I speculate that these apparent differences are primarily related to the type of precipitation and temperature regimes. Rugose microtopography may be related to frequent needle ice and frost-heaving in the temperate zone. Snow constitutes a major portion of the precipitation at higher latitudes and altitudes. The greater the force of falling rain, especially in monsoonal climates, and lack of frost-heaving may keep the microtopography smoother in the lower temperate and subtropical zones. The rugose crusts may be more dynamic and interactive in ecosystem functions, but also more fragile to disturbances.

The relationships of microphyte abundance to soil chemistry are complex, as different groups of microphytes show different correlations. Many mosses and lichens can occur on highly acidic soils, whereas cyanobacteria are generally found on saline soils and perhaps even require sodium for their best growth (Shields and Durrell, 1964; Shubert and Starks, 1980; Starks *et al.*, 1981). The heavy growth of microphytes has often been correlated with high extractable phosphorus (Kleiner and Harper, 1972, 1977; Anderson *et al.*, 1982a) and gypsum (Meyer, 1986; Meyer and Garcia-Moya, 1989). Filamentous cyanobacteria require a high soil calcium concentration. They also respond positively to P additions, e.g. after fire (Paul and Clark, 1989). The groups with N-fixing capability seem to thrive best under low fertility status (Starks *et al.*, 1981).

Some micronutrients are probably limiting at times. For instance, Starks and Shubert (1982) reported a positive correlation between available iron and both the richness and biomass of algae on newly colonized mine spoils in North Dakota. They also presented evidence that excess cadmium, lithium, manganese and strontium may have hampered the colonization and growth of algae on mine spoils.

C. Tolerance to Burning

Although fires are rare in deserts because of the lack of continuous fine fuel, semi-arid regions experience fire whenever enough fuel and ignition sources converge. With the invasion of Eurasian annuals, wildfire has recently become more common in the semi-arid regions of North America (West,

1988). Several recent studies there have demonstrated that microphytic soil covers can be at least temporarily damaged by wildfire (West and Hassan, 1985; Johansen and Rayburn, 1989).

Rosentreter (1986) and Whisenant (1989) described how fire reduced lichen cover on the shrub steppes of south-western Idaho where deeper soils produce more fuel. Kinnell *et al.* (1990) showed that increased fire frequency in semi-arid woodlands in central New South Wales, Australia reduced the spatial distribution of microphytic cover and that this in turn led to an increased soil erosion hazard. Johansen *et al.* (1984) described fire damage to microphytes on the *Artemisia* steppe of central Utah. Assuming equivalent conditions before the fire, lichen and moss cover and algal densities were substantially lower in abundance in the burned area than in the adjacent unburned area up to 3 years after the wildfire. Five years after the fire, the recovery of all these groups was progressing well but not complete. Unusually wet conditions and protection from disturbance prevailed, however, probably making this rate of recovery unusually rapid. West and Hassan (1985) found after 2 years only partial recovery of the pre-burn moss and lichen soil cover on a similar site in central Utah. Johansen *et al.* (1982) found that fire in sagebrush steppe at another central Utah site reduced algal biomass, but the flora remained remarkably similar. Schulten (1985a) found that fire reduced the total cover of mosses and lichens and altered species composition in the initial stages of recovery in a midwestern USA grassland on sandy soils. Callison *et al.* (1985) showed similar problems where wildfire has altered *Coleogyne*-dominated desert scrub in the Mohave Desert. No significant recovery was observed after 20 years.

D. Tolerance to Animal Disturbance

Many researchers (Marshall, 1972; Mack and Thompson, 1982; Johansen, 1986a, b; Harper and Marble, 1988; Rickard and Vaughn, 1988) have observed apparent reductions of microphytic crusts following livestock trampling. Marble and Harper (1989) feel that microphytic crusts are especially susceptible to this mechanical damage when they are dry. They even recommend excluding livestock use during dry periods to reduce erosion. Few investigators, however, have tried to sort out the degree of change affected by livestock versus other impacts and natural fluctuations. The most common approach has been natural "snap-shot" experiments (Diamond, 1986). Many studies have shown less microphytic cover on the more heavily grazed side of fence line contrasts (Brotherson *et al.*, 1983; Johansen and St. Clair, 1986; Graetz and Tongway, 1986) or less on the grazed outside than in the usually adjacent fenced exclosure or relict stand (Crisp, 1975). Statistical inference has been illegitimately (Hurlbert, 1984) applied to some of these studies.

Rogers and Lange (1971) have directly studied livestock grazing effects on soil microphytes in South Australia via the piosphere concept (gradient of grazing intensity and subsequent changes existing around animal watering places: Andrew, 1988). This approach allows pattern-seeking analyses, but can still result only in correlations that cannot lead directly to conclusions about mechanisms of causation. Apparently, only Andrew and Lange (1986a, b) have so far been able to inventory an area before grazing and then follow the creation of a piosphere. Lichen frequencies nearer the watering point severely declined after only 3 months of sheep presence and continued to decrease over the following 2 years.

Mosses and foliose lichens appear to be more susceptible to livestock trampling than crustose lichens and the microscopic forms. Because crustose lichens like *Collema* and free-living cyanobacteria seem to play the major roles as N-fixers, livestock grazing and other kinds of mechanical disturbances may not reduce that function, if applied in moderation.

Yair and Shachak (1987) mention that isopods and porcupines break through biological crusts covering some hill slopes of the Negev Desert in Israel when burrowing and searching for food. They believe that this reduction of crust accelerates soil erosion because loosely aggregated fine soil particles are made available for transport by wind and shallow flows of water. My observations of continued existence of well-developed microphytic crusts in places like Israel (West, 1986), that have had unrestricted livestock use for centuries, flies in the face of warnings regarding their permanent loss following livestock use, at least in that region.

E. Tolerance to Pollutants, Pesticides and Chemical Amendments

The simple anatomical structure, rapid wicking of water throughout thalli, lack of roots and waxy cuticles, and absorption of water and nutrients through the entire surface of microphytes makes them potentially very sensitive to pollutants. Microphytes on soil surfaces are apparently less sensitive than those on rocks or plant surfaces (Nash, 1974) because they may be protected by splashed or blown soil particles and are only intermittently active. Field and fumigation studies by Marsh and Nash (1979) suggest that the build-up of atmospheric pollutants from coal-fired electric power plants are relatively minor at the soil surfaces of deserts even though emission concentrations can be great. Laboratory studies by Sheridan (1979), Henriksson and Pearson (1981) and Pearson and Rodgers (1982), where some soil surface lichens were fumigated with SO_2 and related substances, have yielded less sanguine results.

Studies of SO_2 on mixed prairie in eastern Montana first showed reduced respiration and severe plasmolysis of the terricolous lichen *Parmelia chloro-*

chroa (Eversman, 1978). Later studies by Lauenroth *et al.* (1984) showed no plasmolysis, but the relative cover of soil lichens decreased as SO_2 inputs increased (Taylor *et al.*, 1980). Pitelka (1988) points out that most research on the effects of air pollutants on plants has tested a few acute levels of pollutants. The response to low chronic levels may be far different.

Photosynthesis-inhibiting herbicides are widely used on rangelands to kill undesirable vascular species, yet rarely have the effects of these herbicides on microphytic organisms been considered. Metting and Rayburn (1979) found that pre-emergent herbicides have a greater adverse affect on algal growth rates than do post-emergent herbicides. This topic needs further investigation for all kinds of microphytes.

We could speculate that because cyanobacteria are sensitive to insecticides (Pipe and Shubert, 1983), the application of materials like malathion to control locusts and grasshoppers may be causing reductions of microphytic crusts. Although soil conditioners and fertilizers are not widely used in wildland ecosystems, slurries and explosives are used in fire control operations. Apparently, there have been no field studies of the responses of soil microphytes to soil amendments.

F. Tolerance to Mechanical Impacts

In recent decades, there has been a considerable increase in vehicular traffic across the deserts and semi-deserts of the world (Cameron, 1978). The widespread use of motorized vehicles to explore for minerals, tend livestock, wage war, hunt game and carry on recreational activities means an increasing percentage of these land surfaces are or will be tracked. If microphytes are functionally important parts of these ecosystems, it behoves us to know how they will respond to mechanical disturbance. Only a very few such studies have yet been undertaken. Braunack (1985) merely mentioned that microphytic crusts were present on his semi-arid Australian site, were disturbed by tracked vehicles, and pointed out the possibilities of their modification leading to increased erosion and invasion of annuals. Webb *et al.* (1988) studied differences in the cover of microphytes on streets of a ghost town in Death Valley National Monument, California. Their linear model and assumption of zero microphyte cover when the town was abandoned about 70 years ago, leads them to estimate that 110 years will be required for this desert shrub site to recover fully.

Mack and Black (1983) found that natural falls of volcanic ash from Mt St. Helens in Washington were deep enough to cause the burial of microphytic crusts. Harris *et al.* (1987) later found that 10 mm of fine volcanic ash was sufficient to destroy lichen crusts, whereas 20 mm were required to eliminate mosses. Introduced annuals such as *Bromus tectorum* then had an easier time in expanding. Enhanced susceptibility of the land to fire then ensues (West, 1988).

Tillage of the soil can also destroy microphytic crust (Jacques, 1984). Terry and Burns (1987) showed that plowing temporarily eliminated the ability of desert soils to fix nitrogen. However, within 1 year, the N-fixing capacity of the soil had recovered to about one-third the rate on the untilled control. Lynn and Cameron (1973) observed that about 10 years were required for algal cover and biomass on once tilled, introduced grass seedlings to be comparable to adjacent, untilled native shrub steppe in Curlew Valley, Utah.

V. INTERACTIONS WITH OTHER ECOSYSTEM COMPONENTS

A. Competitive and Successional Relations with Higher Plants

1. Competition

Those who pay attention to the microphytic component of vegetation can take one of three basic views regarding their competitive role: (1) no interactions, (2) negative or (3) positive interactions prevail.

The first view usually comes from observations that vascular plant cover is usually not significantly greater within protected "controls" (exclosures or relict areas) than it is on adjacent or otherwise paired lands where impacts like livestock grazing have reduced microphytic cover via trampling (Marble and Harper, 1989; Rickard and Vaughn, 1988). However, such comparisons are not completely fair, because other features such as invertebrates (as grazers and in bioturbation), microclimate and other soil properties may not be comparable.

The simplest way to regard the interactions between microphytes and vascular plants is as a 1:1 trade-off. For instance, Schofield (1985) concludes that as grass and forb cover declines with excessive mowing or overgrazing of grasslands, mosses often increase in proportion.

Crisp (1975) studied the recovery of chenopod shrubland and arid woodland at Koonamore, South Australia following removal of livestock from an area that later became a nature reserve. He observed that herbaceous species delined in abundance as trees and microphytes increased. Whether these patterns should be regarded as a zero sum game remains to be seen.

Some observers (Lee, 1977; Savory, 1988; Wolf, 1985) essentially regard microphytes as weeds which compete with vascular plants. McIlvanie (1942) found that seeded grasses were apparently prevented from germinating on a protected Arizona desert grassland soil having well-developed microphytic crusts. However, seeds readily germinated on nearby soils which had been heavily grazed and lacked the microphytic crusts. Hacker (1984, 1987) found inverse correlations of seedling abundance of both tall and chenopod shrubs to microphytic cover in semi-arid Western Australia. He contends that livestock disturbance of microphytic crust can increase the available seed-bed area. The alteration of more or less continuous microphytic crusts can therefore contribute to habitat differentiation. Danin (1978) made similar observations for a sandy area in the Negev Desert of Israel.

Because the availability of "safe sites" is usually more limiting to plant establishment than seed rain or soil reserves (Crawley, 1983), microphytes may influence the establishment of vascular plants. If vascular plants are palatable to livestock, trampling is advocated as a means to reduce the microphytes and supposedly enhance soil moisture, nutrient cycling and regeneration of vascular plants (Savory, 1988). Eldridge (1988), however, found no consistently positive relationship of *Atriplex vesicaria* seedling establishment and survival to microphytic cover on duplex soils of the Mundi Mundi Plain of extreme western New South Wales.

The more common view is that microphytes function in a desert as litter does in more mesic systems (Wagner, 1980). Further, microphytes could enhance the site by increasing water and nutrient input and retention. These and associated microclimatic modifications are commonly thought to lead to enhanced regeneration of higher plants. For instance, Schlatterer and Tisdale (1969) reported that litter of the moss *Tortula ruralis* in southern Idaho retarded germination and early growth of the perennial bunchgrasses *Stipa thurberiana, Sitanion hystrix* and *Agropyron spicatum*. Four weeks after germination, however, growth of the same species was stimulated by the moss litter.

Kleiner and Harper (1977b) observed correlations of greater stand richness of vascular plants with increased richness of microphytes in an ungrazed relict of desert grassland in south-eastern Utah. Meyer (1986) has recently shown that vascular gypsophile and gypsovag species in the Mohave Desert are highly correlated with the cover of microphytes. The reasons for these patterns are not yet apparent, however.

The leachates from leaves of some desert shrubs have been shown to inhibit the N-fixing abilities of microphytes by as much as 80% (Rychert and Skujins, 1974). However, because these shrubs often cover less than 25% of the soil surface, fixation may still proceed in the interspaces.

St. Clair *et al.* (1984) have shown in greenhouse experiments that the establishment of the perennial bunchgrasses *Agropyron elongatum* and *Elymus junceus* were enhanced where microphytic crusts developed. Eckert *et al.* (1986a) found better establishment of one shrub, three native perennial bunchgrasses and an introduced annual grass on microsites where microphytes were more developed than in the bare interspaces in shrub steppe of the Great Basin of the USA. They speculate that crusted surfaces provide more "safe sites" and improved moisture and nutrient status. Harper and Marble (1988) present field data from the same region with less positive trends, especially where mosses were abundant. In field and glasshouse experiments in Arizona, Sylla (1987) compared the emergence of seedlings of introduced warm-season grasses on bare soil, undisturbed soils with heavy microphytic cover and disturbed microphytic cover. He found that emergence was greatest on the disturbed soil with microphytic crust.

The positive correlation between the cover of perennial chenopod shrubs and microphytic crusts found in Australia by Graetz and Tongway (1986) is indicative of some possible synergism. They speculate that the crusts result in spatial concentrations of moisture and nutrients into the top 10 cm of soils, thus favoring seedling establishment.

Mücher *et al.* (1988) showed that the greater dry matter production of the understory of dry woodlands in central New South Wales correlated with greater proportions of the soil covered by microphytic mats. They attributed this to the greater lodgement of seeds, the enhancement of soil moisture and increased seed germination.

Findings of apparent chemical interactions between lichens and seed germination, seedling growth, mycorrhizae, bacterial and fungal growth in other environments alert us to such possibilities in drier environments (Rogers, 1977; Rundel, 1978). The interactions between the microphytic and macrophytic components of vegetation in the field are probably more complicated and indirect than previously thought. The respective differences between micro- and macrophytes at different sites probably make generalizations premature. Thus, all we can do at the moment is highlight a few case studies.

Crisp (1975) has shown how increased amounts of microphytic cover following exclusion of sheep grazing at Koonamore in South Australia have resulted in a sorting out of higher plant species. He explains this through seed characteristics. Small roundish seeds without specialized penetration structures (e.g. *Schismus* and *Salsola*) are prevented from breaking through the crusts. *Stipa*, however, has a hydroscopic awn and setae on the seed which penetrates the microphytic crust. This may explain part of the abundance of *Stipa* on semi-arid sites elsewhere. For instance, Kleiner (1982) found that most perennial grasses decreased as mosses and lichens increased on a previously heavily grazed desert grassland in south-eastern Utah. *Stipa*, however, increased in abundance following the removal of livestock.

Rosentreter (1986) presents an even more complex interaction. A fine-grained patchwork of different plant communities and associated soil patterns on the Snake River Plains of southern Idaho has apparently influenced fire and weed invasion patterns. Patches dominated by lichen crusts produce few fire-supporting fine fuels. Because fires cannot sweep across the landscape, some sites are cooler during a fire. This allows more resprouting of the shrub *Chrysothamnus nauseousus* and its subsequent dominance over *Bromus tectorum*. Lichens can more easily re-invade the burned areas than elsewhere probably because shallow, unburned soil patches serve as their refugia and the deciduous *Chrysothamnus* allows more light to reach the soil in winter than did the original evergreen *Artemisia*.

The dark colors of many microphytes, although probably selected to protect them from ultraviolet radiation (Galun *et al.*, 1982), could cause surface soil temperatures to heat up faster in the spring and thus advance seed germination, seedling growth and other aspects of the phenological progression by vascular plants and microbes. Indeed, Harper and Marble (1988) found that intact dark-colored microphytic crusts in Utah consistently had surface temperatures about 5°C higher than intermingled scalped plots. It is thus possible that there may be some synergism between the micro- and macrophytic phases of a plant community rather than a zero sum game where reduction in one phase is countered by a proportional increase in the other.

2. Successional Patterns

In addition to interactions with vascular plants, microphytes probably interact among themselves and the environment to produce successional sequences. The usual expectation is for relative heights to play a strong role in such series. For instance, Danin and Barbour (1982) describe spatial patterns of bare ground and microphytic communities on the Lissan Marl within the Pleistocene lake basin of the Dead Sea in Israel. Algal, lichen and moss-dominated patterns were related to a progressively more stable micro-relief. To conclude that this micro-toposequence is a chronosequence is logical because of the relative vertical projections of cyanobacteria, algae, lichen and mosses. The relative differences in height are as pronounced as from herbs to trees in a forest sere. Moss rhizoids frequently grow several centimeters deeper into the soil than the other microphytes (Bell, 1983). In addition to overtopping or overgrowth, lichens are known to have algacidal effects through the chemicals they produce (Friedmann and Galun, 1974; Topham, 1977). Lichens can, however, successionally overcome mosses in some circumstances (McWhorter, 1921) and allelochemics may be involved. In general, however, lichens are less able to compete with mosses because of slower growth rates (Kershaw, 1985).

Shubert and Starks (1979) describe an increasing diversity of soil algae with age since the abandonment of coal mine spoils in North Dakota. However, the nearby unmined native grassland had many more species than the oldest mine spoil (45 years). Mosses may prevail in the later stages of succession, in part due to their hydroscopicity. That is, some mosses are apparently able to throw off higher plant litter through hydroscopic twisting (Scott, 1982).

The role of microphytes in plant succession and ecosystem development has only recently been considered. It is not yet clear from the available evidence whether the development of a microphytic crust is a prerequisite to the development of higher plant cover or vice versa, or whether both processes proceed simultaneously. Many more case studies are needed, preferably involving direct inventories of changes over time on permanently identified sites.

The long evolutionary history of microphytes has allowed time for chemical-based defenses and offenses to arise. Mosses may function as "climax" species in patches of microphytes. Some of this successional dominance may be due to greater growth rates, relative height and deeper rhizoids. There is some evidence that desert mosses in the Sonoran Desert inhibit growth of several taxa of bacteria and fungi (Anon. 1963).

Certain soil surface lichens can influence the growth and survival of some forest trees through the production of phytotoxins that inhibit the growth of mycorrhizae (Rice, 1984). We must be alert as to the possibility that such mechanisms may be operating in drier environments.

B. Interactions with Animals

The relatively low phytomass and low rates of productivity of microphytes, in addition to their diminutive stature, inevitably limit interactions with vertebrates. These influences are not unknown, however. For instance, the human famine food "manna" mentioned in the Bible was probably an unattached, erratic or *vagrans* form of a foliose lichen, *Lecanora esculenta*, which drifted into windrows (Donkin, 1981). Even if this is only a myth, wombats of the semi-arid southern plains of Australia have been known to eat windrows of foliose lichens, especially *Chrondriopsis* during drought periods (Rod Wells, pers. comm.). Follmann (1964) also noted that lichens are utilized as food by guanacos in arid portions of northern Chile. Roger Rosentreter (pers. comm.) has observed the apparent use of vagrant forms of lichens by pronghorn antelope and both wild and domestic sheep in central Idaho.

Soil surfaces covered by microphytes may be more protective of seeds being gathered by birds, rodents or ants than on bare ground. This possibility could easily be verified experimentally. Vertebrates could disturb microphytic crusts while searching for food, escaping predators and building homes. For instance, the scratching and pecking behaviors of birds during their searches for food could be a natural and repetitive disturbance to microphytic crusts.

A wider variety of invertebrates than vertebrates depend on microphytic crusts. For instance, Shachak *et al.* (1976) and Shachak (1980) have found an isopod (*Hemilepistus reaumuri*) which depends on the high protein contents of growing algal soil crusts in the Negev Desert of Israel just before mating and egg-laying. Their young are also fed with algal components of the crust. Laboratory experiments have shown that the isopods grow less thriftily when microphytic crusts are not available but thrive when they are present. These wood lice serve as prey for rodents, birds, lizards, spiders and scorpions and thus are important in that food chain. These isopods carry on burrowing activities that expose fine sediments to soil erosion, however (Yair and Shachak, 1987). The denuded trails and fecal deposits they leave may contribute to the erosional process in this region.

Shachak and Steinberger (1980) have also observed two species of snails (*Sphincterochila zonata* and *S. prophetarum*) feeding on soil surface algae only after rainfall in the Negev (20–40 days per year). Rodents, birds and lizards eat the snails. Snakes and birds eat the rodents. These two examples from Israel are the only food chains so far discovered that are based on soil surface microphytes.

Rogers *et al.* (1988) note that adult tenebrionid (darkling) beetles feed on microphytes. These invertebrates are particularly abundant in temperate deserts. Steinberger (1989) found that microphytic soil crusts were essential in the energy and protein budgets of the desert isopod *Hemilepistus reaumuri*.

An even wider variety of invertebrates use microphytic crusts as their habitat, even if they do not apparently eat them (Wallwork, 1962; Wood, 1970). Collembolans and mites are other likely consumers of microphytic crusts because they can be active in both wet and dry conditions (Whitford and Freckman, 1988). Ants have been observed to eat desert moss capsules (Loria and Hernstadt, 1980). Nematodes and protozoa probably graze the microscopic phases of microphytic crusts. Ghabbour *et al.* (1980) found that protozoan grazing of cynanobacteria in laboratory cultures derived from desert soils in Egypt actually stimulated N-fixation because they did not consume the heterocysts. Whitford (1986) speculates that invertebrate consumption and consequent mineralization is required to release microphytic-fixed N for use by higher plants.

C. Role in Soil Formation

Because microphytes add organic matter (Gayel and Shtina, 1974; Kononova, 1975) and nutrients (Kleiner and Harper, 1972, 1977a), and improve crumb structure (Yabukov and Bespalova, 1961; Bond, 1964; Bond and Harris, 1964; Bailey *et al.*, 1973; Schulten, 1985b; Mücher *et al.*, 1988), overall productivity is probably increased and thus soil development increased. Despite these findings, no-one has yet speculated on the comparative rates of soil development with and without microphytes.

Graetz and Tongway (1986) have shown that the loss of microphytes in Australia can lead to considerable changes in a soil's physical structure through reduced silts and clays in the top 10 cm and movement of salinity to deeper depths because of increased infiltration and percolation. Those soils that retained microphytes were, however, much more fertile and had greater cation exchange capacities.

VI. INDICATOR VALUES

A. General

Rosentreter (1986) and Safriel (1987) contend that soil surface microphytes should show less year-to-year fluctuations in abundance than higher plants because they are slower growing, but more persistent. This would supposedly make microphytes ideal indicators of long-term climatic and other environmental trends. I can reason just the opposite. Because they are small and fragile to both physical and chemical damage, soil surface microphytes may easily be destroyed, particularly in the temperate zone. Furthermore, they have enormous intra-seasonal variation in abundance and conspicuousness, especially the microscopic components and ephemeral mosses. Microphytes have apparently low thresholds of tolerance to some impacts such as fire and flooding. They can easily be covered by dust and rainwash, and they absorb pollutants readily. They may well have been largely lost over entire regions due to livestock grazing, fire, pollutants and volcanic deposits (Mack, 1981; Harris *et al.*, 1987; Rickard and Vaughn, 1988). It is impossible to scale absence in the way degree of presence and vigor can be used. The apparent lack of "weedy" thallophytes probably limits their indicator values to only moderately altered landscapes.

B. Land Stability

Because microphytes are of such small stature and apparently fragile to mechanical forces, it is often assumed that they can be used to judge the stability of landscapes (e.g. Young *et al.*, 1986; Safriel, 1987; Tongway and Smith, 1989; Harper, 1989). Although this is logical, there have apparently been few experimental studies to find out how much removal or deposition of wind- and water-moved sediments they can tolerate. One exception is the observation by Danin and Yaalon (1981) that microphytes have been able to stay ahead of up to 20 cm of loess deposited on former beaches of the Dead Sea since it began retreating 12 000 to 10 000 years B.P.

Placodial soil lichens of arid regions develop extraordinarily long and strong strands of rhizines or rhizodial hyphae (Vogel, 1955; Poelt and Baumgartner, 1964). Although these rhizines have no apparent role in the water relations of lichens (Poelt and Baumgartner, 1964), they might result in more soil stability than crusts dominated only by algae. The impact of animals and vehicles on microphytic crusts has not yet received the attention it deserves.

Skujins (1984) points out that the stability of microphytic crusts is strongly related to soil texture. Sands have much less stable crusts than where silts and clays prevail at the soil surface.

Measuring soil erosion rates and land deterioration directly on semi-arid to arid rangelands is difficult. Thus even hydrologists recommend that the monitoring of vegetation change is the most feasible way of detecting deterioration (Blackburn *et al.*, 1986). If microphytes were added to the vascular vegetation being monitored it might help improve the sensitivity of the monitoring efforts.

Several investigators (Otterman, 1974, 1977; Robinove *et al.*, 1981; Otterman and Fraser, 1976; Otterman *et al.*, 1975, 1978; Balling, 1988) have noted increased albedos from desert landscapes being impacted by man. They have usually attributed these changes to reductions in the vascular plant cover and its litter and thus increased exposure of soil. Otterman (1981) later recognized that the growth of algae on the soil surface might have been contributing to the development of enormous differences in reflectance between opposite sides of the Israel–Egypt border following the 1973 peace accord for the Sinai. I have personally verified (West, 1986) that the microphytic cover is now well developed on the Israeli side where the land has been little used by grazing animals and fuel gatherers of late. Whenever albedo is regionally changed in global circulation models, so are the predictions of precipitation, temperature and movement of the atmosphere (Mooney *et al.*, 1987).

Graetz and Tongway (1986) recognized that microphytes were major contributors to signatures of satellite imagery in Australia. There can be considerable confoundments in satellite images from microphytes interacting with antecedent soil moisture. The contribution of microphytic cover to these changes in reflectance should be examined more quantitatively, because this is an important issue in the process of desertification (Charney, 1977; West, 1986).

C. Range Conditions

On the assumption that the presence of microphytes means freedom from disturbance and that freedom from disturbance means successional advancement, the relative abundance of microphytes is being used to judge the condition (successional status or health) of rangelands (Dunne, 1989). For instance, Eckert *et al.* (1986a) advocate the use of the proportion of land covered by various soil surface types as a way to judge range condition. Some of the types of land cover are distinguished by the presence of moss and lichens. They have shown elsewhere (Eckert *et al.*, 1986b, 1987) that there are complex interactions between the different soil covers and soil moisture inputs and losses, seed-bed conditions, and the differential survival of various vascular plant species. Livestock grazing, particularly trampling, simultaneously influences vascular and microphytic vegetation as well as the soil. Moderate trampling is shown by Eckert *et al.* (1987) to favor the establishment of some vascular species on some types of soil microtopography. However, how much of this is due to alteration by the microphytes cannot as yet be determined.

Other USA scientists also incorporate microphytic crusts as desirable indicators of soil and range condition. For instance, Anderson *et al.* (1982b) consider that more crust implies a higher (later seral) range condition. National Park Service personnel in the south-west USA generally feel that a greater cover of microphytic crusts is a positive indicator of the return to pristine-like environments that they are trying to achieve (Norman Henderson, National Park Service, pers. comm., 1984). Environmentalists are starting to use the abundance of microphytic crusts as an informal gauge of suitability for wilderness designation (Gary McFarlane, Utah Wilderness Alliance, pers. comm., 1984).

Harrington *et al.* (1984) and Tongway and Smith (1989) imply that a lack of soil microphytes indicates a poorer range condition for several types of vegetation in the semi-arid parts of Australia. Apparently, their logic is that because crusts are immediately impacted by animal hooves and return slowly after disturbance, the abundance of microphytes can be used as a gauge of incipient degradation.

Rogers and Lange (1971) contend that there are no weedy soil lichens, i.e. no taxa apparently increase rapidly following disturbance and replace those which have been destroyed. This view may apply to mosses and liverworts as well, but I have observed apparent increases in the proportion of soil covered by cyanobacteria following light, extensive grazing.

Soviet and South African range ecologists regard very abundant microphytic crusts as a sign of desertification or degraded range condition. For instance, Antonova (1981, p. 307) cites instances in the central Kara-Kum Desert of the USSR where:

> tracts of land are now worsening as a result of understocking. This results in the emergence of a superficial crust of mosses and lichens. Where this is the case, range productivity decreases by 20 to 40%. Thus, desertification in the Kara Kum Desert is induced not only due to excessive influence of the anthropogenic factor, but also for the lack of livestock grazing. This demands recognition of the need for uniform, moderate use of the whole area of the desert range.

Nechaeva (1981) reinforces this view of what is apparently being taught to students in the USSR.

Alan Savory (1988), originally a wildlife biologist from Zimbabwe, reasons that an abundance of soil microphytic crust indicates earlier stages of succession. If the vascular plants are to succeed, disturbance of the crust by ungulate hooves is necessary. He regards relict areas and protected areas such as national parks as having unnatural and incomplete ecosystems where succession is prevented from going forward by lack of enough ungulates and predators to create the critical "hoof action".

The rate of succession or recovery of crusts following disturbance is poorly understood. Anderson *et al.* (1982b) compared microphytic cover in livestock exclosures of different age in the semi-deserts of Utah and developed a curve to estimate recovery. They concluded that at least 14–18 years are needed to develop an asymptotic level of cover. Kleiner (1982, 1983) compared the recovery of vegetation in lightly and heavily grazed with relict desert grasslands in south-eastern Utah over 10–11 years and noted dramatic increases in the microphytic portion of the community, especially mosses, after livestock had been removed.

Most of the above is based on comparisons in space, not the result of following succession in one place over time which is a much more definitive way to study succession (Goodall, 1977). Therefore, I agree with Smith (1986) that while microphytes can potentially indicate some aspects of land stability, condition and their rate of recovery, we need a more complete understanding of the interactions before we incorporate them in our routine assessments. Whether the concepts of "decreaser", "increaser" and "invader" or "weediness" can be applied to microphytes requires more study of successional trajectories where turnover in microphytes are also monitored. It may also be useful to categorize the microphytes by life strategies (During, 1979).

D. Vegetation Type Definition

Reichert (1936) long ago suggested that microphytes could be used to indicate changes in effective environments. He suggested that the presence of the soil surface lichen (*Diploschistes*) indicated a steppe vegetation type, whereas it was absent in the desert types of Israel.

Looman (1964a, b) pointed out that inclusion of thallophytes along with the vascular flora could reduce similarity coefficients between stands and thereby increase the ability to discriminate between them. When more is known of the autecology of microphytes we will be better able to utilize them as indicators of various kinds of environments and degrees of disturbance and recovery. Progress on autecological understanding is already being made in the case of soil algae (Shubert and Starks, 1985).

VII. CONCLUSIONS

Microphytic crusts may have been more abundant before disturbances of arid to semi-arid environments became intensified. Nevertheless, they have been noticed by some researchers on all continents and remarked upon as to their possible functions in such ecosystems.

Plant ecologists have rarely recorded microphytes as part of their vegetation inventories. If microphytes are mentioned, they are commonly considered only qualitatively in text or at most grouped in broad form and/or taxonomic classes in tables of data. This coarseness of description is apparently due to: (1) unfamiliarity with non-vascular plants and difficulties in identifying the component taxa *in situ*; (2) lack of standardized sampling procedures, especially the unrealistic laboratory culturing approaches to identification; (3) the variable conspicuousness of these organisms (compared to vascular plants), dependent on antecedent conditions; and (4) patchiness on small scales of time and space. Obviously, what is needed to correct this lack of attention toward microphytes are more quantitative expressions of their structure via non-destructive *in situ* means of identification, at least to broad taxonomic, structural or functional groups. If we wait for all the taxonomic problems to be resolved so that we can proceed with traditional modes of investigation, many of these communities may be lost. The testing of remote sensing approaches on a microscale is suggested to break this impasse.

Many functional roles have been attributed to the microphytes of desert and semi-deserts. Opinions about their importance in energy transfer, moisture and nutrient economies, response to disturbances and land stability vary widely. Generalities are being eagerly sought by land managers who want to know how to regard microphytes as indicators of ecosystem productivity, fragility and stability. Ecologists should avoid strongly worded pronouncements until many more results from well-designed field experiments are available. Even then broad generalities may be impossible because (1) the crusts may vary in taxonomic composition, almost as widely as vascular plant communities, and (2) the interactions with physical and chemical crusts may be what is more important. For instance, much mucilage and some carbonate is produced where cyanobacteria dominate. If such microphytes overlay a vesicular layer, the soil will be especially impermeable to water. Some kinds of microphytic vegetation may impede the establishment of desirable vascular plants, moisture input and storage, etc., whereas other kinds of microphytic crusts, in concert with particular kinds of soil, may enhance recovery of site productivity and stability. More carefully designed field experiments are needed to obtain more definitive answers. Although broad generalizations will probably be illusive, there are regions and particular circumstances within regions where soil surface microphytes are abundant and possess probable key functions. We thus can no longer avoid addressing these issues.

The repeated and intuitive belief in the protective cover and organic and nutritive inputs via microphytes on desert soils has been so great that some researchers are attempting to accelerate their recovery on disturbances, such as fires (St. Clair *et al.*, 1986), mines (Ashley and Rushforth, 1984) and mill wastes (Dean *et al.*, 1973). I would prefer to wait for experimental data demonstrating the functional need for microphytes before such efforts are expanded.

In summary, our present knowledge of microphytic crusts, particularly their functions and indicator values, is very limited. Land managers should use extant knowledge with caution when addressing needs for monitoring and rehabilitation in arid to semi-arid environments. I hope this chapter stimulates the interest of many others to begin more experimental research, both basic and applied, dealing with microphytic soil crusts.

ACKNOWLEDGMENTS

I wish to thank the National Science Foundation for travel grants to the USSR and Australia. The Lady Davis Trust supported a semester at Hebrew University, Jerusalem, Israel. These trips, along with much travel around western North America, allowed me to observe microphytic crusts in a wide variety of circumstances and find out about research that would have probably gone undetected by usual library search. Thanks also go to Malyn Brower in diligently tracking down obscure references. This chapter has been approved by the Director of the Utah Agricultural Experiment Station as Journal Paper No. 3808.

REFERENCES

Aayad, M. A. and Ghabbour, S. I. (1985). Hot deserts of Egypt and the Sudan. In: *Hot Deserts and Arid Shrublands*, Vol 12B: *Ecosystems of the World* (Ed. by M. Evenari, I. Noy-Meir and D. W. Goodall), pp. 149–202. Elsevier, Amsterdam.

Ahb El Rahman, A. A. (1985). The deserts of the Arabian Peninsula. In: *Hot Deserts and Arid Shrublands*, Vol. 12B: *Ecosystems of the World* (Ed. by M. Evenari, I. Noy-Meir and D. W. Goodall), pp. 29–54. Elsevier, Amsterdam.

Alexander, M. (1969). Microbiological problems of the arid zone. In: *Global Impacts of Applied Microbiology* (Ed. by E. Gaden), Vol. II, pp. 285–291. Interscience Publishers, New York.

Alexander, R. W. and Calvo, A. (in press). Lichens and soil development in some Spanish badlands. In: *Vegetation and Geomorphology* (Ed. by J. Thornes). Brit. Geomorph Res. Group.

Anderson, D. C., Harper, K. T. and Holmgren, R. C. (1982a). Factors influencing the development of cryptogamic soil crusts in Utah deserts. *J. Range Mgmt* **35,** 180–185.

Anderson, D. C., Harper, K. T. and Rushforth, S. R. (1982b). Recovery of cryptogamic soil crusts from grazing on Utah winter ranges. *J. Range Mgmt* **35,** 355–359.

Andrew, M. H. (1988). Grazing impact in relation to livestock watering points. *Trends Res. Ecol. Evol.* **3,** 336–339.

Andrew, M. H. and Lange, R. T. (1986a). Development of a new piosphere in arid chenopod shrubland grazed by sheep. I. Changes to the soil surface. *Austr. J. Ecol.* **11,** 395–409.

Andrew, M. H. and Lange, R. T. (1986b). Development of a new piosphere in arid chenopod shrubland grazed by sheep. II. Changes in the vegetation. *Austr. J. Ecol.* **11,** 411–424.

Anon. (1963). Mosses inhibit growth of bacteria and fungi. *Nat. Conserv. News.* **13,** 12.

Antonova, K. G. (1981). Influence of reservation management upon range productivity. In: *Rangeland Ecology, Management and Productivity: Collection of Instructional Materials of the International Training Course* (Ed. by N. T. Nechaeva), Vol. 2, pp. 291–310. UNEP/USSR Publications and Information Support Project, Moscow.

Ashley, J. and Rushforth, S. R. (1984). Growth of soil algae on topsoil and processed oil shale from the Uintah Basin, Utah, USA. *Reclam. Reveg. Res.* **3,** 49–63.

Ashley, J., Rushforth, S. R. and Johansen, J. R. (1985). Soil algae of cryptogamic crusts from the Uintah Basin, Utah, U.S.A. *Grt Basin Nat.* **45,** 432–442.

Bailey, D., Mazurak, A. P. and Rosowski, J. R. (1973). Aggregation of soil particles by algae. *J. Phycol.* **9,** 99–101.

Bailey, R. H. (1976). Ecological aspects of dispersal and establishment in lichens. In: *Lichenology: Progress and Problems* (Ed. by D. H. Brown, D. L. Hawksworth and R. H. Bailey), pp. 215–247. Academic Press, London.

Balling, R. C. Jr (1988). The climatic impact of a Sonoran vegetation discontinuity. *Climatic Change* **13,** 99–109.

Bell, G. (1983). Mosses and their relationship to an arid environment. *Austr. System. Bot. Soc. Newsl.* 11–13.

Blackburn, W. H. (1975). Factors influencing infiltration rates and sediment production of semi-arid rangelands in Nevada. *Water Resources Res.* **11,** 929–937.

Blackburn, W. H., Thurow, T. L. and Taylor, C. A. Jr (1986). Soil erosion on rangeland. In: *Use of Cover, Soils and Weather Data in Rangeland Monitoring* (Ed. by E. L. Smith *et al.*), *Symposium Proceedings*, 12 February 1986, pp. 31–39. Kissimmee, Florida. Society for Range Management, Denver, Colorado.

Bolyshev, N. N. (1962). Role of algae in soil formation. *Soviet Soil Sci.* **1964,** 630–635.

Bond, R. D. (1964). The influence of the microflora on the physical properties of soils. II. Field studies in water repellent sands. *Austr. J. Soil Res.* **2,** 123–131.

Bond, R. D. and Harris, J. R. (1964). The influence of the microflora on physical properties of soils. 1. Effects associated with filamentous algae and fungi. *Austr. J. Soil Res.* **2,** 111–122.

Booth, W. E. (1941). Algae as pioneers in plant succession and their importance in erosion control. *Ecology* **22,** 38–46.

Box, E. O. (1981). *Macroclimate and Plant Forms: An Introduction to Predictive Modeling in Phytogeography.* W. Junk, The Hague, 258 pp.

Branson, F. A., Gifford, G. F., Renard, K. G. and Hadley, R. F. (1981). *Rangeland Hydrology*, 2nd edn. Society for Range Management, Denver, Colorado, 340 pp.

Braunack, M. V. (1985). The effect of tracked vehicles on soil strength and microrelief of a calcareous earth (Gcl. 12) north of Woomera, South Australia. *Austr. Rangeland J.* **7,** 17–21.

Brazel, A. J. and Nickling, W. G. (1987). Dust storms and their relation to moisture in the Sonoran-Mojave desert region of the southwestern United States. *J. Environ. Mgmt.* **24,** 279–291.

Breckle, S. W. (1983). The temperate deserts and semi-deserts of Afghanistan and Iran. In: *Temperate Deserts and Semi-deserts*, Vol. 5: *Ecosystems of the World* (Ed. by N. E. West), pp. 271–299. Elsevier, Amsterdam.

Brotherson, J. D. and Rushforth, S. R. (1983). Influence of cryptogamic crusts on moisture relationships of a soil in Navajo National Monument, Arizona. *Grt Basin Nat.* **43,** 73–79.

Brotherson, J. D., Rushforth, S. R. and Johansen, J. R. (1983). Effects of long-term grazing on cryptogamic crust cover in Navajo National Monument, Arizona. *J. Range Mgmt* **35,** 579–581.

Bull, W. B. (1981). Soils, geology and hydrology of deserts. In: *Water in Desert Ecosystems* (Ed. by D. D. Evans and J. L. Thames), pp. 42–58. Dowden, Hutchinson and Ross, Stroudsburg, Penn.

Burns, S. J. (1983). Nitrogen fixation and transformations in cryptogamic soil crusts as affected by disturbance. M.S. thesis, Brigham Young University, Provo, Utah, 93 pp.

Callison, J., Brotherson, J. D. and Bowns, J. E. (1985). The effects of fire on the blackbrush (*Coleogyne ramosissima*) community of southwestern Utah. *J. Range Mgmt* **38,** 535–538.

Cameron, R. E. (1972). A comparison of soil microbial systems in hot, cold and polar desert regions. In: *Ecophysiological Foundation of Ecosystem Productivity in the Arid Zone* (Ed. by L. E. Rodin), pp. 185–192. Nauka, Leningrad.

Cameron, R. E. (1978). The perplexity of desert preservation in a threatening world. In: *Earth-care: Global Protection of Natural Areas* (Ed. by E. A. Schofield), pp. 411–443. Westview Press, Boulder, Colorado.

Cameron, R. E. and Blank, G. B. (1966). Desert algae: Soil crusts and diaphanous substrata as algal habitats. *JPL Technical Paper* No. 32–971, 1–41, Jet Propulsion Laboratory, California Institute of Technology, Pasadena, Calif.

Cameron, R. E., and Fuller, W. H. (1960). Nitrogen fixation by some soil algae in Arizona soils. *Soil Sci. Soc. Am. Proc.* **24,** 353–356.

Campbell, S. E. (1979). Soil stabilization by prokaryotic desert crusts: Implications for Precambrian land biota. *Origins of Life* **9,** 335–348.

Campbell, S. E., Seeler, J.-S. and Glolubic, S. (1989). Desert crust formation and soil stabilization. *Arid Soil Res. Rehab.* **3,** 217–228.

Charney, J. G. (1977). A biophysical feedback mechanism in arid lands. In: *Arid Zone Development: Potentials and Problems* (Ed. by Y. Mundlak and S. F. Singer), pp. 181–188. Ballinger, Cambridge, Mass.

Cooke, R. U. and Warren, A. (1973). *Geomorphology of Deserts.* University of California Press, Berkeley, Calif., 353 pp.

Coxson, D. S. and Kershaw, K. A. (1983a). Nitrogenase activity during chinook snowmelt sequences by *Nostoc commune* in *Stipa–Bouteloua* grassland. *Can. J. Microb.* **29,** 938–944.

Coxson, D. S. and Kershaw, K. A. (1983b). Rehydration response of nitrogenase activity and carbon fixation in terrestrial *Nostoc commune* from *Stipa–Bouteloua* grassland. *Can. J. Bot.* **61,** 2658–2668.

Coxson, D. S. and Kershaw, K. A. (1983c). The pattern of *in situ* summer nitrogenase

activity in terrestrial *Nostoc commune* from *Stipa–Bouteloua* grassland, southern Alberta. *Can. J. Bot.* **61,** 2686–2693.
Crawford, C. S. and Gosz, J. (1982). Desert ecosystems: Their resources in space and time. *Environ. Conserv.* **9,** 181–195.
Crawford, C. S. and Gosz, J. R. (1986). Dynamics of desert resources and ecosystem processes. In: *Ecosystem Theory and Application* (Ed. by N. Polunin), pp. 63–88. John Wiley, New York.
Crawley, M. J. (1983). *Herbivory: The Dynamics of Animal–Plant Interactions.* Blackwell, Oxford, 437 pp.
Crisp, M. D. (1975). Long term change in arid zone vegetation at Koonamore, South Australia. Ph.D. thesis, University of South Australia, Adelaide, 373 pp.
Danin, A. (1978). Plant species diversity and plant succession in a sandy area in the Northern Negev. *Flora* **167,** 409–422.
Danin, A. (1983). *Desert Vegetation of Israel and Sinai.* Cana, Jerusalem, 148 pp.
Danin, A. and Barbour, M. G. (1982). Microsuccession of cryptogams and phanerogams in the Dead Sea area, Israel. *Flora* **172,** 173–179.
Danin, A. and Yaalon, D. H. (1980). Trapping of silt and clay by lichens and bryophytes in the desert environment of the Dead Sea region. In: *Seminar On Approaches and Methods in Paleoclimatic Research with Emphasis on Aridic Areas.* Bat Sheva Foundation, Jerusalem, pp. 32.
Dean, K. C., Havens, R., Harper, K. T. and Rosenbaum, J. B. (1973). Vegetative stabilization of mill mineral wastes. In: *Ecology and Reclamation of Devastated Land* (Ed. by R. Hutnik and G. Davis), Vol. 2, pp. 119–136. Gordon and Breach, London.
Diamond, J. (1986). Overview: Laboratory experiments, field experiments and natural experiments. In: *Community Ecology* (Ed. by J. Diamond and T. J. Case), pp. 3–32. Harper and Row, New York.
Donkin, R. A. (1981). The manna lichen *Lecanora esculenta. Anthropos* **76,** 562–572.
Dregne, H. E. (1968). Surface materials of desert environments. In: *Deserts of the World: An Appraisal of Research into their Physical and Biological Environments* (Ed. by W. G. McGinnies, B. G. Goldman and P. Paylore), pp. 287–377. University of Arizona Press, Tucson, Arizona.
Dulieu, D., Gaston, A. and Darley, J. (1977). La degradation des pâturages de la région N'Djamena (République du Tchad) en relation avec la présence de cyanophycées psamnophiles. *Rev. Elev. Med. Vet. Pays Trop.* **30,** 181–190.
Dunne, J. (1989). Cryptogamic crusts in arid ecosystems. *Rangelands* **11,** 180 182.
During, H. J. (1979). Life strategies of bryophytes: A preliminary review. *Lindbergia* **5,** 2–18.
Eckert, R. E. Jr, Wood, M. K., Blackburn, W. H., Peterson, F. F., Stephens, J. L. and Meurisse, M. S. (1978). Effects of surface soil morphology on improvement and management of some arid and semi-arid rangelands. In: *Proceedings of the First International Rangeland Congress* (Ed. by D. N. Hyder), pp. 299–302. Society for Range Management, Denver, Colorado.
Eckert, R. E. Jr, Peterson, F. F., Meurisse, M. S. and Stephens, J. L. (1986a). Effects of soil-surface morphology on emergence and survival of seedlings in big sagebrush communities. *J. Range Mgmt* **39,** 414–420.
Eckert, R. E. Jr, Peterson, F. F. and Belton, J. T. (1986b). Relation between ecological range condition and proportion of soil surface types. *J. Range Mgmt* **39,** 409–414.
Eckert, R. E. Jr, Peterson, F. F. and Emmerich, F. L. (1987). A study of factors influencing secondary succession in the sagebrush (*Artemisia* spp. L.) type. In:

Proc. Symp. on Seed and Seedbed Ecology of Rangeland Plants (Ed. by G. W. Frasier and R. A. Evans), pp. 149–168. USDA Agricultural Research Service, Tucson, Arizona.

Eldridge, D. J. (1989). The population dynamics of the perennial chenopod shrub *Atriplex vesicaria* in semiarid western New South Wales. M.Sc. thesis, Macquarie University, North Ryde, New South Wales, 115 pp.

Epstein, E. and Grant, W. J. (1973). Soil crust formation as affected by raindrop impact. In: *Physical Aspects of Soil Water and Salts in Ecosystems* (Ed. by A. Hadas *et al.*), pp. 195–201. Springer-Verlag, New York.

Eskew, D. L. and Ting, I. P. (1978). Nitrogen fixation by legumes and blue-green algae–lichen crusts in a Colorado Desert environment. *Am. J. Bot.* **65,** 850–856.

Evenari, M. (1985). The desert environment. In: *Hot Deserts and Arid Shrublands,* Vol. 12A: *Ecosystems of the World* (Ed. by M. Evenari, I. Noy-Meir and D. W. Goodall), pp. 1–22. Elsevier, Amsterdam.

Eversman, S. (1978). Effects of low-level SO_2 on *Usnea hirta* and *Parmelia chlorochroa*. *Bryologist* **81,** 368–377.

Faust, W. F. (1971). Blue-green algal effects on some hydrologic processes at the soil surface. In: *Hydrology and Water Resources in Arizona and the Southwest.* Proc. 1971 meetings Ariz. Sec. Am. Water Res. Assn and Hydrol. Sec., pp. 99–105. Arizona Academy of Sciences, Tempe, Arizona.

Finkel, H. J. (1986). *Semiarid Soil and Water Conservation.* CRC Press, Boca Raton, Florida, 298 pp.

Fletcher, J. E. (1960). Some effects of plant growth on infiltration in the southwest. In: *Water Yield in Relation to Environment in the Southwestern U.S.,* pp. 51–63. American Association for the Advancement of Science, Washington, D. C.

Fletcher, J. E. and Martin, W. P. (1948). Some effects of algae and molds in the raincrust of desert soils. *Ecology* **29,** 95–100.

Follmann, G. (1964). Nebelflechten als Futterflanzen der Kusstenguancos. *Die Naturwissenschaften* **51,** 19–20.

Friedmann, E. I. and Galun, M. (1974). Desert algae, lichens, and fungi. In: *Desert Biology* (Ed. by G. W. Brown), Vol. 2, pp. 165–212. Academic Press, London.

Fuller, W. H. (1974). Desert soils. In: *Desert Biology* (Ed. by G. W. Brown), Vol. 2, pp. 31–101. Academic Press, London.

Galun, M., Burbrick, P. and Garty, J. (1982). Structural and metabolic diversity of two desert lichen populations. *J. Hattori Bot. Lab.* **53,** 321–324.

Gayel, A. G. and Shtina, E. A. (1974). Algae on the sands of arid regions and their role in soil formation. *Soviet Soil Sci.* **6,** 311–319.

Ghabbour, S. I., El-Ayouty, E. Y., Khadr, M. S. and El-Ton, A. M. S. (1980). Grazing by microfauna and productivity of heterocystous nitrogen-fixing blue-green algae in desert soils. *Oikos* **34,** 209–218.

Gifford, G. F. (1972). Infiltration rate and sediment production on a plowed big sagebrush site. *J. Range Mgmt* **25,** 53–55.

Gifford, R. O. and Thran, D. F. (1971). Strength of silica cementation. In: *Soil Crusts* (Ed. by J. W. Cary and D. D. Evans), pp. 28–30. *Technical Bulletin* No. 214, University of Arizona Agricultural Experimental Station, Tucson, Arizona.

Gillette, D. A. (1981). Production of dust that may be carried great distances. In: *Desert Dust: Origin, Characteristics and Effects on Man* (Ed. by T. L. Pewe), pp. 11–26. *GSA Special Paper* No. 186, Geological Society of America, Washington D.C.

Gillette, D. A., Adams, J., Endo, A., Smith, D. and Kihl, R. (1980). Threshold velocities for input of soil particles into the air by desert soils. *J. Geophys. Res.* **85,** 5621–5630.

Gillette, D. A., Adams, J., Muhls, D. and Kihl, R. (1982). Threshold friction velocities and rupture moduli for crusted desert soils for input of soil particles into air. *J. Geophys. Res.* **87,** 9003–9015.

Goodall, D. W. (1977). Dynamic changes in ecosystems and their study: The roles of induction and deduction. *J. Environ. Mgmt* **5,** 309–317.

Goudie, A. S. (1978). Dust storms and their geomorphological implications. *J. Arid Environ.* **1,** 291–310.

Goudie, A. S. (1983). Duststorms in space and time. *Prog. Phys. Geogr.* **7,** 502–530.

Graetz, R. D. and Tongway, D. J. (1986). Influence of grazing management on vegetation, soil structure, nutrient distribution and the infiltration of applied rainfall in a semi-arid chenopod shrubland. *Austr. J. Ecol.* **11,** 347–360.

Graf, W. (1986). Fluvial erosion and federal public policy in the Navaho nation. *Phys. Geogr.* **7,** 97–115.

Gupta, R. K. (1985). The Thar Desert. In: *Hot Deserts and Arid Shrublands*, Vol. 12B: *Ecosystems of the World* (Ed. by M. Evenari, I. Noy-Meir and D. W. Goodall), pp. 55–99. Elsevier, Amsterdam.

Hacker, R. B. (1984). Vegetation dynamics in a grazed mulga shrubland community. I. The mid-storey shrubs. *Austr. J. Bot.* **32,** 239–249.

Hacker, R. B. (1986). Effects of grazing on chemical and physical properties of an earthy sand in the Western Australian mulga zone. *Austr. Rangeland J.* **8,** 11–17.

Hacker, R. B. (1987). Species responses to grazing and environmental factors in an arid halophytic shrubland community. *Austr. J. Bot.* **35,** 135–150.

Harper, K. T. (1989). The role of nonvascular plants in long-term stability of vascular plant communities of deserts. *Bull. Ecol. Soc. Am.* **70,** 134.

Harper, K. T. and Marble, J. R. (1988). A role for nonvascular plants in management of arid and semiarid rangelands. In: *Application of Plant Sciences to Rangeland Management and Inventory.* (Ed. by P. T. Tueller), pp. 135–169. Martinus Nijhoff/ W. Junk, Amsterdam.

Harrington, G. N., Wilson, A. D. and Young, M. D. (Eds) (1984). *Management of Australia's Rangelands.* CSIRO, Melbourne, 354 pp.

Harris, E., Mack, R. N. and Ku, M. S. B. (1987). Death of steppe cryptogams under the ash from Mount St. Helens. *Am. J. Bot.* **74,** 1249–1253.

Henriksson, E. and Pearson, L. C. (1981). Nitrogen fixation rate and chlorophyll content of the lichen *Peltigera canina* exposed to sulfur dioxide. *Am. J. Bot.* **68,** 680–684.

Hillel, D. (1982). *Negev: Land, Water, Life in a Desert Environment.* Praeger, New York, 292 pp.

Houerou, H. N. Le (1984). Rain use efficiency: A unifying concept in arid land ecology. *J. Arid Environ.* **7,** 213–247.

Houerou, H. N. Le (1985). The desert and arid zones of northern Africa. *Hot Deserts and Arid Shrublands*, Vol. 12B: *Ecosystems of the World* (Ed. by M. Evenari, I. Noy-Meir and D. W. Goodall), pp 101–148. Elsevier, Amsterdam.

Hugie, V. K. and Passey, H. B. (1964). Soil surface patterns of some semiarid soils in northern Utah, southern Idaho and northeastern Nevada. *Soil Sci. Soc. Am. Proc.* **28,** 786–792.

Hurlbert, S. H. (1984). Pseudoreplication and the design of ecological field experiments. *Ecol. Monogr.* **54,** 187–211.

Imeson, A. C. and Verstraten, J. M. (1989). The microaggregation and erodibility of some semi-arid and Mediterranean soils. *Catena* **14**(suppl.), 11–24.

Isichei, A. O. (1980). Nitrogen fixation by blue-green algae soil crusts in Nigerian savanna. In: *Nitrogen Cycling in West African Ecosystems* (Ed. by T. Rosswall), pp. 191–199. NFR, Stockholm.

Jackson, E. A. (1958). Soils and hydrology at Yudnapinna Station, South Australia. *CSIRO Soils Land Use Ser. Bull.* **24,** 1–66.

Jacques, I. D. (1984). Self-revegetation of a sagebrush–brunchgrass community after surface blading by bulldozer. M.S. thesis, Washington State University, Pullman, Wash., 44 pp.

Jefferies, D. L., Link, S. O. and Klopatek, J. M. (1989). CO_2 fluxes of cryptogamic crusts in response to resaturation. *Bull. Ecol. Soc. Am.* **70,** 156.

Johansen, J. R. (1984). Response of soil algae to a hundred-year storm in the Great Basin Desert, Utah. *Phykos* **23,** 51–54.

Johansen, J. R. (1986a). Importance of cryptogamic soil crusts to arid rangelands: Implications for short duration grazing. In: *Short Duration Grazing: Proc. of a Short Course* (Ed. by J. A. Tiedeman), pp. 127–136. Washington State University, Pullman, Wash.

Johansen, J. R. (1986b). Soil algae and range management. *Appl. Phycol. Forum* **3**(1), 1–2.

Johansen, J. R., and Rayburn, W. R. (1989). Effects of rangefire on soil cryptogamic crusts. In: *Prescribed Fire in the Intermountain Region: Forest Site Preparation and Range Improvement* (Ed. by D. M. Baugartner *et al.*), pp. 161–164. Washington State University, Pullman, Wash.

Johansen, J. R. and Rushforth, S. R. (1985). Cryptogamic crusts: Seasonal variation in algae populations in the Tintic Mountains, Juab County, Utah, U.S.A. *Grt Basin Nat.* **45,** 14–21.

Johansen, J. R. and St. Clair, L. L. (1986). Cryptogamic soil crusts: Recovery from grazing near Camp Floyd State Park, Utah, USA. *Grt Basin Nat.* **46,** 632–640.

Johansen, J. R., Javakul, A. and Rushforth, S. R. (1982). Effects of burning on the algae communities of a high desert soil near Wallsburg, Utah. *J. Range Mgmt* **35,** 598–600.

Johansen, J. J., St. Clair, L. L., Webb, B. L. and Nebeker, G. T. (1984). Recovery patterns of cryptogamic soil crusts in desert rangeland following fire disturbance. *Bryologist* **87,** 238–243.

Johnson, C. W. and Blackburn, W. H. (1987). Factors contributing to sagebrush rangeland soil loss. *Am. Soc. Agric. Engin. Paper* No. 87-2546.

Johnson, C. W. and Gordon, N. D. (1986). Runoff and erosion from rainfall simulator plots on sagebrush rangeland. *Am. Soc. Agric. Engin. Paper* No. 86-2021, 14 pp.

Johnson, C. W., Savabi, M. R. and Loomis, S. A. (1984). Rangeland erosion measurements for the Universal Soil Loss Equation. *Trans. ASAE* **27,** 1313–1320.

Johnson, C. W., Gordon, N. W. and Hanson, C. L. (1985). Northwest rangeland sediment yield analysis by MUSLE. *Trans. ASAE* **28,** 1889–1895.

Kaiser, P. (1983). The role of soil micro-organisms in savanna ecosystems. In: *Tropical Savannas*, Vol. 13: *Ecosystems of the World* (Ed. by F. Bourliere), pp. 541–557. Elsevier, Amsterdam.

Kappen, L., Lange, O. L., Schulze, E. D., Evenari, M. and Buschlom, U. (1975). Primary productivity of lower plants (lichens) and its physiological basis. In: *Photosynthesis and Productivity in Different Environments* (Ed. by J. P. Cooper), pp. 133–143. Cambridge University Press, Cambridge.

Kershaw, K. A. (1985). *Physiological Ecology of Lichens.* Cambridge University Press, Cambridge, 293 pp.

Kinnell, P. I. A., Chartres, C. J. and Watson, C. L. (1990). Edaphic effects of prescribed burning on a red earth soil in semi-arid woodlands: The effect on erosion susceptibility. *Austr. J. Soil Res.* **28** (in press).

Kleiner, E. F. (1982). Eleven year vegetational comparison in an arid grassland. *Am. Phil. Soc.* **125,** 520–526.

Kleiner, E. F. (1983). Successional trends in an ungrazed, arid grassland over a decade. *J. Range Mgmt* **36,** 114–118.

Kleiner, E. F. and Harper, K. T. (1972). Environment and community organization in grasslands of Canyonlands National Park. *Ecology* **53,** 229–309.

Kleiner, E. F. and Harper, K. T. (1977a). Soil properties in relation to cryptogamic ground cover in Canyonlands National Park. *J. Range Mgmt* **30,** 202–205.

Kleiner, E. F. and Harper, K. T. (1977b). Occurrence of four major perennial grasses in relation to edaphic factors in a pristine community. *J. Range Mgmt* **30,** 286–289.

Klubek, B. and Skujins, J. (1980). Heterotrophic N_2-fixation in arid soil crusts. *Soil Biol. Biochem.* **12,** 229–236.

Klubek, B., Eberhardt, P. J. and Skujins, J. (1978). Ammonia volatilization from Great Basin Desert soils. In: *Nitrogen in Desert Ecosystems* (Ed. by N. E. West and J. Skujins), pp. 107–129. US/IBP Synthesis Series Vol. 9. Dowden, Hutchinson and Ross, Stroudsburg, Penn.

Kononova, M. M. (1975). Humus of virgin and cultivated soils. In: *Soil Components* (Ed. by L. Geisking), Vol. 1, pp. 475–526. Springer-Verlag, Berlin.

Kovda, V. (1980). *Land Aridization and Drought Control.* Westview Press, Boulder, Colorado, 277 pp.

Kovda, V. A., Samoilova, E. M., Charley, J. L. and Skujins, J. (1979). Soil processes in arid lands. In: *Arid-land Ecosystems: Structure, Functioning and Management* (Ed. by D. W. Goodall, R. A. Perry and K. M. W. Howes); Vol. 1, pp. 439–470. Cambridge University Press, Cambridge.

Lauenroth, W. K., Milchunas, D. G. and Dodd, J. L. (1984). Responses of the vegetation. In: *The Effects of SO_2 on a Grassland: A Case Study in the Northern Great Plains of the United States* (Ed. by W. K. Lauenroth and E. M. Preston), pp. 97–136. Springer-Verlag, New York.

Lee, K. E. (1977). Physical effects of herbivores on arid and semiarid rangeland ecosystems. In: *The Impact of Herbivores on Arid and Semiarid Rangelands.* Proc. 2nd US/Australian Rangeland Panel, pp. 173–186. Australian Rangeland Society, Perth.

Levi, Y., Berner, T. and Cohen, Y. (1981). CO_2 exchange and growth rate of the loess soil crusts algae in the Negev Desert of Israel. In: *Developments in Arid Zone Ecology and Environmental Quality* (Ed. by H. Shuval), pp. 43–48. Balaban ISS, Philadelphia, Penn.

Levy, G., Shainberg, I. and Morin, J. (1986). Factors affecting the stability of soil crusts in subsequent storms. *Soil Soc. Am. J.* **50,** 196–201.

Loftis, S. G. and Kurtz, E. B. (1980). Field studies of inorganic nitrogen added to semiarid soils by rainfall and blue-green algae. *Soil Sci.* **129,** 150–155.

Looman, J. (1946a). Ecology of lichen and bryophyte communities in Saskatchewan. *Ecology* **45,** 481–491.

Looman, J. (1946b). The distribution of some lichen communities in the Prairie Provinces and adjacent parts of the Great Plains. *Bryologist* **67,** 209–224.

Loope, W. L. and Gifford, G. F. (1972). Influence of a soil microfloral crust on select properties of soils under pinyon–juniper in southeastern Utah. *J. Soil Water Conserv.* **27,** 164–167.

Loria, M. and J. Herrnstadt. (1980). Moss capsules as food of the harvest ant. *Messor. Bryologist* **83,** 524–525.

Lusby, G. C. (1979). Effects of grazing on runoff and sediment yield from desert rangeland at Badger Wash in Western Colorado, 1953–73. *U.S. Geol. Surv. Water Supply Paper* No. 1532–I. US Government Printing Office, Washington, D. C.

Lynn, R. I. and Cameron, R. E. (1973). Role of algae in crust formation and nitrogen cycling in desert soils. *US/IBP Desert Biome Res. Memo* No. 73–40(3): 2.3.45.-1-26. Utah State University, Logan, Utah.

McIlvanie, S. K. (1942). Grass seedling establishment and productivity – over grazed and protected range soils. *Ecology* **23,** 228–231.

McWhorter, F. P. (1921). Destruction of mosses by lichens. *Bot. Gaz.* **72,** 321–325.

MacCracken, J. G., Alexander, L. E. and Uresk, D. W. (1983). An important lichen of southeastern Montana rangelands. *J. Range Mgmt* **36,** 35–37.

MacGregor, A. N. (1972). Gaseous losses of nitrogen from freshly wetted desert soils. *Soil Sci. Soc. Am. Proc.* **36,** 594–596.

MacGregor, A. N. and Johnson, D. E. (1971). Capacity of desert algal crusts to fix atmospheric nitrogen. *Soil Sci. Soc. Am. Proc.* **35,** 843–844.

MacMahon, J. A. and Wagner, F. H., (1985). The Mojave, Sonoran and Chihuahuan Deserts of North America. In: *Hot Deserts and Arid Shrublands*, Vol. 12A: *Ecosystems of the World* (Ed. by M. Evenari, I. Noy-Meir and D. W. Goodall), pp. 105–202. Elsevier, Amsterdam.

Mack, R. N. (1981). Invasion of *Bromus tectorum* L. into western North America: An ecological chronicle. *Agro-Ecosyst.* **7,** 145–165.

Mack, R. N. and Black, R. A. (1983). The effect of the Mt. St. Helens ashfall on the xero-phytic vegetation of eastern Washington. *Bull. Ecol. Soc. Am.* **64,** 129 (abstract).

Mack, R. N. and Thompson, J. N. (1982). Evolution in steppe with few large, hooved mammals. *Am. Nat.* **119,** 757–773.

Malcolm, C. V. (1972). Establishing shrubs in saline environments. *Technical Bulletin* No. 21. Department of Agriculture, Western Australia, 141 pp.

Marathe, K. N. and Anantani, Y. S. (1975). Algal crusts as nitrogen sources in some arid soils of India. *J. Univ. Bombay* (*Science*), **71,** 44–49.

Marble, J. R. and Harper, K. T. (1989). Effect of timing of grazing on soil-surface cryptogamic communities in a Great Basin low shrub desert: A preliminary report. *Grt Basin Nat.* **49,** 104–107.

Mares, M. A., Morello, J. and Goldstern, G. (1985). The Monte Desert and other subtropical semiarid biomes of Argentina, with comments on their relation to North American arid areas. In: *Hot Deserts and Arid Shrublands*, Vol. 12A: *Ecosystems of the World* (Ed. by M. Evenari, I. Noy-Meir and D. W. Goodall), pp. 203–238. Elsevier, Amsterdam.

Margulis, L., Chase, D. and Guerrero, R. (1986). Microbial communities. *BioScience* **36,** 160–170.

Marsh, J. E. and Nash, T. H. (1979). Lichens in relation to the Four Corners Power Plant in New Mexico. *Bryologist* **82,** 20–28.

Marshall, J. K. (1972). Principles of soil erosion and its prevention. In: *The Use of Trees and Shrubs in the Dry Country of Australia* (Ed. by N. Hall), pp. 90–107. Australian Government Publishing Service, Canberra.

Marshall, J. K. (1973). Drought, land use and soil erosion. In: *The Environmental, Economic and Social Significance of Drought* (Ed. by J. Lovett), pp. 55–77. Angus and Robertson, Sydney.

Mayland, H. F. and McIntosh, T. H. (1966). Availability of biologically fixed atmospheric nitrogen 15 to higher plants. *Nature* **209,** 421–422.

Mayland, H. F., McIntosh, T. H. and Fuller, W. H. (1966). Fixation of isotopic nitrogen on a semiarid soil by algal crust organisms. *Soil Sci. Soc. Am. Proc.* **30,** 56–60.

Metting, B. (1981). The systematics and ecology of soil algae. *Bot. Rev.* **47,** 195–312.
Metting, B. and Rayburn, W. R. (1979). Effects of the pre-emergence herbicide diallate and the post-emergence herbicide MCPA on the growth of some soil algae. *Phycologia* **18,** 269–272.
Meyer, S. E. (1986). The ecology of gypsophile endemism in the eastern Mojave Desert. *Ecology* **67,** 1303–1313.
Meyer, S. E. and Garcia-Moya, E. (1989). Plant community patterns and soil moisture regime in gypsum grasslands of north central Mexico. *J. Arid Environ.* **16,** 147–155.
Mooney, H. A., Vitousek, P. M. and Matson, P. A. (1987). Exchange of materials between terrestrial ecosystems and the atmosphere. *Science* **238,** 929–932.
Mücher, H. J., Chartres, C. J., Tongway, D. J. and Greene, R. S. B. (1988). Micromorphology and significance of the surface crusts of soils in rangelands near Cobar, Australia. *Geoderma* **42,** 227–244.
Nash, T. H. (1974). Lichens of the Page environs as potential indications of air pollution. *J. Arizona Acad. Sci.* **9,** 97–101.
Nash, T. H. and Moser, T. J. (1982). Vegetational and physiological patterns of lichens in North American deserts. *J. Hattori Bot. Lab.* **53,** 331–336.
Nash, T. H., White, S. L. and Marsh, J. E. (1977). Lichen and moss distribution and biomass in hot desert ecosystems. *Bryologist* **80,** 470–479.
Nash, T. H., Nebeker, G. T., Mosher, T. J. and Reeves, T. (1979). Lichen vegetational gradients in relation to the Pacific coast of Baja California: The maritime influence. *Madroño* **26,** 149–163.
Nechaeva, N. T. (1981). Indicators of range degradation. In: *Rangeland Ecology, Management and Productivity: Collection of Instructional Materials of the International Training Course* (Ed. by N. T. Nechaeva), Vol. 1, pp. 98–123. UNEP/USSR Publications and Information Support Project, Moscow.
Nickling, W. G. (1984). The stabilizing role of bonding agents on the entrainment of sediment by wind. *Sedimentology* **31,** 111–117.
Novichkova-Ivanova, L. N. (1972). Soil algae of Middle Asian deserts. In: *Ecophysiological Foundation of Ecosystem Productivity in the Arid Zone* (Ed. by L. E Rodin), pp. 180–182. Acad. Sci., Leningrad.
Noy-Meir, I. (1973). Desert ecosystems: Environment and producers. *Ann. Rev. Ecol. Syst.* **4,** 25–551.
Orshan, G. (1985). The deserts of the Middle East. In: *Hot Desert and Arid Shrublands*, Vol. 12B: *Ecosystems of the World* (Ed. by M. Evenari, I. Noy-Meir and D. W. Goodall), pp. 1–28. Elsevier, Amsterdam.
Osborn, B. (1952). Range conditions influence water uptake. *J. Soil Water Conserv.* **7,** 128–132.
Otterman, J. (1974). Baring high albedo soils by overgrazing: A hypothesized desertification mechanism. *Science* **186,** 153–533.
Otterman, J. (1977). Anthropogenic impact on the albedo of the earth. *Clim. Change* **1,** 137–155.
Otterman, J. (1981). Satellite and field studies of man's impact on the surface in arid regions. *Tellus* **33,** 68–77.
Otterman, J. and Fraser, R. S. (1976). Earth–atmosphere system and surface reflectivities in arid regions from Landsat (MSS data). *Remote Sens. Environ.* **5,** 247–266.
Otterman, J., Waisel, Y. and Rosenberg, E. (1975). Western Negev and Sinai ecosystems: A comparative study of vegetation, albedo and temperatures. *Agroecosystems* **2,** 47–60.
Otterman, J., Waisel, Y. and Imber, A. (1978). Rangeland recovery in an exclosure:

Satellite and ground observations. In: *The Contribution of Space Observations to Global Food Information System* (Ed. by E. A. Godby and J. Otterman), pp. 23–26. Pergamon Press, New York.

Paul, E. A. and Clark, F. E. (1989). *Soil Microbiology and Biochemistry*. Academic Press, London, 273 pp.

Pearson, L. C. and Rodgers, G. A. (1982). Air pollution damage to cell membranes in lichens. III. Field experiments. *Phyton* **22,** 329–337.

Penning deVries, F. W. T. and Djiteye, M. A. (Eds) (1982). *La productive des Pâturages Saheliens*. PUDOC Centre for Agricultural Publishing and Documentation, Wageningen, 525 pp.

Petrov, M. P. (1979). Land-use of semi-deserts in the U.S.S.R. In: *Management of Semi-arid Ecosystems*. (Ed. by B. H. Walker), pp. 301–327. Elsevier, Amsterdam.

Pewe, T. L. (1981). Desert dust: An overview. In: *Desert Dust: Origin, Characteristics and Effects on Man* (Ed. by T. L. Pewe), pp 1–10. *GSA Special Paper* No. 186, Geological Society of America, Washington, D.C.

Pipe, A. E. and Shubert, L. E. (1983). The use of algae as indicators of soil fertility. In: *Algae as Ecological Indicators* (Ed. by L. E. Shubert), pp. 213–233. Academic Press, London.

Pitelka, L. F. (1988). Evolutionary responses of plants to anthropogenic pollutants. *Trends Res. Ecol. Evol.* **3,** 233–236.

Poelt, M. J. and Baumgartner, H. (1964). Uber Rhizenenstrange bei placodialen Flechten. *Oesterr. Bot.* **110,** 194–269.

Pye, K. (1987). *Aeolian Dust and Dust Deposits*. Academic Press, London, 334 pp.

Rauh, W. (1985). The Peruvian–Chilean Deserts. In: *Hot Deserts and Arid Shrublands*, Vol 12A: *Ecosystems of the World* (Ed. by M. Evenari, I. Noy-Meir and D. W. Goodall), pp. 239–268. Elsevier, Amsterdam.

Reichert, I. (1936). Steppe and desert in the light of lichen vegetation. *Proc. Linn. Soc. (Lond.)* **149,** 19–23.

Rice, E. L. (1984). *Allelopathy*, 2nd edn. Academic Press, London, 422 pp.

Richards, B. N. (1974). *Introduction to the Soil Ecosystem*. Longman, London, 266 pp.

Rickard, W. H. and Vaughn, B. E. (1988). Plant community characteristics. In: *Shrub–Steppe: Balance and Change in a Semi-arid Terrestrial Ecosystem* (Ed. By W. H. Rickard, L. E. Rogers, B. E. Vaughn and S. F. Leibetrau), pp. 109–179. Elsevier, Oxford.

Robinove, C. J., Chavez, P. S., Gehring D, and Holmgren, R. (1981). Arid land monitoring using LANDSAT albedo difference images. *Remote Sens. Environ.* **11,** 133–156.

Rogers, L. E., Fitzner, R. E., Cadwell, L. L. and Vaughn, B. E. (1988). Terrestrial animal habitats and population responses. In: *Shrub–Steppe: Balance and Change in a Semi-arid Terrestrial Ecosystem* (Ed. by W. H. Rickard, L. E. Rogers, B. E. Vaughn and S. F. Liebetrau), pp. 181–256. Elsevier, Oxford.

Rogers, R. W. (1972). Soil surface lichens in arid and subarid southeastern Australia. III. The relationship between distribution and environment. *Austr. J. Bot.* **20,** 301–316.

Rogers, R. W. (1977). Lichens of hot arid and semi-arid lands. In: *Lichen Ecology* (Ed. by M. R. D. Steward), pp. 211–252. Academic Press, London.

Rogers, R. W. (1982). Lichens of arid Australia. *J. Hattori Bot. Lab.* **53,** 351–355.

Rogers, R. W. (1989). Blue-green algae in southern Australian rangeland soils. *Austr. Ranges. J.* **11,** 67–73.

Rogers, R. W. and Lange, R. T. (1971). Lichen populations on arid soil crusts around sheep watering places in South Australia. *Oikos* **22,** 93–100.

Rogers, R. W., Lange, R. T. and Nicholas, D. J. D. (1966). Nitrogen fixation by lichens of arid soil crusts. *Nature* **209,** 96–97.

Rosentreter, R. (1986). Compositional patterns within a rabbitbrush (*Chrysothamnus*) community of the Idaho Snake River Plain. *Proceedings of a Symposium on the Biology of Artemisia and Chrysothamnus*, 9–13 July 1984, Provo, Utah, pp. 273–277. *General Technical Report* No. INT-200, USDA Forest Service, Intermountain Research Station, Ogden, Utah.

Rozanov, A. N. (1951). *Serozems of Central Asia* (translated from Russian). Israel Program for Scientific Translations, Jerusalem (O.T.S. 60-21834), 541 pp.

Rundel, P. W. (1978). The ecological role of secondary lichen substances. *Biochem. Syst. Ecol.* **6,** 157–170.

Rushforth, S. R. and Brotherson, J. D. (1982). Cryptogamic soil crusts in the deserts of North America. *Amer. Biol. Teacher* **44,** 472–475.

Rychert, R. C. and Skujins, J. (1974). Nitrogen fixation by blue-green algae–lichen crusts in the Great Basin Desert. *Soil Sci. Soc. Am. Proc.* **38,** 768–771.

Rychert, R. C., Skujins, J., Sorenson, D. and Porcella, D. (1978). Nitrogen fixation by lichens and free-living microorganisms in deserts. In: *Nitrogen in Desert Ecosystems* (Ed. by N. E. West and J. Skujins), pp. 20–30. Dowden, Hutchinson and Ross, Stroudsburg, Penn.

St. Clair, L. L., Webb, B. L., Johansen, J. R. and Nebeker, G. T. (1984). Cryptogamic soil crusts: Enhancement of seedling establishment in disturbed and undisturbed areas. *Reclam. Reveg. Res.* **3,** 129–136.

St. Clair, L. L., Johansen, J. R. and Webb, B. L. (1986). Rapid stabilization of fire-disturbed sites using a soil crust slurry: Inoculation studies. *Reclam. Reveg. Res.* **4,** 261–269.

Safriel, U. N. (1987). The stability of the Negev Desert ecosystems: Why and how to investigate it. In: *Progress in Desert Research* (Ed. by L. Berkofsy and M. G. Wurtele), pp. 133–144. Rowan and Littlefield, Totowa, N. J.

Satterlund, D. R. (1972). *Wildland Watershed Management.* Ronald Press, New York, 370 pp.

Savory, A. (1988). *Holistic Resource Management.* Island Press, Covelo, Calif., 564 pp.

Scherer, S., Ernst, A., Chen, T.-W. and Boger, P. (1984). Rewetting of drought resistant blue-green algae: Time course of water uptake and reappearance of respiration, photosynthesis and nitrogen fixation. *Oecologia (Berl.)* **62,** 418–423.

Schlatterer, E. F. and Tisdale, E. W. (1969). Effects of litter of *Artemisia, Chrysothamnus* and *Tortula* on germination and growth of three perennial grasses. *Ecology* **50,** 869–873.

Schofield, W. B. (1985). *Introduction to Bryology.* Macmillan, New York, 431 pp.

Schulten, J. A. (1985a). The effects of burning on the soil lichen community of a sand prairie. *Bryologist* **72,** 110–114.

Schulten, J. A. (1985b). Soil aggregation by cryptogams of a sand prairie. *Am. J. Bot.* **72,** 1657–1661.

Schwabe, G. H. (1960). Zur autotrophen Vegetation in ariden Boden, Blaualgen und Lebensraum IV. *Osterr. Bot.* **107,** 281–309.

Scott, B. A. M. (1982). Ecology of bryophytes in arid regions. In: *Bryophyte Ecology* (Ed. by A. J. E. Smith), pp. 105–122. Chapman and Hall, New York.

Shachak, M. (1980). Energy allocation and life history strategy of the desert isopod, *Hemilepistus reaumuri. Oecologia (Berl.)* **45,** 404–413.

Shachak, M. and Steinberger, Y. (1980). An algae–desert snail food chain: Energy flow and soil turnover. *Oecologia (Berl.)* **46,** 402–411.

Shachak, M., Chapman, E. A. and Steinberger, Y. (1976). Feeding, energy flow and soil turnover in the desert isopod, *Hemilepistus reaumuri. Oecologia* **24,** 57–69.

Shainberg, I. and Singer, M. J. (1985). Effect of electrolytic properties of depositional crust. *Soil Sci. Soc. Am. J.* **49,** 1260–1263.

Sheridan, R. P. (1979). Impact of emissions from coal-fired electrical generating facilities on N_2-fixing lichens. *Bryologist* **82,** 54–58.

Shields, L. M. (1957). Algal and lichen floras in relation to nitrogen content of certain volcanic and arid range soil. *Ecology* **38,** 661–663.

Shields, L. M. and Drouet, F. (1962). Distributions of terrestrial algae within the Nevada test site. *Am. J. Bot.* **48,** 547–554.

Shields, L. M. and Durrell, L. W. (1964). Algae in relation to soil fertility. *Bot. Rev.* **30,** 92–128.

Shields, L. M., Mitchell, C. and Drouet, F. (1957). Algae- and lichen-stabilized surface crusts as soil nitrogen sources. *Am. J. Bot.* **44,** 489–498.

Shmida, A., Evenari, M. and Noy-Meir, I. (1986). Hot desert ecosystems: An integrated view. In: *Hot Deserts and Arid Shrublands*, Vol. 12B: *Ecosystems of the World* (Ed. by M. Evenari, I. Noy-Meir and D. W. Goodall), pp. 379–387. Elsevier, Amsterdam.

Shubert, L. D. and Starks, T. L. (1979). Algal succession on orphaned coal mine spoils. In: *Ecology and Coal Resource Development* (Ed. by M. Wali), pp. 661–669. Pergamon Press, New York.

Shubert, L. E. and Starks, T. L. (1980). Soil–algal relationships from surface mined soils. *Br. Phycol. J.* **15,** 417–428.

Shubert, L. E. and Starks, T. L. (1985). Diagnostic aspects of algal ecology in disturbed lands. In: *Soil Reclamation Processes: Microbiological Analyses and Applications* (Ed. by R. L. Tate III and D. A. Klein), pp. 83–106. Marcel Dekker, New York.

Sinclair, P. C. (1976). Vertical transport of desert particulates by dust devils and clear thermals. In: *Atmosphere Surface Exchange of Particulate and Gaseous Pollutants* (Ed. by R. J. Engelmann and G. A. Schmel), pp. 497–526. Nat. Tech. Inf. Service CONF-740921, Richland, Wash.

Skarpe, C. and Henriksson, E. (1987). Nitrogen fixation by cyanobacterial crusts and by associative-symbiotic bacteria in western Kalahari, Botswana. *Arid Soil Res. Rehab.* **1,** 55–60.

Skujins, J. (1984). Microbial ecology of desert soils. In: *Advances in Microbiol Ecology* (Ed. by C. C. Marshall), Vol. 7 pp. 49–91. Plenum Press, New York.

Skujins, J. (1981). Nitrogen cycling in arid ecosystems. *Ecol. Bull. (Stockholm)* **33,** 477–491.

Skujins, J. and Klubek, B. (1978a). Nitrogen fixation and cycling by blue-green algae–lichen crusts in arid rangeland soils. *Ecol. Bull (Stockholm)* **26,** 164–171.

Skujins, J. and Klubek, B. (1978b). Nitrogen fixation and denitrification in arid soil cryptogamic crust microenvironments. In: *Environmental Biogeochemistry and Geomicrobiology*, Vol. 2: *The Terrestrial Environment* (Ed. By W. Krumbein), pp. 543–552. Ann Arbor, Science, Ann Arbor, Michigan.

Smith, E. L. (1986). Soil condition as a factor in assessing range condition. In: *Use of Cover, Soils and Weather Data in Rangeland Monitoring* (Ed. by E. L. Smith *et al.*), pp. 25–29. *Symposium Proceedings*, 12 February 1986, Kissimmee, Florida. Society for Range Management, Denver, Colorado.

Smith, S. D. and Nobel, P. S. (1986). Deserts. In: *Topics in Photosynthesis* (Ed. by N. R. Baker and S. F. Long), Vol. 7, pp. 13–62. Elsevier, Amsterdam.

Snyder, J. M. and Wullstein, L. H. (1973). The role of desert cryptogams in nitrogen fixation. *Am Midl. Nat.* **90,** 257–265.

Soriano, A. (1983). Deserts and semi-deserts of Patagonia. In: *Temperate Deserts and Semi-deserts*, Vol. 5: *Ecosystems of the World* (Ed. by N. E. West), pp. 423–260. Elsevier, Amsterdam.

Sprent, J. I. (1985). Nitrogen fixation in arid environments. In: *Plants for Arid Lands* (Ed. by G. E. Wickens, J. R. Goodin and D. V. Field), pp 215–229. George Allen and Unwin, London.

Stanley, R. J. (1983). Soils and vegetation: An assessment of current status. In: *What Future for Australia's Arid Land?* (Ed. by J. Messer and G. Mosely), pp. 8–18. Australian Conservation Foundation, Canberra.

Starks, T. L. and Shubert, L. E. (1982). Colonization and succession of algae and soil–algal interactions associated with disturbed areas. *J. Phycol.* **18,** 199–207.

Starks, T. L., Shubert, L. E. and Trainor, F. R. (1981). Ecology of soil algae: A review. *Phycologia* **20,** 65–80.

Steinberger, Y. (1989). Energy and protein budgets in the desert isopod *Hemilepistus reaumuri. Acta Oecologia* **10,** 117–134.

Sylla, D. (1987). Effect of microphytic crust on emergence of range grasses. M.Sc. thesis, University of Arizona, Tucson, Arizona, 87 pp.

Taylor, J. E., Leininger, W. C. and Hoard, M. W. (1980). Plant community structure on ZAPS. In: *The Bioenvironmental Impact of a Coal-fired Power Plant* (Ed. by E. M. Preston and D. W. O'Guinn), pp. 216–234. *Fifth Interim Report*, Corvallis Environmental Research Laboratory, US Environmental Protection Agency.

Tchoupopnou, E. (1989). Splash from microphytic soil crusts following simulated rain. M.Sc. thesis, Utah State University, Logan, Utah, 63 pp.

Terry, R. E. and Burns, S. J. (1987). Nitrogen fixation in cryptogamic soil crusts as affected by disturbance. In: *Proceedings of the Pinyon–Juniper Conference* (Ed. by R. E. Everett), pp. 369–372. *General Technical Report* No. INT-215. USDA Forest Service, Intermountain Research Station, Ogden, Utah.

Thompson, J. W. and Iltis, H. (1968). A fog-induced lichen community in the coastal desert of southern Peru. *Bryologist* **71,** 31–34.

Tongway, D. J. and Smith, E. L. (1989). Soil surface features as indicators of rangeland site production. *Austr. Rangeland J.* **11,** 15–20.

Topham, P. B. (1977). Colonization, growth, succession and competition. In: *Lichen Ecology* (Ed. by M. R. D. Seaward), pp. 31–68. Academic Press, London.

Tsarchitzky, J., Banin, A., Morin, J. and Chen, Y. (1984). Nature, formation and effects of soil crusts formed by water drop impact. *Geoderma* **33,** 135–155.

Tsoar, H. and Pye, K. (1987). Dust transport and the question of desert loess formation. *Sedimentology* **34,** 139–153.

Tsoar, H. and Tyge Moller, J. (1986). The role of vegetation in the formation of linear sand dunes. In *Aeolian Geomorphology* (Ed. by W. G. Nickling), pp. 75–95. Allen and Unwin, Boston, Mass.

Turner, G. T. (1971). Soil and grazing influences on a salt-desert shrub range in western Colorado. *J. Range Mgmt* **24,** 31–37.

Uehara, G. and Jones, R. C. (1974). Bonding mechanisms for soil crusts. Part I. Particle surfaces and cementing agents. In: *Soil Crusts* (Ed. by J. W. Cary and D. D. Evans), pp. 17–28. *Technical Bulletin* No. 214, University of Arizona Agricultural Experimental Station, Tucson, Arizona.

Vogel, S. (1955). Niedere "Fensterpflanzen" in der sudaforkanischen Wuste. *Bertr. Biol. Pflanz.* **31,** 45–135.

Wagner, F. H. (1980). Integrating and control mechanisms in arid and semi-arid ecosystems–considerations for impact assessment. In: *Proceedings of a Symposium on the Biological Evaluation of Environmental Impact*, pp. 145–158. USDA Fish and Wildlife Service, Washington, D.C., FWS/OBS-80/26.

Walker, B. H. (1979). Game ranching in Africa. In: *Management of Semi-arid Ecosystems* (Ed. by B. H. Walker), pp. 55–81. Elsevier, Amsterdam.

Wallwork, J. A. (1962). Distribution patterns and population dynamics of the microarthropods of a desert soil in southern California. *J. Anim. Ecol.* **41,** 291–310.

Walter, H. (1973). *Vegetation of the Earth in Relation to Climate and Eco-physiological Conditions.* Springer-Verlag, New York.

Walter, H. (1985). The Namib Desert. In: *Hot Deserts and Arid Shrublands,* Vol. 12B: *Ecosystems of the World* (Ed. by M. Evenari, I. Noy-Meir and D. W. Goodall), pp. 245–282. Elsevier, Amsterdam.

Walter, H. and Box, E. O. (1983). Continential deserts and semi-deserts of Eurasia. In: *Temperate Deserts and Semi-deserts,* Vol. 5: *Ecosystems of the World* (Ed. by N. E. West), pp. 3–270. Elsevier, Amsterdam.

Watson, A. (1979). Gypsum crusts in deserts. *J. Arid Environ.* **2,** 3–20.

Weaver, R. W. (1986). Measurement of biological dinitrogen fixation in the field. In: *Field Measurement of Dinitrogen Fixation and Denitrification* (Ed. by R. D. Hauck and R. W. Weaver), pp. 1–9. *Special Publication* No. 18, Soil Science Society of America, Madison, Wisconsin.

Webb, R. H., Steiger, J. W. and Newman, E. B. (1988). The response of vegetation to disturbance in Death Valley National Monument, California. *US Geological Survey Bull.* **1793,** 103 pp.

Werger, M. J. A. (1986). The Karoo and Southern Kalahari. In: *Hot Deserts and Arid Shrublands,* Vol. 12B: *Ecosystems of the World* (Ed. by M. Evenari, I. Noy-Meir and D. W. Goodall), pp. 283–360. Elsevier, Amsterdam.

West, N. E. (1983). North American temperate deserts and semi-deserts. In: *Temperate Deserts and Semi-deserts,* Vol. 5: *Ecosystems of the World* (Ed. by N. E. West), pp. 321–422. Elsevier, Amsterdam.

West, N. E. (1985). Shortcomings of plant frequency-based methods for range condition and trend. In: *Selected Papers presented at the 38th Annual Meeting of the Society for Range Management,* pp. 87–90. Society for Range Management, Denver, Colorado.

West, N. E. (1986). Desertification or xerification? *Nature* **321,** 562–563.

West, N. E. (1988). Intermountain deserts, shrub steppes, and woodlands. In: *North American Terrestrial Vegetation* (Ed. by M. G. Barbour and W. D. Billings), pp. 201–320. Cambridge University Press, New York.

West, N. E. (1989). Spatial pattern–functional interactions in shrub-dominated plant communities. In: *The Biology and Utilization of Shrubs* (Ed. by C. M. McKell), pp. 283–305. Academic Press, London.

West, N. E. and Hassan, M. A. (1985). Recovery of sagebrush-grass vegetation following wildfire. *J. Range Mgmt* **38,** 131–134.

West, N. E. and Skujins, J. (1977). The nitrogen cycle in North American cold winter semi-desert ecosystems. *Oecol. Plant.* **12,** 45–53.

Westerman, R. L. and Tucker, T. C. (1978). Denitrification in desert soils. In: *Nitrogen in Desert Ecosystems* (Ed. by N. E. West and J. Skujins), pp. 75–106. IBP Synthesis Vol. 9. Dowden, Hutchinson and Ross, Stroudsberg, Penn.

Westoby, M. (1987). Soil erosion, as a landscape ecology phenomenon. *Trends Ecol. Evol.* **2,** 321–322.

Whisenant, S. G. (1989). Changing fire frequencies on Idaho's Snake River Plains: Ecological and management implications. *Sixth Wildland Shrub Symposium,* Las Vegas, Nevada, p. 1 (abstract).

Whitford, W. G. (1986). Decomposition and nutrient cycling in deserts. In: *Pattern and Process in Desert Ecosystems* (Ed. by W. G. Whitford), pp. 93–117. University of New Mexico Press, Albuquerque, New Mexico.

Whitford, W. G. and Freckman, D. W. (1988). The role of soil biota in soil processes in the Chihuahuan Desert. In: *Arid Lands: Today and Tomorrow*. (Ed. by E. E. Whitehead *et al.*), pp. 1063–1073. Westview Press, Boulder, Colorado.

Wolf, E. C. (1985). Managing rangeland. In: *The State of the World in 1985* (Ed. by L. Brown *et al.*), pp. 62–77. Worldwatch Institute, Washington, D.C.

Wood, J. C., Wood, M. K. and Tromble, J. M. (1987). Important factors influencing infiltration and sediment production on arid lands in New Mexico. *J. Arid Environ.* **12,** 111–118.

Wood, M. K. (1988). Rangeland vegetation–hydrologic interactions. In: *Vegetation Science Applications for Rangeland Analysis and Management* (Ed. by P. T. Tueller), pp. 469–491. Kluwer, Dordrecht.

Wood, T. G. (1970). Micro-arthropods from soils of the arid zone in southern Australia. *Search* **1,** 75–76.

Woodmansee, R. G., Vallis, J. and Mott, J. J. (1981). Grassland nitrogen. In: *Terrestrial Nitrogen Cycles* (Ed. by F. E. Clark and T. Rosswall), pp. 443–462. *Ecol. Bull. (Stockholm)* **33.**

Yabukov, T. F. and Bespalova, R. Y. (1961). Soil formation processes during the invasion of sands by plants in the northern deserts of the Caspian region. *Soviet Soil Sci.* **6,** 651 658.

Yair, A. and Shachak, M. (1987). Studies in watershed ecology of an arid area. In: *Progress in Desert Research* (Ed. by L. Berkofsky and M. G. Wurtele), pp. 145–193 Rowman Littlefield, Totawa, N.J.

Young, J. A., Evans, R. A., Roundy, B. A. and Brown, J. A. (1986). Dynamic landforms and plant communities in a pluvial lake basin. *Grt Basin Nat.* **46,** 1–21.

Ecology of Mushroom-feeding Drosophilidae

STEVEN P. COURTNEY, TRAVIS T. KIBOTA and
THOMAS A. SINGLETON

1. INTRODUCTION

Rapid generation times and polytene chromosomes have made fruit-flies of the genus *Drosophila* ubiquitous tools for genetic research. Some laboratory-based ecological studies have also been carried out, often yielding conclusions of general interest (e.g. Ayala *et al.*, 1973; Gilpin *et al.*, 1986; Mueller, 1987). However, the field ecology of the genus is largely unexplored due to a single unfortunate circumstance: the breeding sites of many species are either

ADVANCES IN ECOLOGICAL RESEARCH VOL. 20
ISBN 0-12-013920-0

inaccessible, hard to find, or simply unknown. For instance, *D. pseudoobscura* and related species have been extensively investigated by population geneticists, but only recently have the larvae been found in their normal host – oak acorns parasitized by other insects. Moreover, those species which are accessible are often now so firmly domesticated that we can have little idea of what their ecology was like before humans developed such extensive *Drosophila* habitats as orchards and garbage cans. However, there are a few species which occupy habitats which are natural, accessible and suitable for ecological experimentation. An extensive body of work has, for instance, developed on the ecology of drosophilids associated with cacti and related substances. Particularly strong have been studies of host choice, dispersal and associations with microflora. Perhaps more especially suited to ecological studies are the many taxa which feed on mushrooms and other fungi. These species are rapidly becoming important tools for those interested in such basic ecological questions as community composition (e.g. Shorrocks and his associates) and the evolution of interactions.

In 1975, Gilbert and Singer brought together most of the existing information on butterfly ecology. Their synthesis was the spark for much of the subsequent exploitation of butterflies for studies of host use, population structure, etc. Mycophagous drosophilids are in an analogous situation to that of butterflies in the early 1970s: much is known, but the literature was scattered and hard to penetrate. Our purpose here is not just to present what has already been discovered about these interesting insects, but also to emphasize the ideal nature of these systems for many ecological and evolutionary investigations. We have little doubt that increasing attention will be focused on these flies, as more and more researchers realize their great potential.

Mushrooms are discrete, easily studied hosts and, notoriously, often endowed with an array of highly toxic chemicals. It is probably unsurprising, therefore, that considerable attention has been given to host choice in mycophagous insects. Host use by mushroom feeders is often compared to the much larger literature on phytophagous insects, in the hope that general principles of host use will emerge. In part, this hope is justified: certain of the basic principles of host use were indeed first studied in mycophagous drosophilids (Jaenike, 1978). Mushrooms are, however, quite unlike plants as hosts in many important ways. One of the additional functions of this chapter will be to determine the extent to which results obtained with mycophagous drosophilids will be generalizable to other communities of insects and hosts.

II. GENERAL ASPECTS

A. Mycophagy within Drosophilidae

Mycophagy has arisen many times within the Drosophilidae (Throckmorton, 1975). A number of factors probably favor this association. First, fungal tissue is highly nutritious (see below) and does not contain the structural materials which make plants so difficult to digest (particularly cellulose). In addition, macrofungi are often associated with decomposing organisms, the natural foods of many drosophilids. It is likely that mycophagy has evolved from detritivory in this taxon as in many others (Throckmorton, 1975; Hammond and Lawrence, 1989; Lawrence, 1989). Whatever the reasons for the association, there are many taxa within Drosophilidae which have (apparently independently) evolved use of mushrooms as their primary larval resource. Some of the more important mycophagous groups are shown in Fig. 1.

Several points must be made about Fig. 1. First, *Drosophila* itself is a huge genus, and is certainly polyphyletic. Some well-defined taxa have been split from it, and are recognized at the generic level (e.g. *Zygothrica, Mycodrosophila* and *Scaptomyza*), whereas others remain lumped within (notably the many species of the subgenus *Hirtodrosophila*). Secondly, many taxa across the whole family are mycophagous. In some lineages, this association appears to be conservative, with most taxa using mushroom hosts. Nearly all *Hirtodrosophila*, for instance, are mycophagous, except for a few rare instances, such as a Central American species bred from tree frog eggs (Gimaldi, 1987). *Mycodrosophila* (as the name suggests) is usually found in association with fungi. In other lineages, mycophagy is more labile. In *Zygothrica*, for instance, despite a strong association of adults with fungi, fewer than 10% of species feed there as larvae. In the major radiations within *Drosophila* itself, some groups, such as the *immigrans* radiation, contain species breeding in an array of many different hosts. The evolutionary instability of host associations is underscored in the *D. quinaria* species group where several host shifts have occurred at the species or population level (Fig. 2). The essential information to be gleaned from Fig. 2 is that host shifts are apparently rather easily accomplished in comparison to other insects (particularly many phytophagous ones).

A further point to be drawn from Fig. 1 is that we know far more about a few mycophagous drosophilids (particularly the *quinaria* and *tripunctata* species groups) than about some others. There are for instance, some 70 species of the "fungus-feeder" group in the endemic Hawaiian fauna; these are almost entirely unexplored from an ecological viewpoint, as are the many tropical species in *Leucophenga, D. (Hirtodrosophila)* and *Mycodrosophila*. It is hoped that attention will soon be focused on these less well-known

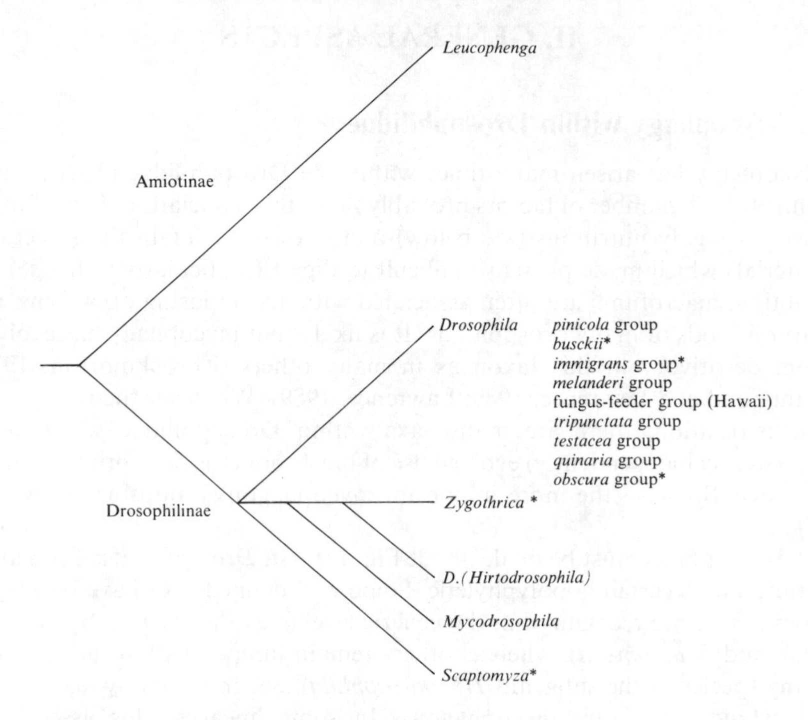

Fig. 1. Mycophagous Drosophilidae. A partial phylogeny showing only the major taxa associated with mushroom hosts. Mycophagy is distributed throughout the family. Within *Drosophila*, a number of species groups are primarily mushroom-feeders, while some others are generalists, feeding facultatively on a range of hosts (including fungi), (indicated by an asterisk). Some mycophagous taxa of uncertain status, e.g. *Samoaia*, are not shown. Note that *Drosophila* is polyphyletic. *D. (Hirtodrosophila)* may warrant generic status. Based (in part) on Throckmorton (1975) and Grimaldi (1987).

groups. It will be instructive to know whether the principles applying to temperate zone relatives also determine the ecology of these tropical taxa. Lastly, it should be emphasized that some species are polyphagous to the degree that fungi represent only a part of the host range. The cosmopolitan species *D. busckii, D. immigrans* and *Scaptomyza pallida* are all often reared from fungi, and they may be locally more numerous than strict mycophages (Dely-Draskovits and Papp, 1973; Shorrocks, 1977).

B. On the Value of Mushrooms

"Vegetable beef-steak" (Crisan and Sands, 1978), a monicker which may reflect the historical confusion over fungal taxonomy, is nevertheless an appropriate description of mushroom nutritional quality. The essential

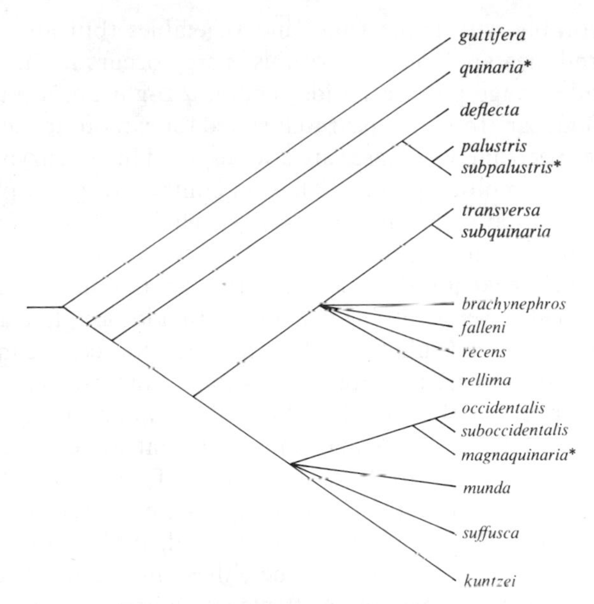

Fig. 2. The *D. quinaria* species complex. *D. guttifera* is a sister group, which is mycophagous. All species in the *quinaria* group are associated with mushrooms, except for those indicated by an asterisk, which use skunk cabbage or other Araceae, and the *deflecta/palustris* lineage which use decaying water plants. In addition, *D. recens* is primarily mycophagous but is sometimes found on Araceae. *D. subquinaria* is a mushroom-feeder which has colonized *Marah oreganus* (Cucurbitaceae) in Oregon. Not shown (of uncertain position) are: *D. limbata* (European: from mushrooms, Araceae, Cucurbitaceae), *D. phalerata* (European: mycophagous, primarily on *Phallus*) and some Japanese species, including *D. nigromaculata* (fruit-feeder). Also in this complex are the mycophagous *D. tenebrosa* and *D. innubila* from North America, and close to *D. suffusca*. Constructed from characters in published keys.

components of insect diets appear to occur at necessary levels in most mushrooms (see Martin, 1979, for a review). Insects, like other heterotrophs, require several basic dietary macronutrients for growth and development: a carbon supply, nitrogenous compounds, lipids and water. In addition, other compounds such as choline, the B-vitamins, polyunsaturated fatty acids, sterols and minerals such as potassium and phosphorus are needed at somewhat lower levels (House, 1974). Some compounds which are required by other organisms — the fat-soluble vitamins A, D, E and K — are not required by insects (Dadd, 1985).

Macrofungi appear to provide an excellent source of energy. With a caloric content ranging from 0·1 to 0·7 calories per gram of fresh weight, mushrooms compare favorably with many fruits and vegetables (but are poorer than seeds, nuts and meats). A majority of this energy occurs in the form of the structural and storage polysaccharides chitin, α-D-glucan, β-D-(1.3)-glucan and β-D-(1,6)-glucan. Insects appear well-suited for exploiting the mushroom carbohydrate spectrum. Chitinases are a widespread insect enzyme playing a crucial role in the molting process. Thus, in contrast to green plants which contain sugars that are difficult to digest (e.g. cellulose, lignins and pectins), mushrooms offer a relatively accessible carbon source for insects.

Proteins, which make up 20–40% of the dry weight of mushrooms, contain all the amino acids essential in the insect diet. In addition, free amino acids are also available for consumption. These nitrogen sources are more readily utilized by insects than are plant proteins because tannins, which reduce the digestibility of proteins (Feeny, 1968, 1970; Bernays and Chapman, 1978), are not known in fungi. Urea and ammonia contribute to the nitrogen content of macrofungi (up to 7·5% dry weight: Tyler *et al.*, 1965). These products may be undigestible (Cochran, 1985), as the correct enzymes are not commonly found in insect digestive tracts (Bursell, 1967). However, even in the absence of the correct insect enzymes, these nitrogen sources may be utilized because the fungi themselves often contain breakdown enzymes (e.g. urease). When the fungal tissue is consumed, the barriers between urea and fungal urease are destroyed and the urea may be digested.

The other major nutrient requirements (lipids and water) of insects are also met by macrofungi (Lilly, 1965; Kukor and Martin, 1987). The micronutrients choline, the B-vitamins and polyunsaturated fatty acids are also present at satisfactory levels. The main fungal sterol is ergosterol (a ^{28}C chain with a methyl group at the C-24 position). The ability to dealkylate ^{28}C and ^{29}C sterols at the C-24 position is widespread in insects. Thus, this major micronutrient requirement is also met by mushrooms. Macrofungi, then, seem to offer an assemblage of abundant and readily available nutrients to mycophages. It should be noted, however, that some mycophagous species require specific yeast and bacterial microflora rather than the fungal tissue itself (see below).

An aspect of mushroom chemistry that must not be overlooked is the potent toxicity of at least some mushrooms. The highly poisonous "death cap" (*Amanita*) mushrooms are infamous for their destruction of human lives. α-Amanitin is but one of a myriad of secondary metabolites which may be toxic to mycophages. As with theories of plant chemistry, mycotoxins are thought to be produced as intermediates in basic metabolic pathways (Turner, 1971). Metabolites which increase fungal fitness by decreasing the efficiency of consumers will be selected, i.e. there will be evolution toward increased toxicity. Major theories of plant–herbivore interactions — the theory of co-evolution (Ehrlich and Raven, 1964) and the plant apparency hypothesis (Feeny, 1976; Rhoades and Cates, 1976) — are based on these principles. It has been suggested, however, that fungus–insect interactions do not behave in similar ways (Hanski, 1989). Most mycophagous insects feed on mushrooms which are in various stages of decay. Thus, the fungi have performed most of their reproductive activities by the time that insects begin to consume their tissue. Because there is not much effect of insect mycophagy on mushroom fitness, it is unlikely that mycotoxins evolve in response to insects. Instead, it is probable that a high proportion of fungal defences (such as amanitin) have evolved to thwart vertebrate mycophagy (which is widespread, e.g. Lacy, 1984a), although other enemies such as parasitic bacteria and fungi may also have effects. Whatever the selection favoring fungal defences, these will have consequences for insect mycophages, which have evolved means to tolerate or circumvent toxins (Kukor and Martin, 1987; Hanski, 1989). However, these interactions are not reciprocal in effect. An important distinction between mycophagy and phytophagy is that there is little opportunity for co-evolution (whether pairwise or diffuse) in mycophagous insects and their hosts.

C. Predictability

Mushrooms vary in predictability. The effects of this predictability on mycophagous *Drosophila* have been considered in studies of host ecology and genetics (Jaenike, 1978; Jaenike and Selander, 1979; Lacy 1982, 1983, 1984a, b). Jaenike (1978) proposed that the unpredictable nature of mushrooms accounts for the high degree of polyphagy seen in mycophagous *Drosophila*. A broad correlation, where stenophagy is associated with predictable hosts, is seen in Japan (Kimura *et al.*, 1978), and the eastern (Lacy, 1984b) and western USA (Courtney, pers. observation). Lacy (1983) has shown that polyphagous species are more genetically variable, in agreement with the niche-width-variation hypothesis. Most of these studies emphasize the *unpredictability* of the mushroom habitat, especially when compared to the *predictability* (Feeny, 1976; Rhoades and Cates, 1976) of plants.

Are mushrooms generally more unpredictable hosts than plants, as many of these authors suggest? In order to answer this question, one must consider both the spatial and temporal components of predictability. Some mushrooms are the fruiting bodies of long-term mycorrhizal associations with plants and therefore are quite spatially predictable. On the other hand, some plant species can be very unpredictable spatially (Strong *et al.*, 1984). Temporally, fungal fruiting bodies exhibit a wide range in variation, from short-lived caps of *Coprinus* which may only last 3 days (Lacy, 1984b), to perennial polypores which can last for years (Hanski, 1989). In the Oregon Cascades, *Ramaria formosa* may be very abundant after short periods of rainfall, yet apparently absent after 2 weeks of dry weather (Courtney and

Singleton, pers. observation). Year-to-year variation can be quite substantial in many fungal species. British mushroom communities are highly sensitive to rainfall (Shorrocks and Charlesworth, 1980), while yields of macrofungi in Finnish Lapland are dependent upon the annual summer temperature (Ohenoja and Metsanheimo, 1982; Ohenoja and Koistinen, 1984). Year-to-year variation in predictability also occurs in plants (e.g. some desert annuals). Courtney and Chew (1987), for instance, record cruciferous host-plants which on average vary in abundance by an order of magnitude between years. Such plants are certainly more sporadic in their occurrence than some of the perennial fungi. For fungi in general it is probable that the temporal component of variance in abundance is high relative to the spatial component (a result of the large and immediate influence of temperature and rainfall). The opposite may be true in plants, especially perennials. We feel that it is an oversimplification to describe fungi as unpredictable, and emphasize the need to quantify variation in time and space before generalizing to particular systems.

III. LARVAL ECOLOGY

A. Host Effects

Mushrooms appear to represent a dietary alternative of high value (see above) and are thought to be homogeneous both within an individual fruiting body and across taxa (Hanski, 1989). Additionally, the co-evolutionary dynamics which would constrain the host range of mycophagous insects appear to be rare or non-existent. One might expect, therefore, that mycophagous larvae should have large host ranges. There is some indication, for example, that larvae of *D. suboccidentalis* can develop as well on skunk cabbage or tomato as on their natural host, *Ramaria formosa* (Kibota, pers. observation). Unfortunately, not much rigorous work has focused on the effects of different larval substrates on growth and development. In phytophagous insects, by contrast, it is well known that the larval host range is often much broader than that of adults (Wiklund, 1975). It would be interesting to discover whether mushroom feeding imposes special physiological requirements or if, as is seen in plant-feeders, host utilization is constrained not by larval fitness but by adult behavior.

The state of decay is known to be important in larval performance. Kimura (1980a) found that larvae of several *Drosophila* spp. prefer decayed over fresh mushrooms. This is true even in the species (*Hirtodrosophila*) which exploit early stages of mushrooms in the wild. This preference may result from many factors. One factor which has been observed in our laboratory is that larvae seem unable to burrow into some fresh mushrooms and are thus not able to feed properly. Klopfenstein (1971) noticed this same occurrence in another insect, *Hadraule blaisdelli*; hard conks of the fungus *Fomes everhartii* are impenetrable by the adult insects but larvae will readily feed on a powderized version of the fungus.

Yeast microflora may also be critical in host utilization by larvae. Kearney and Shorrocks (1981) found that three species of *Drosophila* exhibit differential fitness in the presence of yeasts associated with different host substrates. The association of *Drosophila* with specific yeast communities is well known in the cactophilic species. These associations may have resulted from the saprophagous habit of the ancestors of many mycophagous drosophilids (see above).

Lastly, and possibly most well known, are the tests by Jaenike and co-workers (Jaenike *et al.*, 1983; Jaenike, 1985a) on the effects of α-amanitin on larval fitness. This research showed that mycophagous *Drosophila* spp. are better able to tolerate the RNA polymerase toxin, α-amanitin, than are non-mycophagous species. The ability to tolerate amanitin, however, does not lead to an increased ability to utilize mushrooms. An interesting question that arises from this research is simply why has amanitin-tolerance evolved in mycophagous insects which are, in the majority of cases, highly polyphagous. Amanitin-containing mushrooms make up a very small fraction of the mycophage diet and selection for amanitin-tolerance would appear to be negligible.

B. Intraspecific Interactions

Fungal fruiting bodies, as mushrooms, are relatively small and discrete, and rather short-lived when compared to the hosts of many phytophagous insects. It might be anticipated that few strong interactions would be found in such peripheral habitats. However, mushrooms differ from plants in two other characteristics; they are generally good food, and are relatively homogeneous material. Hence all of the fruiting body is often available to insects or other fungivores. A large number of insect taxa do indeed use fungal food (Wheeler and Blackwell, 1984; Wilding *et al.*, 1989). There seems to be as much opportunity for ecological interactions between these and other groups as there is for communities associated with plants.

Intraspecific competition in fungivorous *Drosophila* has been suggested by some authors (Grimaldi and Jaenike, 1984; Grimaldi, 1985). This does seem a real possibility in light. The high densities of larvae which are sometimes seen in single fruiting bodies (Lachaise and Tsacas 1983). Often, large numbers of eggs are laid by a few or even single females (Jaenike and Selander, 1979). It is clear that females remain in the vicinity of a mushroom patch for several days, adding more eggs as they become available for oviposition. In a number of species, intraspecific aggregation also occurs.

D. phalerata adults show a clear association with low-density patches of the host *Phallus* (Shorrocks and Charlesworth, 1982), while *D. suboccidentalis* is typically found on just a few individual fruiting bodies of *Ramaria formosa* in Oregon. In these circumstances, the number of larvae developing in a mushroom may vary widely, as indeed seems to be the case (Grimaldi and Jaenike, 1984). It is also to be expected that the larval population will be composed of insects at all stages of development as new individuals are added. We therefore expect intraspecific competition to follow the scramble pattern, as limited resources are consumed, but for the effect to be felt unevenly among competitors; late arrivals will develop poorly, eggs laid early will suffer little detrimental effect. Note that this will be true even for individuals laid as eggs by the same female. The complete consumption of fungi by Diptera does occur regularly; on Long Island, New York, *Russula* and *Tricholoma* were sometimes completely consumed by the larvae of *D. falleni* and *D. testacea* (S. P. Courtney, pers. observation). Lachaise and Tsacas (1983) report occasionally high local densities of *Leucophenga* and other species in tropical Africa. Antagonistic effects are also sometimes seen. In laboratory cultures, eggs are often buried by burrowing larvae, and die from asphyxiation. The surface of a mushroom in an advanced state of decay is almost liquid, and eggs are sometimes pushed under the surface here too. These deaths do not appear to be the result of any directed action by the larvae, but are the accidental consequence of normal feeding behavior.

It is by no means clear, however, that intraspecific competition is an important process at the population level, and that it sets a limit to population growth. The sporadic local depletion of hosts is commonplace in many phytophagous insects which show no signs of being food-limited (e.g. Courtney and Courtney, 1982). Grimaldi and Jaenike (1984) and Worthen (1989a) have performed experiments which demonstrate that there is a potential for individual suppression by conspecifics and congeners, but there is currently no evidence available on the relative importance of competition, or indeed any larval mortality factor, relative to other population processes, such as the factors affecting host detection by adults. In this circumstance, strong advocacy of competition in mycophagous drosophilids seems premature.

Not all larval–larval interactions are antagonistic. Microbial competitors are often seen in host fungi; *Drosophila* larvae often graze these microorganisms as well as the host tissue. If rather few larvae are present, the competitors may reproduce rapidly and smother them. A sufficiently large number of larvae is necessary to prevent microbial growth, for any drosophilid to emerge. For instance, in *D. suboccidentalis*, there is evidence for an Allee effect, where survivorship of larvae at first rises with larval density, until the effects of competition outweigh the effects of facilitation (Fig. 3). In this case, the larvae are grazing on fungal hyphae which would otherwise

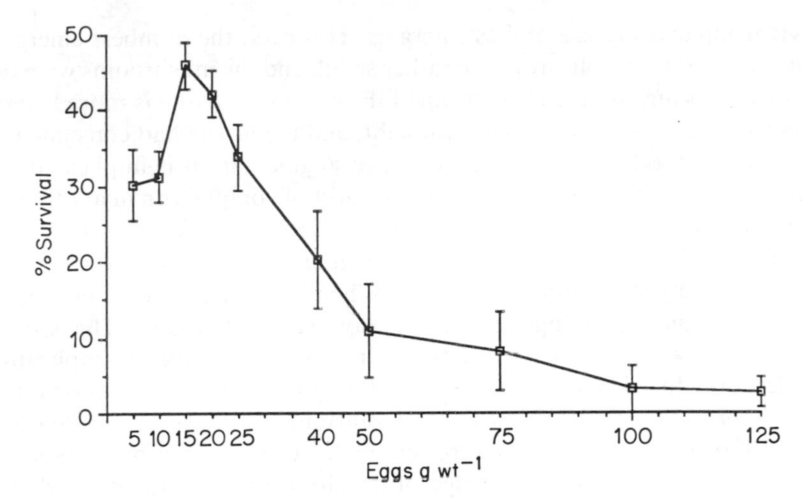

Fig. 3. Survivorship (to pupation) of *D. suboccidentalis* larvae as a function of initial density (eggs per g of *Ramaria* host ±1 s.e.). Females were allowed to lay eggs for 24 h, on 0·5–1·0 g pieces of mushroom, which were kept over moistened insect medium, until all surviving larvae had emerged and pupated. Survivorship increases with density at low levels, because larval feeding inhibits the growth of fungal hyphae, which compete for host material. As larval density is further increased, competition for food among larvae outweighs facilitation, and survivorship declines

dominate the mushroom. This effect is commonly seen in laboratory cultures, and is probably present in the wild. However, detecting such intraspecific facilitation would require careful manipulation of the numbers of eggs present on mushrooms, which has yet to be attempted in any study.

C. Interspecific Interactions

Interspecific competition, between mycophagous drosophilids, and between these and other mushroom feeders, has been much discussed (Spieth and Heed, 1975; Shorrocks and Charlesworth, 1980; Grimaldi and Jaenike, 1984; Grimaldi, 1985; Worthen and McGuire, 1988). Often, several species of dipteran are reared from a single fruiting body. Is this evidence of competition? Spieth and Heed (1975) and Grimaldi (1985) appear to think so, and have discussed factors which might limit the effects of competition and allow co-existence. Clearly, occasional instances of complete resource consumption will bring about competition among all larvae in that host. As before, however, we do not have much information on how important a process this will be at the population level. Grimaldi and Jaenike (1984), in an elegant series of experiments, showed that augmenting the food available to the Diptera naturally present in collections of mushrooms increased both the survivorship and the size of flies emerging. However, the numbers emerging from many of their replicates were rather small, and the mushrooms were not reported as being completely consumed. Either fungal tissue is more heterogeneous than has been previously thought, and the larvae had consumed all the available food, or some other effect of augmentation is implicated. By contrast, Worthen (1989a) has shown the effect of competition in another set of experiments where ovipositing adults were also subject to predation. When no predators were present, *D. tripunctata* dominated *D. falleni* and *D. putrida* in emergence from mushrooms. When a rove beetle was allowed to feed *ad lib.* on egg-laying adults, this dominance was lost. The size of emergent flies also increased when predators were present, clearly implicating an effect of reduced larval competition, although differential predation on *D. tripunctata* adults cannot be ruled out. Unfortunately, however, the densities used in Worthen's experiments were very high (105 adults per bottle) and are certainly at the extreme high range of densities observed in the field. We cannot yet be assured that interspecific competition, although it may well occur, is an important contributor to either larval mortality or to population fluctuations (Worthen, 1989b). This is essentially the same position reached in studies of phytophagous insects where the mere presence of competition is still a controversial subject (e.g. Lawton and Strong, 1981). We need studies which will document the patterns of larval mortality, including its spatial and temporal variance, and which will also determine whether larval factors play a significant role in population dynamics. Large year-to-year fluctuations of the resource mushrooms, in response to climatic rather than fungivore effects, suggests that host availability will be an important influence on drosophilids and other fungivores (Hanski, 1989), as is also seen in many phytophagous insects with unpredictable resources (Courtney and Duggan, 1983). Muona and Lumme (1981) have, for instance, noted the *D. phalerata* and *D. transversa* are far more variable in abundance than non-mycophagous *Drosophila* spp. in the same habitats. In one year (1974), *D. testacea* and *D. phalerata*, normally common, were both extinct at their study sites. Such large fluctuations in abundance, especially when seen simultaneously in two species, suggests that host availability is more important than interspecific competition in setting the abundance of either population.

It is probable that competition with non-drosophilid fungivores is more important than competition within the family. Because many other Diptera breed within fungi (see Hanski, 1989, for a review), and these are often much larger than *Drosophila*, these may be competitive dominants. However, other organisms are more likely to be major competitors with all Diptera in fungi. Decay fungi and bacteria consume the majority of fruiting bodies in the field, and are probably responsible for the rapid removal of much drosophilid food. Essentially similar interactions occur between Diptera and decomposing organisms on other substrates, such as corpses (Wilson and Fudge, 1984).

It is to be expected, therefore, that drosophilids would be sensitive to the state of decay of mushrooms, because highly decayed hosts will not be suitable food for very long. This does appear to be the case in many communities (see Section III.B; Kimura, 1980a; Kimura and Toda, 1989). In this respect, fungivorous insects may differ from phytophagous insects which interact with detritivores to a much lower degree.

Interactions with other organisms have received rather little attention. It is known that several parasitoid species are found on mycophagous drosophilids, but their importance in larval mortality is debatable. Grimaldi (1985) and some others in both Europe and the USA have noted rather low parasitization rates (<3%), but Janssen *et al.* (1988) and Janssen (1989) found much higher levels of attack. Part of the reason for such disagreement may lie in the different sampling methods used by different authors. Vet (1983) notes that the development of the eucoilid parasitoid, *Leptopilina clavipes*, is some 26 days when reared in *D. phalerata*. This is considerably longer than the development of the fly larvae, and will certainly exceed the duration of many of the fungal fruiting bodies. A sampling method which records the emergence of insects from fungi and associated soil (as is routinely practiced) will miss most parasitoids unless sampling is continued long after flies have ceased emerging. For this reason, we have greater confidence in the estimates of Janssen *et al.* (1988) than in the more casual, lower estimates obtained by ourselves and others.

Janssen *et al.* (1988) record six species of parasitoids attacking mycophagous *Drosophila* (primarily *D. phalerata* and *D. kuntzei*, plus some facultative species such as *D. immigrans*). Of these, three (*L. clavipes*, *Kleidotoma bicolor* and *Phaenocarpa tacita*) are apparently restricted to such hosts. Parasitization rates vary with season, reaching the highest levels (up to 50%) in mid-summer, which represents an early generation of the succeeding generations of *D. phalerata* (Shorrocks and Charlesworth, 1982). Parasitization rates also vary strongly with fungal or other host identity. No parasitoids were reared from larvae growing in *Polyporus squamosus*, which appears to be a refuge for *Drosophila* from parasitoid attack. Parasitism was usually high on other hosts, notably *Phallus impudicus*, which may be the primary host for *D. phalerata*. Interestingly, high attack rates were also seen for larvae found in slime-fluxes on oak, sycamore or buckthorn (*Hyppophae rhamnoides*), or in rotting material of *Heracleum mantegazzianum*; none of these are normal substrates for either *D. phalerata* or *D. kuntzei*. A number of authors have suggested that host shifts may be favored if the new host is "enemy-free space" (Lawton and Strong, 1981; Bernays and Graham, 1988). In this community, larvae feeding in abnormal habitats appear to have (if anything) higher losses to parasitoids, suggesting that escape from these enemies will not favor the use of novel hosts.

Vet (1983, 1985) has studied the behavior of several drosophilid parasitoids in some depth. *L. clavipes* appears to develop normally in several species (*D. phalerata, D. transversa* and *D. subobscura*, but not *D. melanogaster*). One parasitoid normally emerges from each larva and the female wasps appear to avoid superparasitism. Parasitoids are strongly attracted to the odor of decaying (but not rotten) mushrooms, which is the host stage where drosophilid larvae will be most abundant. However, they are not attracted to the odor of *D. phalerata* larvae. Females have a strong attraction to mushroom odors, but can be conditioned (at emergence) to respond also to yeasts and fruits (mushrooms are still preferred). There is also a much stronger associative learning of habitat cues during oviposition which can modify preference for mushrooms. *L. clavipes* can be trained to search for *D. phalerata* in any habitat where it has previously found this host. Vet interprets this behavior as being optimal in searching for a polyphagous host, with seasonal fluctuating abundances in alternative habits. *K. bicolor* similarly fails to detect *D. phalerata* larvae by olfactory cues (Vet, 1985).

The studies of Janssen *et al.* (1988), Jannsen (1989) and Vet (1983, 1985) suggest that parasitism by Hymenoptera may play an unexpected and large role in population dynamics of drosophilids. More studies in many more localities are badly needed. Some points appear worth making at this point, however. First, these parasitoids appear able to attack all drosophilids in mushrooms, and may act in a density-dependent fashion, attacking most severely the most abundant host. Secondly, other *Leptopilina* spp. are most successful when attacking crowded *Drosophila* larvae; crowding appears to lower the resistance of fly larvae to attack (Bouletreau and Wajaburg, 1986). If similar responses are found in species attacking mycophagous drosophilids, then this will be a strong density-dependent response. Thirdly, the attack rate upon drosophilids in different fungi are highly variable, and appear sufficiently strong and predictable that they may influence the evolution of host choice. Lastly, the attack rates seen by Janssen *et al.* (1988) are still only moderate in comparison to those seen in many parasitoids of phytophages. Further work is necessary before we can be sure that even these factors have much effect on population or community dynamics.

Parasitism by nematodes has recently received some attention (Jaenike, 1985a, 1988a; Montague and Jaenike, 1985; Kimura and Toda, 1989). Nematodes can be quite common in many mushrooms and are regularly found in stocks in our laboratory. They sterilize insects which are carrying them and also use them as vectors to new fungal fruiting bodies. Parasitism levels are higher in flies at more decayed mushrooms, resulting in some species (particularly *D. testacea*) having higher levels of attack than others (Kimura and Toda, 1989). Similarly, those species of fungi which decay most rapidly (such as *Coprinus*) yield many parasitized flies. Other fungi show much lower levels of association with nematodes. *Amanita* spp. (which contain amanitin toxins) and *Pleurotus* spp. (whose hyphae may predate nematodes) appear to yield few parasite-bearing flies. Jaenike (1985a) and Kimura and Toda (1989) argue that these hosts are included in the diet of many drosophilids because they are in effect enemy-free space. Note that some Japanese *D. (Hirtodrosophila)* appear to specialize on *Pleurotus*, perhaps suggesting that nematode parasitism has selected for this specialization. Note also, however, that the levels of parasitism by nematodes are rarely high enough to have significant effects on the population dynamics of the flies. However, nematode parasitism may affect choice among males by females (Jaenike 1988a).

Worthen (1988) has recently described an unusual interaction between mycophagous drosophilids and slugs which also graze on mushrooms. The number of adult flies emerging from slug-damaged mushrooms was higher than that from undamaged fungi. A number of explanations for this facilitation effect are possible. Worthen favors an increased level of oviposition by flies on damaged mushrooms which have open wounds (attractive sites for egg-laying) and a more textured surface. Alternative explanations include increased olfactory detection of damaged mushrooms, and an increased rate of decay following slug damage. Slugs are common on mushrooms throughout North America and typically affect a high proportion of the fruiting bodies of certain species, such as *Pleurotus* and *Russula*. It will be interesting to see whether this facilitation effect is common in many communities. Note, however, that Worthen's data imply that more flies can be reared from many mushrooms than often do emerge. This interpretation is not consistent with larval competition being simultaneously prevalent. One possibility that should be explored is slug inhibition of microbial competitors of Diptera.

We have left until now discussion of perhaps the most important mortality factor acting upon larvae–predation. Some generalist insect predators, including beetles and ants, do attack larvae in mushrooms. One of us (S.P.C.) has observed a single *Tricholoma* fungus, in an advanced state of decay, which contained hundreds of *D. falleni* larvae, and where ants were removing larvae at the rate of one every minute or so. In this case, the predator had detected and attacked a locally high density of flies and appeared to be having a major impact on the population in the host. Such events are probably quite opportunistic and rare, however. The majority of predation is probably due to mammalian fungivores which may consume the majority of fruiting bodies in an area. Lacy (1984b), for instance, recorded that skunks, mice and deer took many of his study mushrooms. We have recorded essentially similar results for chipmunks, squirrels and other vertebrates in western North America. One estimate of mammal consumption of *Ramaria formosa*, the major host of *D. suboccidentalis*, suggests that 55% of fungi are consumed by mammals before any flies could complete development (S.P. Courtney, pers. observation). It is important to note here that the majority of fungal toxins appear to be inactive against specialized fungivorous insects (Jaenike *et al.*, 1983) but to have evolved instead in response to mammal predation of the fruiting body. Squirrels (a major predator of fungi in many forests) appear to be the only mammals immune to the effects of amanitin. All these observations suggest that mammal predation of fungi, and their associated dipteran larvae, may have important consequences for population dynamics, by both reducing host abundance and by killing fly larvae. This incidental predation has yet to receive any systematic investigation.

IV. ADULT ECOLOGY

A. Feeding

Adults feed on a variety of substances ranging from yeasts, bacteria and other microorganisms that grow on fungal material, to fungal spores, to fresh fungal material. Ancestrally, *Drosophila* are believed to have been saprophagous (Throckmorton, 1975), and therefore exploitation of the fungal environment probably began with decaying fungi and the associated microbes.

In phytophagous insects, adults and larvae often, but not always feed on different hosts. In mycophagous drosophilids, adults and larvae usually feed at the same fruiting bodies. However, there may still be some differences in resource use at a finer scale. Adults usually feed (and lay eggs) on host material that is in a fresher state than that fed on by larvae. This may be associated with a difference in preference behavior between adults and larvae (e.g. *Hirtodrosophila*: Kimura, 1980a). Some other studies show adult drosophilids feeding on different material than do their larvae. For instance, Carson *et al.* (1956) found different yeasts in the gut of adult flies than the yeasts isolated from their breeding sites, thus implying different food sources for adults and their larvae. Adult *D. trivittata* (now known to be three species) and *D. sexvittata* probably feed on fungal spores, whereas their larvae feed on fungal hyphae (Kimura, 1976). Lacy (1984a) has emphasized the spatial component of niche partitioning between adults and larvae by pointing out the simple fact that adults restrict feeding to the surface of the sporophore while larvae tunnel through the fungal material. In a few mycophagous species, adults feed at non-fungal sites. Flies of the *quinaria* species group are often found at fruit baits (worldwide); in Japanese *D. brachynephros*, adults are routinely found on wild fruits and tree sap in the autumn (Watabe *et al.*, 1985). *Leucophenga* adults may usually feed on sap (Okada, 1962; Lachaise, 1975; Lachaise and Tsacas, 1983).

Kimura (1980a) has suggested that mushroom material is not necessarily good for adult survival. Females kept over yeast or banana media survive better than those kept over mushrooms. Kimura suggests that adults are poisoned by toxic decomposition products. In North American taxa, survivorship (over a 48-h period) on *Amanita muscaria* was low in four species of the *quinaria* species complex, ranging from 60% in *D. magnaquinaria* to 80% in *quinaria* and *suboccidentalis* (S. P. Courtney, unpublished data). Spieth and Heed (1975) note that *D. flavopinicola* adults cannot be maintained solely on *Agaricus* and must have access to a carbohydrate-rich food source. Sturtevant (1942) suggested that *D. pinicola* are sap feeders, as he observed adults on moist spots on the trunks of *Sequoiadendron giganteum*. More work is needed on the role of nutrition in adult survival.

Although many feeding habits of mycophagous *Drosophila* have been identified and many pressures for the observed patterns have been proposed, it is still unclear whether the distribution of fungi, competition (interspecific, intraspecific, or adult/larval), predation, or a combination of these pressures has shaped the adult feeding habit of mycophagous *Drosophila* (see Section IV.H).

B. Mating

Mating occurs at or near fungal sites in many mycophagous drosophilids. Typically, adult males sit on top of mushrooms "waiting" for females (see Fig. 4). When females fly to the mushroom to feed or oviposit, copulation occurs. Males may also be found on nearby vegetation if many males are present. Spieth and Heed (1975) observed this behavior in *D. melanderi*. Males of this species defend areas on top of mushroom caps and, consequently only one or two males per mushroom are ever found. However, there are a number of variations on this typical encounter behavior.

Two notable variations occur in *quinaria* group flies in Oregon (*D. suboccidentalis* and *D. magnaquinaria*). In *D. suboccidentalis*, the fungal host is *Ramaria formosa*, a structurally complex mushroom. Females oviposit low on the fungal fruiting bodies in the field, and they also show a tendency to oviposit on the underside of hosts in the laboratory. In the field, *D. suboccidentalis* males usually position themselves on the side or base of the fungal fruiting bodies (see Fig. 4). This positioning probably helps males sight females, as vision would be obstructed if males were positioned in the convolutions on top of the mushroom. When many males are present, nearby vegetation and the mushroom top are also used as perch sites. A similar situation appears to occur in *D. flavopinicola* (Spieth and Heed, 1975). *D. magnaquinaria* feeds and breeds on *Lysichitum americanum* (skunk cabbage). Encounter sites are located in rolled spathes of the flowers (where most feeding and egg laying occurs in the spring when flowers are abundant), and in rolled leaves which are sheltering sites (see Fig. 4). Flies (both males and females) are less frequently found on open leaves (Kibota, unpublished). These differences among species appear to reflect differences in availability of

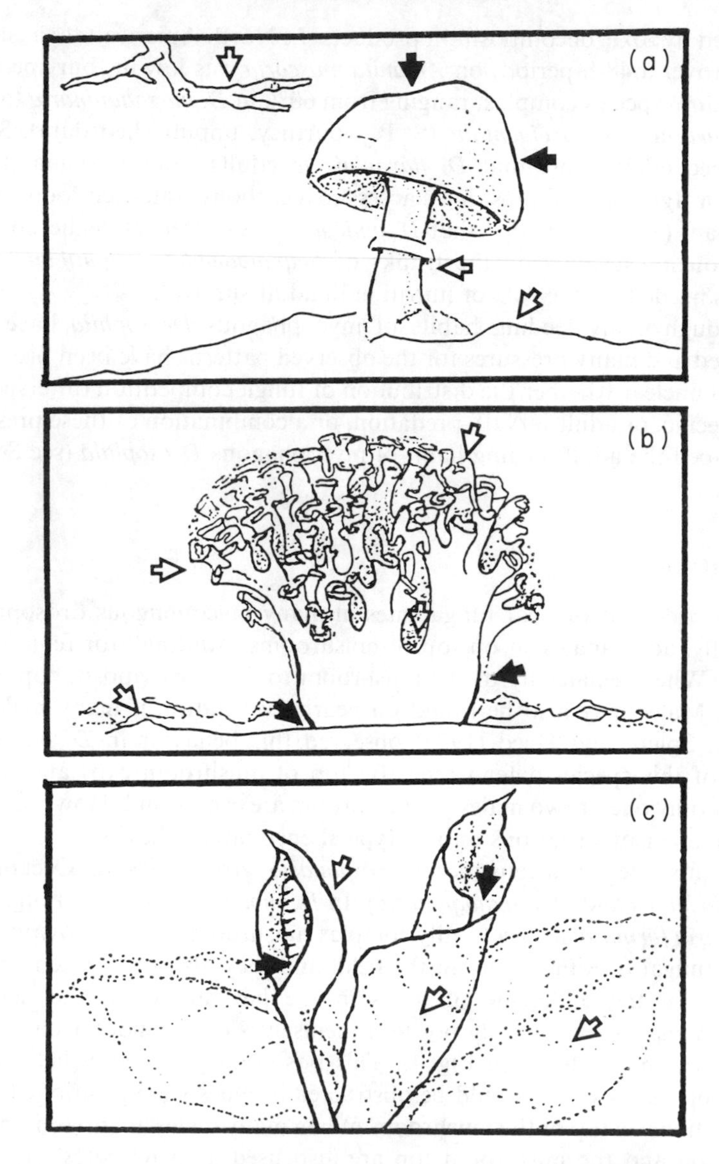

Fig. 4. Mating sites in three members of the *D. quinaria* complex. Primary (closed arrow) and subsidiary (open arrow) perch sites of males. At low male densities, only primary sites are occupied. Secondary sites are used at higher densities. (a) *D. rellima* (and many other species, including those in other genera). Males perch on the upper surface of a mushroom host, and respond visually to incoming flies of either sex. These are inspected on arrival at the host, and male–male chases and male–female courtships are frequent. At high male densities, some individuals perch on the stipe, on the ground, or (often) on nearby vegetation. (b) *D. suboccidentalis* on its host, *Ramaria formosa* (Basidiomycetes) in Oregon, USA. Males perch at or under the base

females at different parts of the host. Parker (1978) discusses the evolution of encounter-site conventions associated with resources used by females. These models appear equally applicable to phytophagous and mycophagous species.

Some researchers have described mating behavior that occurs on the light-colored undersides of bracket fungi that are not used for breeding. In Australia, some members of the subgenus *Hirtodrosophila* lek on bracket fungi, but breed on other soft fleshy fungi (Brock and Parsons, 1982; Parsons, 1977, 1982). In Peru and Panama, males of the closely related genus *Zygothrica* also use bracket fungi as mating display grounds, while females lay eggs on other substrates (Grimaldi, 1987). *Zygothrica* and *Hirtodrosophila* are sister groups of *Mycodrosophila* (see Section I), where mating and oviposition occur on the same host, as in most mycophagous Drosophilidae. This observation suggests that host use and mating site may co-evolve for long periods of time but that the correlation may be broken at the macro-evolutionary scale. This pattern is consistent with the hypothesis that sexual selection may drive host selection at micro-evolutionary scales (Colwell, 1986; Courtney *et al.*, unpublished), but that at larger taxonomic levels the genetic correlation between the behavior of the sexes changes and there is no necessary association of mating site and oviposition site.

Sexual dimorphism is present in some mycophagous or partially mycophagous drosophilids. Males of some species in the genus *Chymomyza* and *Zygothrica* have broadened heads (hypercephaly) and other unique physical characteristics (such as femoral spines). Grimaldi (1986, 1987) describes slashing of forelegs, head butting and jousting behaviors that accompany these unique physical characteristics. It is believed that territoriality is probably a prerequisite for the evolution of these dimorphisms (Grimaldi, 1987). Territoriality also appears in the sexually monomorphic *D. melanderi* (Spieth and Heed, 1975). Fungal encounter sites, especially the white undersides of bracket fungi, provide discrete and often visually distinguishable areas for territorial behavior.

Sex-ratio differences dependent on the age of mushroom-baited traps have been noted in Oregon (Muttulingham, pers. comm.). Three-day-old traps produce male-biased sex ratios, whereas 6-day-old traps produce female-biased sex ratios in *D. suboccidentalis, D. rellima* and *D. testacea*. In the laboratory, male defecation on the host stimulates female egg laying (see

---

of the mushroom, where females crawl to lay eggs. At higher densities, they may perch on the nearby soil or on top of the host. (c) *D. magnaquinaria* (sibling species of *D. suboccidentalis*) on its host, *Lysichitum americanum* (western skunk cabbage). Males perch in one of two distinct areas – inside the flowers, or in the coiled roll of unopened leaves. Females feed and oviposit in flowers, and shelter in leaf-rolls. At very high densities, males perch outside flowers, or on open leaves.

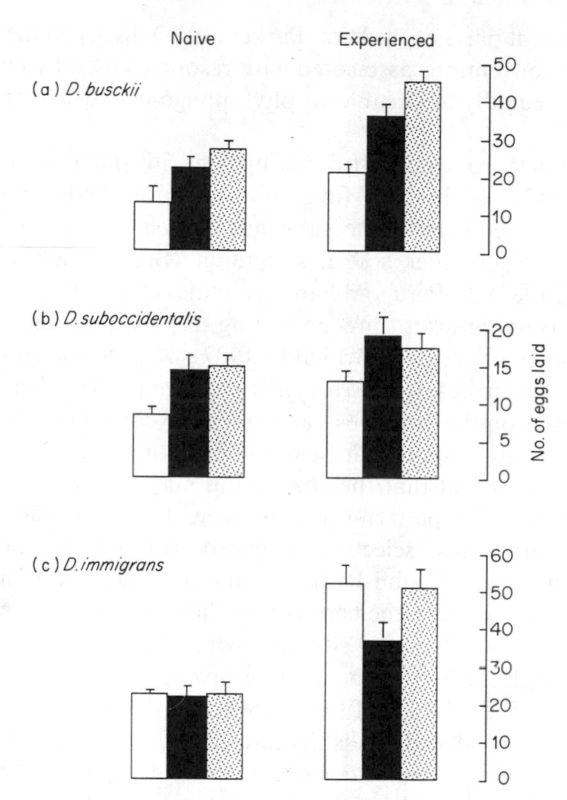

Fig. 5. The number of eggs laid (± 2 s.e.) by females of three drosophilids, in 24-h trials, on 1 g of *Agaricus* mushroom treated in one of three ways: fresh (open figures); decayed at room temperature for 24 h (closed figures); exposed to male defecation and regurgitation for 24 h (stippled figures). Females of two classes were used: naive females, which had not previously experienced hosts; experienced females, which had fed (but not oviposited) on *Agaricus*. In all three species there is a strong effect of experience (increased oviposition on *Agaricus*) and no interaction with host state (two-way ANOVA). In *D. busckii* (facultative mushroom-feeder) and *D. suboccidentalis* (mushroom specialist), host decay state significantly affects oviposition. In *D. busckii*, male-exposed hosts are the most attractive to females.

Fig. 5). It may be that males are attracted to pre-oviposition substrates and alter them by means of introduced microflora to a point where the hosts are more attractive to females. This opens up the possibility that host choice is partly determined by males.

C. Overwintering and Diapause

Many mycophagous *Drosophila* overwinter as diapausing adults. Overwintered adults can often be recognized in the following spring by their much darker abdominal pigmentation (Sabath *et al.*, 1973; Watabe, 1979). Diapause is often triggered by changes in daylength (Geyspits and Simonenko, 1970; Lakovaara *et al.*, 1972; Lumme *et al.*, 1974, 1978, 1979; Begon, 1982; Charlesworth and Shorrocks, 1980; Muona and Lumme, 1981; Lumme and Lakovaara, 1983), although there may be interactions with temperature. Photoperiodic effects may be mediated through juvenile hormone production (Kambysellis and Heed, 1974). In *D. phalerata, D. (H.) cameraria* and some other species, high temperatures can abolish photoperiodic induction of dormancy (Geyspits and Simonenko, 1970; Charlesworth and Shorrocks, 1980). Reproductive activity can be rapidly reversed in such species by increasing daylength and temperatures. In *D. transversa*, temperature appears to be the controlling factor in the onset of reproduction in spring (Muona and Lumme, 1981; Lumme and Lakovaara, 1983). A number of authors have studied diapause and phenology, and have attempted to explain phenology in terms of resource availability and temperature (Lumme *et al.*, 1978; Shorrocks and Charlesworth, 1980; Toda *et al.*, 1986). Often, there are fewer generations per year in areas with lower temperatures (Kimura *et al.*, 1978).

In *D. phalerata*, there are suggestions of aestivation with flies entering non-reproductive condition in response to long daylengths (Geyspits and Simonenko, 1970; Muona and Lumme, 1981). This may reflect seasonal shortages of mushrooms in continental dry summers; it is interesting that northern and western populations do not show evidence of aestivation. In North America, western species similarly do not show aestivation: fungi are common in midsummer in Oregon. It would be instructive to study the several *quinaria* group species of Arizona and New Mexico, which experience much drier summer conditions.

Kimura (1980b) describes the overwintering sites of some mycophagous species of *Drosophila*. *D. trivittata* and *D. sexvittata* were found on tree-trunks, while *D. alboralis* were located in cliff-shelters and tree holes. It is interesting to note that females of *D. alboralis* were inseminated, whereas most *Drosophila* are believed to enter diapause in a sexually immature state. Kimura suggests that because *D. alboralis* is a rather rare species (especially in spring), mate location may be infrequent enough to make copulation upon encounter advantageous, regardless of ovarian development. It is possible that such a strategy could prove advantageous in other mycophagous species that experience periodic rarity due to fluctuations in host abundance.

D. Adult Survivorship

A few studies have looked at adult survival in natural populations of *Drosophila*. These studies consider active adults during the breeding season (Charlesworth and Shorrocks, 1980, have shown that survivorship decreases dramatically during oviposition period), not overwintering inactive flies (see above). Rosewell and Shorrocks (1987) summarize studies from the literature that use Fisher and Ford's (1947) method to estimate survival rates from capture–recapture data. They then use these survival rates to estimate adult life-expectancy. The resulting estimates range from 1·3 to 6·2 days "with a most representative value of 2·8 days". They compare these figures to those of Birch *et al.* (1963), who estimated average adult survival rates in the laboratory of 0·98. This survival rate results in a life-expectancy of 45 days. This estimate agrees well with estimates of survivorship and fecundity in some semi-natural or natural populations (Rocha Pite *et al.*, 1985; Avelar *et al.*, 1987; Matos *et al.*, 1987; Singleton, unpublished). Rosewell and Shorrocks (1987) point out the inadequacy of referring to laboratory estimates of adult survival when studying the importance of factors such as fecundity rates in wild *Drosophila* populations.

Although this point is well taken, we believe that it is premature to assume that the adult life-expectancy of mycophagous *Drosophila* is of the order of 2·8 days (see Worthen, 1989b). One problem with the capture–recapture method in determining adult survival is that this technique fails to distinguish between mortality and emigration. Another problem with the existing field data is that it is limited to a handful of species of limited taxonomic diversity. Within the *quinaria* group, *D. limbata* appears to be relatively short-lived (Hummel *et al.*, 1979), but *D. suboccidentalis* have longer recapture times and many adults of the closely related *D. magnaquinaria* have been recaptured up to 40 days after marking (T. A. Singleton, unpublished data). Although Singleton's recapture techniques do not allow for the rigorous estimation of survival rates, the life-expectancy of *quinaria* group flies must either be longer or far more variable than an estimate of 2·8 days would indicate. Further research on the survival rate and life-expectancy of *Drosophila* is needed before meaningful general conclusions can be drawn about differences in the survival of species using different substrates.

Rosewell and Shorrocks (1987) have argued that as drosophilids have short life-expectancies in the field, the period immediately after eclosion is critical to the ecology of the species. If some drosophilids are in fact longer-lived, this conclusion does not hold. Given that many species have a pre-reproductive adult stage (when eggs are maturing) which may last several days (5 days in *D. phalerata*; Shorrocks and Rosewell 1986), current estimates of the potential reproduction in populations are likely to be significant underestimates.

There have been some incidental observations of mortality factors. In Oregon, mycophagous flies have been seen tangled in spider webs or with wings stuck to mushrooms, probably as a result of rain (S. P. Courtney, pers. observation). In culture, many adults die when their wings become stuck to the sides of vials. Desiccation may also be an important mortality factor in mycophagous species that are subject to fluctuations in climatic conditions.

Predation by staphylinids (Grimaldi, 1987; Worthen, 1989a) and by mantids, reduviids and anurans (Grimaldi, 1987) has been witnessed in the wild. Worthen (1989a) shows that predation by the rove beetle *Ontholestes cingulatus* facilitates co-existence of the mycophagous *D. tripunctata, D. putrida* and *D. falleni* in the laboratory. However, we know of no studies that estimate the magnitude of these varying factors in the wild. It is yet to be determined which factors are most important in adult mortality in natural populations.

E. Movement

Ever since Timofeeff-Ressovsky and Timofeeff-Ressovsky's (1940) dispersal experiments on *D. melanogaster* and *D. funebris* and the experiments of Dobzhansky and Wright (1943, 1947) on *D. pseudoobscura*, workers have tried to quantify movement in *Drosophila*. Various marking (genetic markers, nail polish and UV dust) and recapturing (baited arrays and aspiration from natural hosts) techniques have been employed. Dispersal events have ranged from less than 1 m in over 2 weeks in *D. magnaquinaria* (Singleton, unpublished) to 15 km in 15 h in *D. pseudoobscura* (Coyne *et al.*, 1982). Criticisms of many of these studies have been plentiful and varied. Early studies used laboratory-reared flies that were often of mutant stock. Many studies rely upon releasing unrealistically large numbers of flies from a central point in order to assure adequate recapture rates. Often, trap configurations fail to take into account the greater area that must be sampled with increasing distance from release point by placing traps in linear or cross-shaped arrays. Dobzhansky (1973) pointed out the difference between active dispersal and passive transport. Begon (1976b) emphasized the need to distinguish between dispersion (a description of distribution in space) and active dispersal (the "normal" movement of an insect under its own power in its life-time). In this section, we are concerned primarily with active dispersal. Finally, baited traps themselves have been criticized on the grounds that they fail to mimic natural host distribution and ignore interactions between traps and otherwise attractive areas. Arrays of traps inherently produce biased dispersal estimates that are directly dependent upon the spacing of traps (Johnston and Heed, 1975). As a result, the techniques for measuring dispersal and the resulting estimates are so varied that it is difficult to draw any general conclusions from the existing data. Dispersal rates must be, at least to a certain extent, system-specific. Much care needs to be taken to consider the particular ecology of the species in question in order to incorporate a realistic experimental design for estimating dispersal. Such is the case in Johnston and Heed's (1976) dispersal experiments on cactophilic *D. nigrospiracula*. In this study, host distribution was taken into consideration and marking and recapture were made at natural substrates.

In comparison to the information on general *Drosophila* dispersal, information on dispersal in mycophagous *Drosophila* is sparse. The few studies that do exist have used UV dusts to mark individuals. Researchers have failed to find the effects of UV dust on mortality (Crumpacker, 1974; Moth and Barker, 1975; but see Montague, 1989) and have found only slight effects on dispersal behavior (Turelli *et al.*, 1986) in the genus *Drosophila*. Our results also show no measurable effects of dusting. Montague (1985), for example, measured dispersal in *D. falleni* with a baited array consisting of traps placed at 21-m intervals (the average distance between naturally occurring hosts was measured to be 50–150 m). The resulting dispersal rates ranged from 9·54 to 16·43 m per day, a distance not only lower than the natural host distribution, but also lower than the distance between traps in the array. In a similar study, Worthen (1989b) measured dispersal in a number of *Drosophila* in New Jersey using an array of traps at 20, 28 and >60 m from the release points. The placement of the traps was designed to mimic natural host distribution. Of the recaptures, 53·7% were at release sites, 43·2% at 20 m and the remainder at 28 and >60 m. T. A. Singleton (unpublished data) also used an array attempting to mimic natural host distribution to measure dispersal in *D. suboccidentalis* and found similar results: flies dispersed 6·7 m per day, a lower estimate than Montague's, but the traps were placed closer together than in Montague's study. Of the recaptured flies, 69% were at 5 m, 18% at 10 m and the remainder at 20 and 40 m. These results are all quite similar, especially when compared to the wide range of estimates for frugivorous *Drosophila*. This similarity may indicate a pattern of generally low dispersal rates in mycophagous flies. Mycophagous drosophilids of the Pacific islands exhibit a higher rate of endemism than frugivorous drosophilids, perhaps indicating that low dispersal rates may in fact be associated with mycophagy (Wheeler and Takada, 1964; Wheeler and Kambysellis, 1967). However, low estimates could also reflect the similarity in techniques used by the various researchers. It is interesting to note that *D. magnaquinaria* (a non-mycophagous sibling species of *D. suboccidentalis* that feeds on skunk cabbage) shows extremely low rates of dispersal (Singleton, unpublished).

There are a number of factors that could bias the current estimates of mycophagous *Drosophila* dispersal. Most obvious are the problems associated with using baited arrays, even if sampling problems and natural host distributions are taken into account (see above). In addition, as Worthen (1989b) recognized, individuals that move greater distances will always be harder to find, and will therefore be undersampled. If one is concerned with the genetic consequences of dispersal, these rare long-distance dispersers may actually be most important (Slatkin, 1987). We have certainly observed events that could be better explained if adults disperse further than current estimates indicate. *D. suboccidentalis* occurs at high altitudes in the Oregon

Cascades and populations are known to exist on "islands" of forest isolated by lava flows. On one of these small islands (10 ha, 1 km to nearest "main" island of forest), the fungal host (*Ramaria formosa*) has been found only once in the past 4 years of field observation, yet flies are consistently caught in traps placed on this island (the authors, pers. observation). Does this population exist as an isolated population, or is it maintained by migration from larger neighboring islands? In three preliminary experiments, no marked flies were recaptured on opposite sides of the lava flow, and indeed the flow does appear to be a formidable barrier to dispersal. One possibility is that dispersal occurs only at certain times of the year, e.g. early spring, when much of the habitat in the Cascades is snow-covered. We cannot address this question at present. However, observations in Japan of seasonal migration from areas of lower to higher elevation indicates that seasonal migration is present in other systems (Kimura *et al.*, 1978). The influence of wind (passive transport) is thought to be a potentially important factor in such long-distance dispersal (Dobzhansky, 1973; Kimura *et al.*, 1978).

Even in "mainland" forests, fungi occur in island patches. Due to natural fluctuations in abundance, distance to the nearest host can be magnitudes greater than the current estimates of dispersal distances. It would appear that some individuals must disperse further than studies have reported in order to find breeding sites, and maintain their observed populations. Seasonal migration may also operate in systems with large temporal components of host variability.

These discrepancies between observed patterns and dispersal estimates illustrate the inadequacy of many current dispersal estimates. Electrophoretic data, or "indirect" measures of dispersal, help in determining if populations are isolated (see, e.g. Jaenike and Selander, 1979; Jones *et al.*, 1981). Techniques for measuring dispersal using genetically marked flies that allow observers to trace offspring may help with problems of detecting distant migrants (Dobzhansky and Wright, 1947; Coyne and Milstead, 1987), but there is no obvious way to overcome the increased sampling problem with increased distance. Despite the complications involved in producing accurate dispersal estimates, we maintain that productive work can continue on the question of dispersal in drosophilids by carrying out studies that employ a variety of techniques. Studies involving baited arrays, recapture work at natural hosts, and genetic techniques for measuring isolation, could accomplish this goal. Often, specific factors affecting dispersal can be addressed in addition to simply estimating dispersal rates. To date there are few extensive studies which attempt to assess the role of resource predictability in determining dispersal rates (see Johnston and Heed, 1976). Because mycophagous *Drosophila* experience a wide range in habitat predictability, this goal could easily be accomplished by comparing a number of mycophagous species that utilize hosts of varying predictabilities. In addition, because mycophagous drosophilids are easily captured directly from their natural hosts (which are discrete and form definable patches), we believe that the mushroom–drosophilid system is one of the best-suited systems for studying dispersal. It is hoped that we will stimulate increased effort to overcome sampling problems, so that the important ecological and evolutionary processes of immigration and gene flow may be better understood.

F. Habitat Selection

Studies on adult needs, independent of host sites, are relatively rare. Spieth (1987) found that in the extreme desiccating conditions of the dry season in Blodgett Forest, California, many species of *Drosophila* were found throughout their forest habitat, even in the apparent absence of free water. However, the fungivorous species (*D. occidentalis* and *D. pinicola*) congregated near moist stream beds indicating that mycophagous species may be less desiccation-resistant than the other *Drosophila* species.

Toda (1977, 1985a, b) has studied vertical stratification of *Drosophila* communities in Japan and Canada. In both regions, species were associated with particular heights, largely dependent upon their feeding habits. These results are consistent with the idea that adults remain in, or search for, particular habitats where their hosts are found (Shorrocks and Nigro, 1981). However, in one Japanese study, Toda (1977) found that the distribution of some fungivores changed in the autumn. This is possibly a migration response as flies move from breeding to overwintering sites.

In some non-mycophagous species, intraspecific habitat selection has been shown, although the generality of these results is questionable (Taylor and Powell, 1978; Kekic *et al.*, 1980; Shorrocks and Nigro, 1981; but see Atkinson and Miller, 1980; Turelli *et al.*, 1984; Hey and Houle, 1987). Jaenike (1986a) reports apparent intraspecific variation in attraction to three hosts in *D. falleni* (mycophagous) and in *D. tripunctata* (mycophagous and frugivorous) but not in *D. putrida* (mycophagous). However, his experimental protocol does not rule out micro-habitat effects on host choice. For instance, his data on *D. falleni* could result from summing together flies of different habitat preferences producing a deficiency of "switchers" in a manner akin to the Wahlund effect (Futuyma, 1979). Similarly, his analysis does not distinguish the effect of hunger which is known to be important (Hoffman and Turelli, 1985). If *D. tripunctata* are more attracted to tomatoes when hungry, then we would expect hungry flies to maintain a preference for tomatoes. This emphasizes the difficulty of distinguishing between the ovipositional and nutritional needs of females which both play a role in habitat choice.

G. Clutch Size

Recently, there has been an abundance of theoretical papers concerning optimal clutch size in insects. Most of these papers include the relationship of female behavior to the optimal size of clutch and the number of eggs that will maximize the number of surviving larvae (Charnov and Skinner, 1984, 1988; Parker and Courtney, 1984; Parker and Begon, 1986; Skinner, 1985; Mangel, 1987; Godfray and Ives, 1988). These models then make predictions about clutch size based upon varying attributes of the system being modeled, both environmental (such as distance between oviposition sites) and physiological (such as current egg load of the adult female).

Although many of these models can be readily applied to drosophilids, relatively little empirical data exist on clutch size in these insects. Two studies on clutch size in phytophagous *Drosophila* attempt to explain clutch size differences between species. Toda *et al.* (1984) examined clutch size (as indicated by ovariole number) in herbage feeders and found differences in clutch size between species that were very similar in habitat preference and host utilization. Toda *et al.* attribute these differences to differences in voltinism, with the univoltine species laying a large number of small eggs and the bivoltine species laying a small number of large eggs. Atkinson (1979) compared clutch size (as indicated by ovariole number and mature eggs per ovary) in a number of *Drosophila* in a fruit stand and vegetable market and found that large-bodied species lay large clutches of small eggs, whereas small-bodied species lay small clutches of large eggs, contrary to the predictions of "r"- and "K"-selection theory (Pianka, 1970). One factor that would tend to select for large clutches of small eggs is infrequent breeding sites, but Atkinson did not collect data of this type. Both of these pressures could easily be tested in mycophagous *Drosophila* which vary both in voltinism (Toda *et al.*, 1986) and in frequency of breeding sites (see Section II.C).

In their electrophoretic work on *Drosophila*, Jaenike and Selander (1979) found that flies emerging from individual fungi collected in the wild were the offspring of a few or even one female. These data provide indirect evidence that mycophagous *Drosophila* do cluster eggs in the wild. But how much clutch size variation exists within a species, and what are the important factors in determining the clutch size a particular female lays? In *D. suboccidentalis*, a fungus feeder whose host goes through wide seasonal fluctuations in abundance, clutch size in the laboratory (measured as the number of eggs a female lays before choosing to leave a host) is correlated with egg load. Clutch size also shows significant variation between three different hosts (*Ramaria*, the natural host; *Agaricus; Lysichitum americanum* or skunk cabbage, the natural host of the sibling species *D. magnaquinaria*). It is interesting to note that clutch-size differences between hosts are most pronounced at intermediate egg loads. Females with small egg loads lay small clutches on all three hosts and females with high egg loads lay large clutches on all three hosts (Fig. 6).

Although data are still scarce on clutch size, their importance for the study of mycophagous *Drosophila* should be stressed. One property of most dipteran life-cycles is the relative immobility of larvae relative to the many lepidopteran species which have so far been used more frequently in the study of clutch size (Pilson and Rausher, 1988; Tatar, 1989). Diptera larvae are "stranded" wherever they are laid as eggs, and we expect clutch size to have more obvious effects on fitness. In addition to effects of clutch size on intraspecific competition, lowered resistance to attack by parasitoids (Bouletreau and Wajaburg, 1986) and larval facilitation (Allee effect) will also depend on clutch size. Clutch size may also affect community level interactions such as degree of aggregation, as suggested by Atkinson and Shorrocks

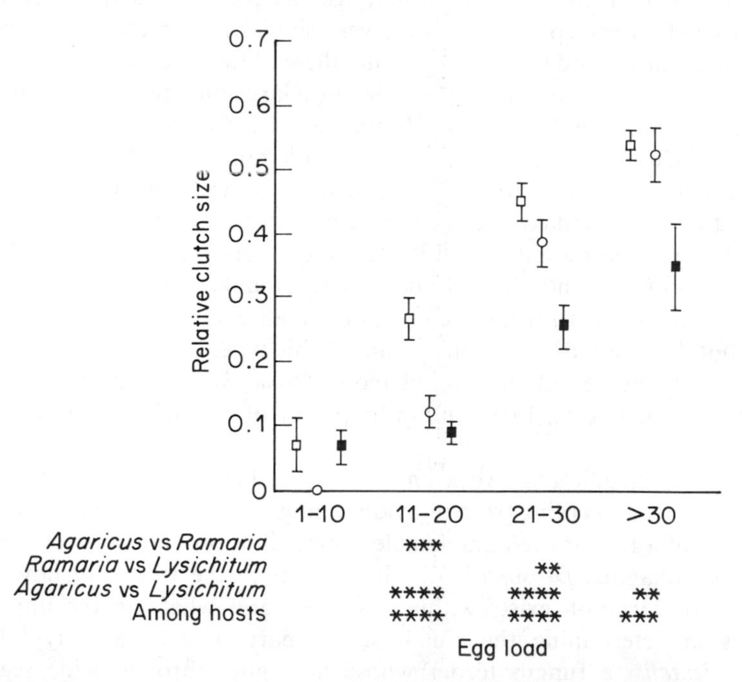

Fig. 6. Clutch size in *D. suboccidentalis*. Females were placed in open-ended vials with a 1-g piece of host (i.e. flies were free to leave the vial but were not allowed to return to the host). The number of eggs laid as a function of the number of eggs females had in the abdomen (females were dissected at the end of the experiment; the number of eggs laid + the number unlaid = the number that females started the experiment with). "Relative clutch size" is the proportion of a female's total egg complement laid at this single host encounter. Hosts differ in the clutch sizes received. The significance of these differences are given below the figure. □, *Agaricus*; ○, *Ramaria*; ■, *Lysichitum*.

(1984; but see Green, 1986). Finally, clutch size has important implications for host choice. It has been suggested that ovipositing females only discriminate between hosts at certain egg loads (Courtney *et al.*, 1989), and this appears to be the case in *D. busckii* (Courtney *et al.*, 1989) and *D. suboccidentalis* (see Fig. 6: Singleton, unpublished). If clutch size is largely determined by egg load, less preferred hosts may then only be used when egg load is high and clutch size is large. This implies that less preferred hosts will typically receive large clutches, with interesting consequences for both the population dynamics of the insect and for the evolutionary future of the interaction. Certainly more work on clutch size is necessary to determine what factors shape clutch size, what interactions are affected by clutch size, and how the resulting combination in turn shapes insect communities.

H. Host Choice

The consequences of host selection (defined here as the process leading to oviposition) bear upon all the work so far discussed. The manner in which a female lays her eggs will determine the substrate her larval offspring must develop on, the habitat that the resulting adult progeny will initially encounter, and to some extent the ecological interactions her offspring will face. Accordingly, much of the emphasis in studies of plant–insect interactions has been placed on the determination of host choice patterns (see reviews by Papaj and Rausher, 1983; Thompson, 1988; Courtney and Kibota, in press). Historically, mycophagous insects have been lumped in with their phytophagous relatives. Thus Jaenike (1978) states: "Individual mushroom species are an unpredictable resource. ... Because of this, and because the inclusion of more host species is unlikely to expose offspring of *D. falleni* to potential toxins that they cannot tolerate, selection will favor an increase in the variety of acceptable host *plants* [our italics]." As we have seen, this assumption of comparability between plants and mushrooms is valid in some instances but not in others.

Some authors have divided the host choice sequence into pre- and post-alighting discrimination phases (Singer, 1982, 1983; Papaj and Rausher, 1983). This distinction has proven useful in light of the evidence that many phytophagous insects show different host preferences between these two stages (Jaenike and Grimaldi, 1983; Rausher, 1984). Mycophagous insects may follow similar patterns. Kimura (1980a) tested adult attraction to various stages of fungal decay (fresh, intermediate and decayed) and oviposition on those stages in eight species of *Drosophila*. In six of the eight fly species, the number of adults attracted to each decay state did not correspond to the hierarchy of oviposition preference. For example, while *D. sexvittata* adults preferred fresh mushrooms to those at intermediate levels of decay and intermediate mushrooms to decayed stages (i.e. fresh > intermediate > decayed), more eggs were laid on intermediate than on fresh stages (i.e. intermediate > fresh, decayed = 0). These asymmetries between long-range attraction and oviposition response are even more pronounced in *D. tripunctata* (Jaenike and Grimaldi, 1983). Two isofemale strains of this fly exhibited opposite oviposition preferences for *Agaricus* and tomato. However, in long-range tests of attraction ("feeding-preference" *sensu* Jaenike and Grimaldi, 1983) the strain which laid eggs on *Agaricus* was more attracted to tomato (and vice versa). Because isofemale lines were used, Jaenike could infer a genetic basis to this variation. Note that in some species there are strong diurnal rhythms to oviposition (Azarian 1968; Goryshin *et al.*, 1973).

Here we will discuss the various sampling methods, which have been employed to study *Drosophila* host usage, within the framework of the pre- and post-alighting behavioral categories. Much of the mycophagous drosophilid literature is simply a report of the numbers of flies collected from various mushroom species. Collection of adults from naturally occurring hosts yields an estimate of the pre-alighting host range of the flies, although there will be some error caused by flies which are attracted to feed rather than to oviposit. The species list is long and includes obligate fungivores (e.g. *Mycodrosophila*), detritivores and frugivores which occasionally utilize fungi (e.g. *D. nigromaculata*), and broad generalists which may often be the dominant species in collections (e.g. *D. busckii*). It is important to remember, however, that this type of sampling method gives only presence/absence information. We cannot infer anything about the interaction between insect and fungi, nor can we formulate any ideas about other factors which may influence these interactions. These objections can be raised to other methods of sampling adult distribution such as sweeping (Lacy, 1984b) and baiting (see above), which give similar results. In a sweep collection of adults, for example, Lacy (1984a) found seven *Drosophila* species in association with between 4 and 14 genera. In addition, many strictly mycophagous species have been captured from banana-baited traps (Toda, 1985a, b; Spieth, 1987), indicating that at the pre-alighting stage these flies are attracted to a very wide range of hosts. We have also collected fungivores from a wide array of baits; *D. rellima, D. testacea* and *D. suboccidentalis* have all been collected from traps baited with tomato, cucumber, *Agaricus bisporus* and *Ramaria formosa* (T. T. Kibota, pers. observation; T. A. Singleton, pers. observation).

Stronger evidence of mycophagy for most species has come from the rearing out of adults from field-collected larval material (e.g. Falcoz, 1921; Riel, 1921; Bonnamour, 1926; Smith, 1956; Buxton, 1960; Shorrocks and Charlesworth, 1982). As with collections of adults from natural substrates, the rearing of their progeny shows a wide range of mushroom usage from very occasional to obligate mycophagy. As stated earlier, the larvae of phytophagous insects generally have a wider host range than the ovipositing adults. Assuming that this is also true in mycophagy, the results of these rearing programs can show evidence of parental post-alighting behavior. For instance, Kimura and Toda (1989) show that *D. (H.) sexvittata* and *D. (H.) alboralis* adults are very frequent visitors to *Pleurotus* fruiting bodies but no emergence is seen from these mushrooms. This difference between adult and larval distribution is evidence for rejection of *Pleurotus* by alighted adults.

What general patterns can be obtained from these data? First, unlike herbivorous insects, polyphagy is very common among fungivores (Lacy, 1984a; Hanski, 1989). This is consistent for all sampling methods. How can this degree of polyphagy be accounted for? Mushrooms, as described earlier, are of high nutritional value and are easily digested. In addition, co-evolution between insects and fungi is thought not to occur; mycotoxic defences should not pose a great evolutionary barrier to the insects. Indeed, many polyphagous species do incorporate highly toxic mushrooms into their diets (Jaenike, 1985a; Kimura and Toda, 1989). Macrofungi, then, appear to present a somewhat homogeneous pool of resources which the mushroom-feeders may exploit. Hanski (1989) has termed this the "quality" explanation for polyphagy in fungivores, and contrasts it with an alternative explanation, the "quantity" hypothesis, which states that mushrooms are highly unpredictable, rendering specialization unlikely. A third hypothesis is that many fungivores, such as drosophilids, may be associated as much with decomposing organisms (e.g. yeasts) as with the fungal tissue itself (Kearney and Shorrocks, 1981; Kearney, 1983). It would be interesting to test the attractiveness of a variety of baits to those flies which prefer the fresher mushroom stages and which, perhaps not coincidentally, have narrower hosts ranges, (e.g. *D.(H.) trivittata* and *D.(H.) sexvittata* (Kimura, 1980b; Kimura and Toda, 1989).

A second general pattern of host use is that those species which do have a narrow host range are *not* specialists on mushrooms having a unique, toxic chemistry (Lacy, 1984a). This is contrary to the expectations of major theories of plant–insect interactions, which are based on the premise that monophagy evolves as a specialization, enabling the insect to use defended, otherwise unavailable resources (Ehrlich and Raven, 1964; Feeny, 1976; Rhoades and Cates, 1976). Fungivores, instead, appear to specialize on those hosts which are most predictable (i.e. long-lasting). Lacy (1984a) reports that *D. duncani* and the two "*M. claytonae*" species are specialists on bracket fungi of the family Polyporaceae (among the most durable of the macrofungi). These flies, as stated above, are usually rare in all types of collections. In Oregon, *D. suboccidentalis* is strongly associated with a single genus of long-lived basidiomycete, *Ramaria*. However, trends toward the use of long-lived fruiting bodies are also seen in the often abundant host generalists. For five of the seven *Drosophila* species in Lacy's (1984a) collections (at both New York and the Great Smoky Mountains), over 50% of the individuals were collected from a single fungal genus (mainly *Russula* or *Polyporus*, both of which are long-lived). Similarly, *D. phalerata* is predominantly found on *Phallus impudicus*, which is highly abundant and predictable in some British woodlands (Shorrocks and Charlesworth, 1982). Predictability, then, seems to be of importance not only in absolute host range but also in relative preferences among hosts within that range.

Another influence on host range is the state of decay, which may also reflect microfloral succession (Starmer, 1981; Oakeshott *et al*., 1989). Kimura (1980a) has shown that many of the Japanese *D. (Hirtodrosophila)* that feed on fresh fungi are more host-specific (most notably on fungi in the family Trocholomataceae) than species of the *immigrans* radiation that prefer decaying fungi. The predictability and abundance of fresh mushrooms should be higher than that of decaying mushrooms, yet as Kimura (1980a) points out, fresh fungal material may be less nutritionally suitable or even less palatable (see above). He suggests that the use of fresh material may result from a combination of increased predictability and a decrease in competition (see Section V). These findings are directly analogous to those for phytophagous insects where, for example, pierid butterflies show a more varied response to the age of the crucifer tissue than to the crucifer species identity (Jones *et al*., 1982).

Taken together, these observations do suggest that host quality, host predictability and microbial associations all affect host selection. In the absence of critical tests, we cannot distinguish between these alternatives or determine whether one is more important than the others in any particular case. However, we believe that the unpredictability of mushrooms has been overstated, and that plants and mushrooms are in fact comparable in this respect. Because the degree of polyphagy observed in mycophagous drosophilids seems large, relative to that seen in many phytophagous insects, we currently favor hypotheses which emphasize resource homogeneity.

Mycophagous drosophilids have been reported to breed in a number of non-fungal habitats. The flowers and fruits of Araceae (*D. recens, D. limbata*. Grimaldi and Jaenike, 1983), cucurbitaceous plants (*D. limbata, D. rellima*. Hummel *et al*., 1979; Singleton, pers. observation), rotting Compositae (*D. phalerata, D. kuntzei*: Shorrocks and Charlesworth, 1980; Janssen *et al*., 1988) slime fluxes (*D. putrida, D. tripunctata, D. phalerata*: Sturtevant, 1921; Carson and Stalker, 1951; Janssen *et al*., 1988) and rotting vegetation (*D. testacea*: Kimura *et al*., 1977) have all been recorded. These observations indicate a number of important points. First, given the otherwise relatively sparse data on drosophilid ecology, these indicate a relatively high rate of oviposition "mistakes" by females, as compared to phytophagous insects, such as butterflies (Gilbert and Singer, 1975). Such "mistakes" are essential to the evolution of host shifts. Secondly, these unusual records are often on rotting substrates, implying that microfloral associations may be critical in determining the probability of host shifts. Thirdly, these "mistakes" often result in successful larvae. This implies that larval nutrition is often not a barrier to host shifts.

Genetic variance in host use has been studied for several mycophagous drosophilids and their relatives. *D. tripunctata* exhibits substantial intra-population variation for oviposition on alternative natural hosts (mushrooms and fruits: Jaenike and Grimaldi, 1983). Surprisingly little between-population variance is seen (Jaenike, 1989); this contrasts with the results often seen in phytophagous insects (Futuyma and Peterson, 1985). However, until further geographic surveys are available, we cannot know whether this pattern is peculiar to *D. tripunctata* or to mycophagous drosophilids in general. In *D. suboccidentalis*, and in the facultative mushroom-feeder *D. busckii*, substantial genetic variation exists for oviposition on novel hosts not normally present in populations (Courtney and Chen, 1988; Courtney and Hard, in press). In these species, as in *D. tripunctata*, the presence of genetic variance for oviposition behavior suggests that rapid evolution of host use could occur. Certain host shifts, such as that to skunk cabbage, do seem to occur repeatedly (see Section I). The larvae of mycophagous species in the *quinaria* group grow well on skunk cabbage (Jaenike, 1985a); the converse is also true (James *et al.*, 1988). Taken together, these results suggest that genetic variance in oviposition behavior is an important factor allowing host shifts in Drosophilidae. One factor maintaining such variation may be pleiotropic interaction with other fitness components (Courtney *et al.*, 1989).

In a number of species of *Drosophila*, adult females change their responses to potential hosts as a result of experience (Jaenike, 1982, 1985b, 1986b, 1988b; Hoffman, 1985). This behavior is potentially important for a number of reasons; it may result in populations tracking abundant resources and in females maximizing individual reproduction and fitness. To date, the only reports on such behavior in mycophagous flies are Jaenike's study of three American species where the results are somewhat equivocal (Jaenike, 1986a), and the date of Kearney (1983) which show that *D. phalerata* and two facultative fungivorous species have the same preferences early and late in adult life (suggesting that experience is unimportant). In our laboratory, experience appears to play a large role in determining how many eggs are laid by the females of these three species when given no-choice tests (Fig. 5).

A number of authors have raised the possibility that choice of diet is affected by interspecific interactions other than with the host. Montague and Jaenike (1985) and Kimura and Toda (1989) have suggested that nematode-parasitism may favor the incorporation of fungi which are defended against nematodes. Other people have suggested that interspecific competition between drosophilids or between these and other insects may limit diet breadth (Jaenike, 1985b; Grimaldi, 1985). While we cannot rule out such interactions being important selective factors in the past, the balance of evidence suggests that parasitism and competition currently do not dramatically reduce larval survival (see Section II C) and hence cannot be important in restricting diet. In phytophagous insects, many alternative factors determining diet breadth have been discussed (Bernays and Graham, 1988, and respondents). One factor, important in many phytophagous insects, which may be important in mycophagous species, is the limited time in which a female has to find hosts (Courtney, 1984). The observed life-spans of mycophagous Drosophilidae are short relative to the overall flight season and availability of hosts. Periods of bad weather, which inhibit flight activity, may reduce female reproduction so far that polyphagy (even incorporation of poor hosts into the diet) is strongly favored (Futuyma, 1983). This suggestion awaits critical testing, which will require the accurate monitoring of female activity.

V. COMMUNITY ECOLOGY

A. Species Interactions

Drosophilids are easily collected at baits or at natural substrates. Species richness and the relative abundance of species are therefore known for a number of communities worldwide. Several hypotheses have been put forward to explain the patterns observed in such data. Some of these have subsequently spawned theories of interactions in communities which are of general interest, far beyond the narrow confines of mycophagous drosophilids (Atkinson and Shorrocks, 1984; Ives and May, 1985). Theoretical work on the effects of aggregation on competitive interactions will continue, even if it should be proven that many mycophagous drosophilid communities (the original inspiration of the idea) do not in fact work in this way.

Extensive collections of drosophilids are reported from various localities, often with sufficient data to determine species abundance relationships. Kimura *et al.* (1977), for instance, list mycophagous flies collected from mushrooms, and general collections from banana baits at the same site. The species abundance curve for mycophagous flies is shown in Fig. 7a. Figures 7b and c show similar data for two collections of drosophilids emerging from fungi in the eastern USA (Lacy, 1984a). Common patterns are seen in both Japan and the USA — a few species are very abundant relative to the others (*D. brachynephros, D. testacea, D. (H.) sexvittata* in Japan; *D. falleni* and *D. testacea* in the USA). Similar findings are reported by Shorrocks and Charlesworth (1980) for the European fauna. This numerical dominance of a few species is all the more surprising when it is considered that many of the rare species in these collections are obligate mycophages, i.e. it is unlikely

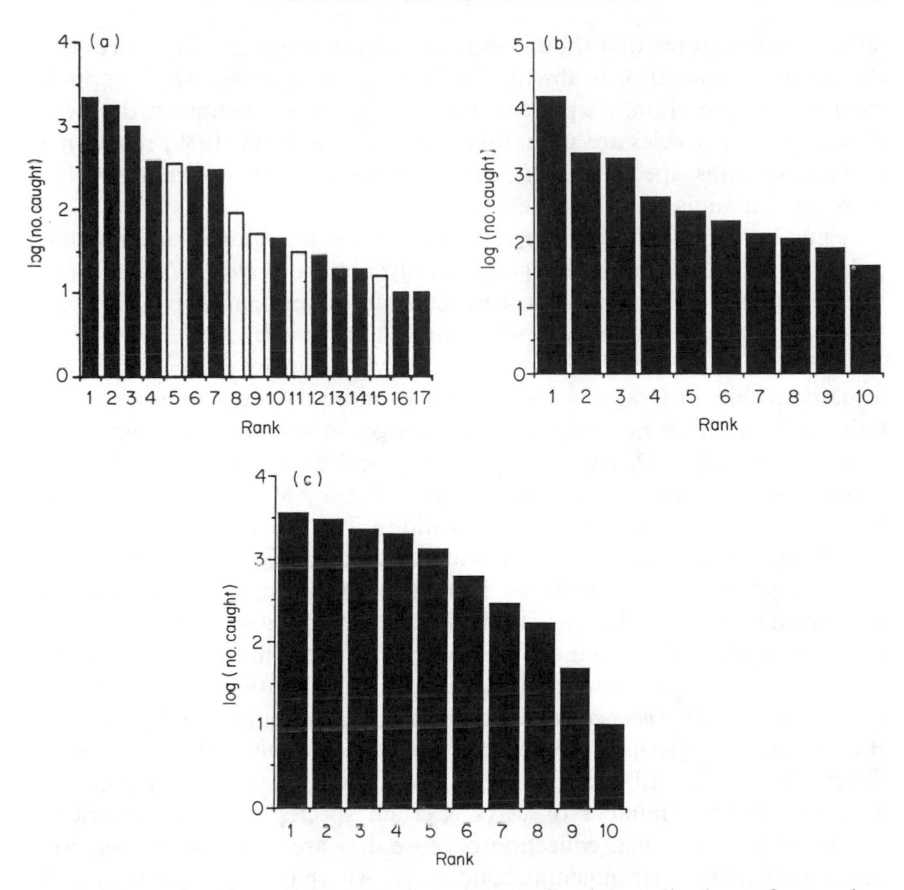

Fig. 7. The rank-order, species abundance curves for three collections of mycophagous drosophilids (■). (a) Captures of adult flies at mushrooms in Japan (Kimura, 1976); (b) larvae reared to adults from hosts in New York State (Lacy, 1984a); (c) larvae reared to adults from hosts in Tennessee (Lacy, 1984a). □, "Non-mycophagous" or facultatively mycophagous species. Lacy (1984a) does not record these in his collections.

that they are present in large numbers but remain undetected. Particularly striking is that the same or related taxa are rare in different geographic locations. *Mycodrosophila* spp. are never common from temperate zone collections, even though several species occur over wide ranges. What factors might explain such species abundance curves? One possibility is that rare species are fugitives from competition and persist by means of dispersal to vacant areas. An alternative is that rare species have more specific, limited resources. Some of the rare species in both Lacy's and Kimura's collections are indeed host specialists. The hypothesis that competition keeps rare species rare requires that those species should increase in abundance when the superior competitor is absent. We know of no data set which supports such an interpretation. At present, the data which exist support the hypothesis that rare species are specialists. Kimura and Toda (1989) report how one rare specialist species, *D. (H.) pirka*, has increased in abundance with an increase in abundance of its sole host.

A number of studies have reported apparent segregation of species along habitat gradients. Vertical segregation of mycophagous *Drosophila* in forests is absent (Toda *et al.*, 1984), but seasonal shifts in dominance of a particular species have been seen in several communities (Spieth, 1987; Toda *et al.*, 1986; Kimura *et al.*, 1977). For instance, S. Muttulingam (pers. comm.) sees a seasonal shift in Oregon with *D. testacea* more common in late autumn. Other authors have noted elevational changes in species abundance. Lacy (1984a) notes that *D. recens, D. testacea* and *D. ordinaria* dominate in collections at higher elevations in the Great Smoky Mountains, whereas *D. putrida* is more common at lower elevations. This elevational pattern may reflect thermal tolerances — *D. recens* and *D. testacea* have northern distributions suggesting cold adaptedness. Similarly, in Oregon, species abundances are strongly affected by elevation (Fig. 8). Here there is a less strong correlation with cold tolerance—*D. testacea*, with the most northerly distribution, is *not* the species which dominates at high elevations. Instead, we see dominance by *D. suboccidetalis*, a species which ranges to Costa Rica. Biogeographic patterns of common European *Drosophila* are discussed by Shorrocks (1977). All these patterns (elevational, seasonal, regional) are interpretable in a number of ways. Certain species may be numerically dominant in a particular collection because they are competitively superior under particular environmental conditions. Alternatively, even in the absence of competition, species' abundances may be strongly influenced by climatic factors, thus producing a similar separation of species. A last alternative is that each species has evolved in response to a unique set of niche characteristics (such as host identity) and is common only when these conditions prevail. At present, all these hypotheses are defendable, but we believe that climatic influences are very strong. Many authors have described the effect of temperature regimes on generation time and the growth of

---

Fig. 8. Changes in composition of the mycophagous *Drosophila* fauna along an elevational gradient in Oregon, USA. (a) Captures of adult flies from mushroom-baited traps (1989) (Muttulingam, pers. comm.); (b) captures of adult flies from mushroom-baited traps (1987–88) (S. P. Courtney, unpublished data). At lower elevations the fauna is diverse. As elevation increases, *D. suboccidentalis*, becomes the dominant (if not the only) species. Note that *D. testacea*, not *D. suboccidentalis* has the most northerly distribution, suggesting that some factor, other than cold-hardiness, is important (see text).

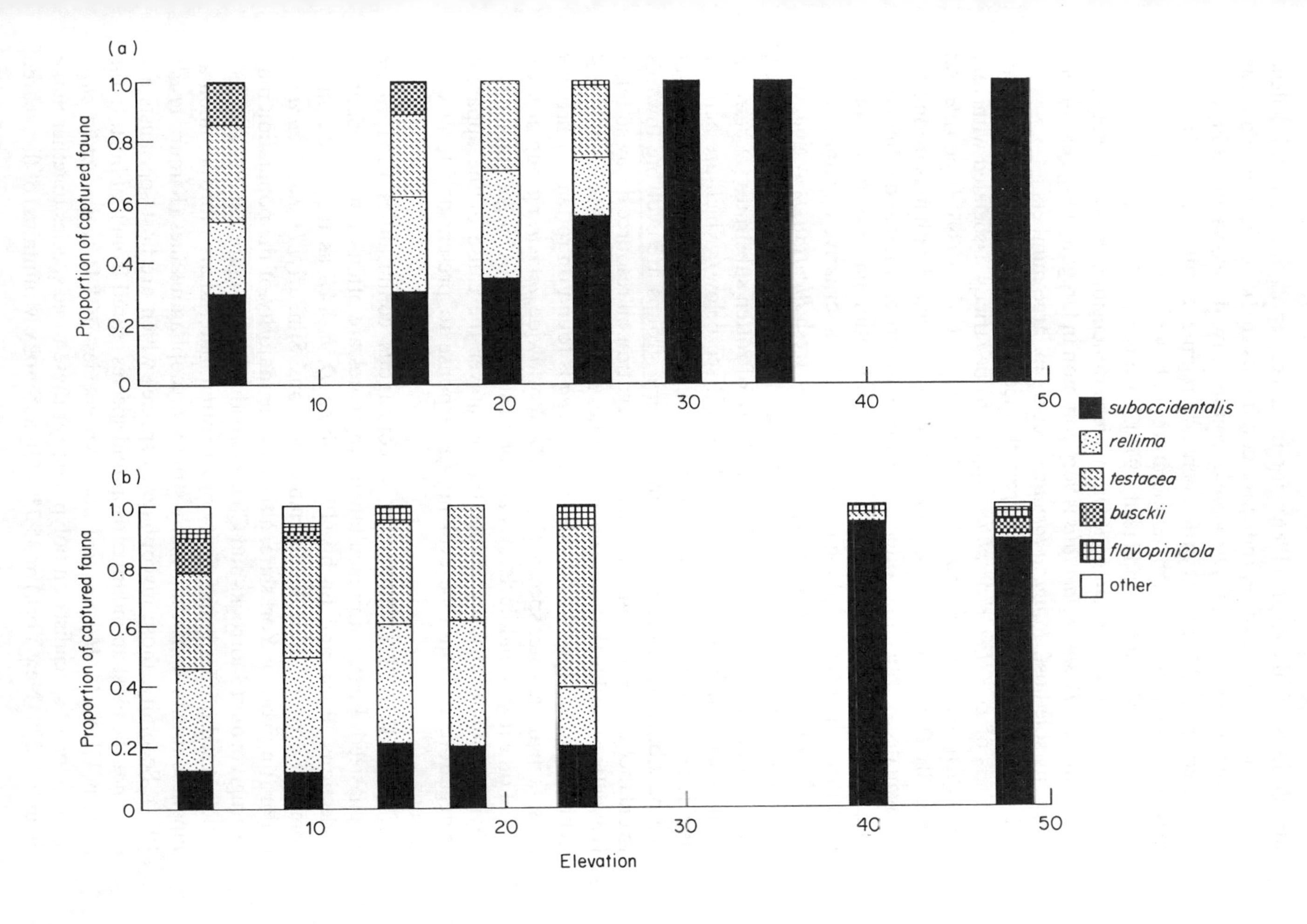
(a)
Proportion of captured fauna
1.0
0.8
0.6
0.4
0.2
0
10
20
30
40
50
(b)
Proportion of captured fauna
1.0
0.8
0.6
0.4
0.2
0
10
20
30
40
50
Elevation
suboccidentalis
rellima
testacea
busckii
flavopinicola
other

populations. Toda *et al.* (1986), for instance, report how climatic effects determine the onset of reproductive conditions in *D. confusa* in two areas. Observed differences in *Drosophila* community development follows as predicted from differences in temperature regime. Similarly, *D. pinicola* is extremely sensitive to high temperatures (above 18°C it becomes sterile: Sturtevant, 1942; Spieth and Heed, 1975). Unsurprisingly, this species occurs at higher elevations in Oregon. Other niche dimensions, however, may also be important. *D. suboccidentalis* is very common in Oregon at high elevations where its sole host, *Ramaria formosa*, occurs in large numbers.

Host decay state has a very large influence on species composition. Some members of *D. (Hirtodrosophila)* appear to be strictly associated with very fresh *Pleurotus* (Kimura *et al.*, 1977; Kimura and Toda, 1989). *D. putrida* and especially *D. testacea* by contrast are strongly associated with mushrooms far along in decay which may often already contain drosophilid and other larvae (e.g. Basden, 1954). Does this imply that competition has been an important organizing influence in these communities? An alternative explanation for specialization by *D. (Hirtodrosophila)* spp. on fresh *Pleurotus* is avoidance of nematodes (which are associated with rotting material). *D. testacea* and *D. putrida* appear to have evolved towards use of rotten substrates (*D. testacea* is also taken from other rotting substrates: see below), which are actively avoided by other mycophagous species (see above). While we cannot rule out the possibility that the *testacea* and *quinaria* groups are evolving towards avoidance of mutual competition (via segregation on resource decay state), it is equally plausible that these characteristics are evolved for reasons independent of competition. Worldwide, *D. testacea* is found on rotten mushrooms as is its nearctic sister species *D. putrida*. Worldwide, *quinaria* group flies are found on early or middle decay states of mushrooms. This biogeographic and phylogenetic consistency argues that whatever the cause of the apparent segregation, it cannot be evolving in response to present-day local conditions.

Two further observations on mycophagous communities need to be emphasized. First, many facultative fungivores are often quite abundant in collections. Burla and Bachli (1968) noted *D. busckii* as the second most common drosophilid in Hungarian collections. Similarly, *D. subobscura* is a major fungivore in Yorkshire, although the majority of the population may be frugivorous (Shorrocks and Charlesworth, 1980). Figure 7 shows that flies which breed primarily in fruit or other rotten material are more common at Japanese mushrooms than many strictly mycophagous flies (Kimura *et al.*, 1977). Secondly, those mycophagous species which are host specialists are often *not* the most abundant drosophilid species reared from that host. In the eastern USA, *D.(H.) duncani* and two species of "*Mycodrosophila claytonae*" are all specialists on polypores; yet these three species together make up only 5·2% (New York) or 35·7% (Great Smoky Mountains) of flies reared from these hosts. Similar conclusions can be drawn from the data of Kimura *et al.* (1977) and Kimura and Toda (1989). Both these patterns are unexpected from orthodox competition theory, where one would anticipate that insects with the more restricted diet would be the stronger competitor on that diet, and would be numerically dominant on it. Competition theory does not seem compatible with the existence of consistently rare specialists which are numerically subordinate on their own hosts.

B. Models of Community Organization in Species Using Ephemeral Resources

Based on their experience with mycophagous and other drosophilids, Shorrocks and his co-workers (Atkinson and Shorrocks, 1981, 1984; Shorrocks *et al.*, 1984; Shorrocks and Rosewell, 1987, 1988), and others (Phillips and Hubbard, 1989), have developed a series of models which seek to understand the co-existence of species which do not differ in resource use. Ives and May (1985) have extended these simulations with an analytical model and confirmed the earlier results: intraspecific aggregation may allow competing species to co-exist provided that interspecific distributions are independent of each other. Basically, this theory states that intraspecific competition will always tend to be more important than interspecific competition and that co-existence is then possible for a large number of species even if they do not differ in any niche dimension. This has been an important insight and is a useful addition to ecological theory. Some criticisms have been labeled at particular interpretations, such as the discussion by Green (1986) on the role of clutch oviposition in causing aggregation. However, these criticisms do not mar the intrinsic interest of the theory. It remains to be proven, however, that these models are indeed applicable to mycophagous communities. Worthen and McGuire (1988), for instance, have shown that, in the eastern USA, many drosophilid species are positively aggregated with each other— this violates an essential assumption of the models. Similarly, Hanski (1989) has argued that any differences in host use patterns will reduce competition between species: in effect, each species will have a refuge on its unique hosts. It is unsurprising that co-existence will be possible in such circumstances. If this is in fact the case, then the models of Atkinson and Shorrocks (1984) do not incorporate the essential features which allow species co-existence. We take this argument a stage further yet, and contend that the evidence for competition itself is still weak. If competition is absent, then co-existence is no puzzle, regardless of patterns of host use or aggregation. The appropriate model for mycophagous drosophilid communities may well be Huston's (1979) non-equilibrial model, where potential competitors rarely reach population levels where they influence each others' abundance, but, instead, fluctuate widely in abundance as a consequence of external environmental conditions. We believe that this more closely resembles the case with many mycophagous flies, where resource abundance is extremely sensitive to climatic conditions (Shorrocks and Charlesworth, 1980; Muona and Lumme, 1981; Hanski, 1989; Worthen and McGuire, in press) and probably insensitive to drosophilid activity. An essential point here is that mycophages (unlike phytophages) are often consuming a post-reproductive host. The flies are unlikely therefore to affect future resource availability. We are unconvinced of the importance or prevalence of competition in these communities, a view originally shared by Shorrocks and Charlesworth (1980). Only if competition is more convincingly demonstrated than at present, will it be necessary to determine whether intraspecific aggregation is the main factor allowing the co-existence of the species-rich mycophagous communities.

VI. UNANSWERED QUESTIONS

The purpose of this chapter has been to synthesize existing data in order to stimulate more research. The mycophagous drosophilids are uniquely appropriate organisms for the study of the interface between ecology and evolution. We hope that our efforts will encourage others to begin studies on this fascinating group. Below we list some of the questions we would like to see answered in the next decade of research.

1. Are repeated host-shifts (e.g. to skunk cabbage or cucumber) really independent events? Do they represent reversions to an ancestral condition?
2. What are the *relative* roles of mushroom tissue and microorganisms in larval nutrition? Are patterns of adult feeding and oviposition congruent?
3. What underlies the ability of some drosophilids to use fungi with severe toxins, such as amanitin?
4. Which factors determine survivorship of larvae at the levels of individual hosts, and of the population as a whole? To what extent do processes at one scale (such as competition in a single mushroom) affect patterns at other scales?
5. Is there a limit to population size, set by resource availability? If so, is this limit imposed by depletion of the resources (as in competition) or by searching constraints?
6. Are Allee effects common? Do they vary with host size and species? Can they occur between larvae of different drosophilids in the same host?
7. Is facilitation by other taxa common? Is this mediated through microfloral communities?
8. Are parasitoids as rare as has been previously suggested? In species-rich communities, do parasitoids attack many different drosophilids?
9. What proportion of *Drosophila* larval populations are consumed incidentally by mammalian fungivores?

10. Are mushrooms sufficiently nutritious to support adult drosophilids? Does nutrition affect survivorship?

11. Does male defecation on hosts lead to increased attraction of females? If so, are there cooperative effects within and between species? Do increased mating opportunities (benefits) outweigh increased mating competition and larval competition (costs)?

12. What is adult survivorship under realistic field conditions?

13. What is the distribution of dispersal distances in most populations? How does this vary with seasonal changes in resources? Do isolated populations receive sufficient gene immigration to swamp local evolutionary changes?

14. Do facultative mycophagous drosophilids play a role in structuring obligate mycophagous communities?

15. Are pre- and post-alighting oviposition preferences congruent? If not, which behavior displays most environmental (e.g. learnt) or genetic variance? Which behavior has evolved most divergently in related taxa?

16. Are mushrooms as homogeneous as currently believed or is this, as with previous generalizations of unpredictability, an oversimplification?

ACKNOWLEDGMENTS

This work was aided by the Mazamas (TAS) grants-in-aid-of-research from Sigma Xi (TTK, TAS), and the Theodore Roosevelt Memorial Fund of the American Museum of Natural History (TAS). April P. Claggett helped with fieldwork and drew Fig. 4. A number of authors sent us reprints and manuscripts. The US Forest Service, The Nature Conservancy, S. Greene and A. McKee have helped with logistical support. We thank all of these for their assistance.

John Jaenike has encouraged all our work with mycophagous drosophilids; we are grateful to him for his generous support.

REFERENCES

Atkinson, W. D. (1979). A comparison of the reproductive strategies of domestic species of *Drosophila*. *J. Anim. Ecol.* **48,** 53–64.

Atkinson, W. D. and Miller, J. A. (1980). Lack of habitat choice in a natural population of *Drosophila subobscura*. *Heredity* **44,** 193–199.

Atkinson, W. D. and Shorrocks, B. (1981). Competition on a divided and ephemeral resource: A simulation model. *J. Anim. Ecol.* **50,** 461–471.

Atkinson, W. D. and Shorrocks, B. (1984). Aggregation of larval Diptera over discrete and ephemeral breeding sites: The implications for coexistence. *Am. Nat.* **124,** 336–351.

Avelar, T., Rocha Pite, M. T. and Matos, M. (1987). Variation of some demographic parameters in *Drosophila simulans, D. subobscura* and *D. phalerata* (Diptera: Drosophilidae) throughout the year in semi-natural conditions in Portugal. *Acta Oecol. Gen.* **8,** 347–356.

Ayala, F. S., Gilpin, M. E. and Ehrenfeld, J. G. (1973). Competition between species: Theoretical models and experimental tests. *Theor. Popul. Biol.* **4,** 331–356.

Azarian, A. G. (1968). The circadian rhythms of flight and oviposition of *Drosophila transversa* and *Drosophila phalerata* in different photoperiodic conditions (in Russian). *Entomol. Obozrenie* **47,** 809–814.

Basden, E. B. (1954). The distribution and biology of Drosophilidae (Diptera) in Scotland, including a new species of *Drosophila. Trans. R. Soc. Edinburgh* **62,** 602–654.

Begon, M. (1976a). Temporal variations in the reproductive condition of *Drosophila obscura* Fallen and *D. subobscura* Collin. *Oecologia* **23,** 31–47.

Begon, M. (1976b). Dispersal, density and microdistribution in *Drosophila subobscura* Collin. *J. Anim. Ecol.* **45,** 441–456.

Bernays, E. A. and Chapman, R. F. (1978). Plant chemistry and acridoid feeding behaviour. In: *Coevolution of Plants and Animals* (Ed. by J. B. Harborne), pp. 100–141. Academic Press, London.

Bernays, E. A. and Graham, M. (1988). On the evolution of host specificity in phytophagous arthropods. *Ecology* **69,** 886–892.

Birch, L. C., Dobzhansky, T., Elliot, P. O. and Lewontin, R. C. (1963). Relative fitness of geographic races of *Drosophila serrata. Evolution* **17,** 72–83.

Bock, I. R. and Parsons, P. A. (1978). Australian endemic Drosophilidae V. Queensland rainforest species associated with fungi with descriptions of six new species and a redescription of *D. pectipennis* ketesz. *Austr. J. Zool.* **26,** 331–347.

Bonnamour, S. (1926). Les insectes parasites des champignons II. clevages et nouvelle liste de Diptères fungicoles. *Ann. Soc. Linn. Lyon,* **72,** 85–93.

Bouletreau, M. and Wajaburg, E. (1986). Comparative response of two sympatric parasitoid cynipids to the genetic and epigenetic variations of the larvae of their host *Drosophila melanogaster. Entomol. exp. appl.* **41,** 107–114.

Burla, H. and Bachli, G. (1968). Beitrag zur Kenntnis der Schweizerischen Dipteran insbesondere Drosophila — Arten, die sich in Fruchtkorper von Hutpilzen entwickelen. *Vierteljahrschr. Nat. Ges. Zurich* **113,** 311–336.

Bursell, E. (1967). The excretion of nitrogen in insects. In: *Advances in Insect Physiology* (Ed. by J. W. L. Beament, J. E. Treherne and V. B. Wigglesworth), Vol. 4, pp. 33–67. Academic Press, London.

Buxton, P. A. (1960). British Diptera associated with fungi III. Flies of all families reared from about 150 species of fungi. *Ent. Mon. Mag.* **96,** 61–94.

Carson, H. L. and Stalker, H. D. (1951). Natural breeding sites for some wild species of *Drosophila* in eastern United States. *Ecology* **32,** 317–330.

Carson, H. L., Knapp, E. P. and Phaff, H. J. (1956). Studies on the ecology of *Drosophila* in the Yellowstone region of California. III. The yeast flora of the natural breeding sites of some species of *Drosophila. Ecology* **37,** 538–544.

Charlesworth, P. and Shorrocks, B. (1980). The reproductive biology and diapause of the British fungal-breeding *Drosophila. Ecol. Entomol.* **5,** 315–326.

Charnov, E. L. and Skinner, S. W. (1984). Evolution of host selection and clutch size in parasitoid wasps. *Fla. Entomol.* **67,** 5–21.

Charnov, E. L. and Skinner, S. W. (1988). Clutch size in parasitoids: The egg production rate as a constraint. *Evol. Ecol.* **2,** 167–174.

Cochran, D. G. (1985). Nitrogenous excretion. In: *Comprehensive Insect Physiology,*

Biochemistry and Pharmacology, Vol. 4, Regulation: Digestion, Nutrition, Excretion (Ed. by G. A. Kerkut and L. I. Gilbert), pp. 467–506. Pergamon Press, Oxford.

Colwell, R. K. (1986). Community biology and sexual selection: Lessons from hummingbird flower mites. In: *Community Ecology* (Ed. by J. Diamond and T. J. Case), pp. 406–424. Harper and Row, New York.

Courtney, S. P. (1984). The evolution of batch oviposition by Lepidoptera and other insects. *Am. Nat.* **123,** 276–281.

Courtney, S. P. and Chen, G. K. (1988). Genetic and environmental variation in oviposition behaviour of the mycophagous *Drosophila suboccidentalis* Spcr. *Funct. Ecol.* **2,** 521–528.

Courtney, S. P. and Chew, F. S. (1987). Coexistence and host use by a large community of pierid butterflies: Habitat is the templet. *Oecologia* **71,** 210–220.

Courtney, S. P. and Courtney, S. (1982). The "edge-effect" in butterfly oviposition causality in *Anthocharis cardamines. Ecol. Entomol.* **7,** 131–137.

Courtney, S. P. and Duggan, A. E. (1983). The population of the orange-tip butterfly, *Anthocharis cardamine*, in Britain. *Ecol. Entomol.* **8,** 271–281.

Courtney, S. P. and Hard, J. (in press). Host acceptance and life-history in *Drosophila busckii*: Tests of the hierarchy-threshold model. *Heredity*.

Courtney, S. P. and Kibota, T. T. (in press). Mother doesn't know best: Selection of hosts by ovipositing insects. In: *Insect–Plant Interactions* (Ed. by E. A. Bernays), Vol. II. CRC Press, Boca Raton, Florida.

Courtney, S. P., Chen, G. K. and Gardner, A. (1989). A general model for individual host selection. *Oikos* **55,** 55–65.

Coyne, J. A. and Milstead, B. (1987). Long-distance migration of *Drosophila*. 3. Dispersal of *D. melanogaster* alleles from a Maryland orchard. *Am. Nat.* **130,** 70–82.

Coyne, J. A., Boussy, I. A., Prout, T., Bryant, S. H., Jones, J. S. and Moore, J. A. (1982). Long-distance migration of *Drosophila*. *Am. Nat.* **119,** 589–595.

Crisan, E. V. and Sands, A. (1978). Nutritional value. In: *The Biology and Cultivation of Edible Mushrooms* (Ed. by S. T. Chang and W. A. Hayes), pp. 137–168. Academic Press, London.

Crumpacker, D. W. (1974). The use of micronized fluorescent dusts to mark adult *Drosophila pseudoobscura. Am. Midl. Nat.* **91,** 118–129.

Dadd, R. H. (1985). Nutrition: Organisms. In: *Comprehensive Insect Physiology, Biochemistry and Pharmacology, Vol. 4, Regulation: Digestion, Nutrition, and Excretion* (Ed. by G. A. Kerkut and L. I. Gilbert), pp. 313–390. Pergamon Press, Oxford.

Dely-Draskovits, A. and Papp, L. (1973). Systematical and ecological investigations of fly pests of mushrooms in Hungary. V. Drosophilidae (Diptera). *Folia Entomol. Hungarica* **26,** 21–29.

Dobzhansky, T. (1973). Active dispersal and passive transport in *Drosophila*. *Evolution* **27,** 565–575.

Dobzhansky, T. and Wright, S. (1943). Genetics of natural populations. X. Dispersion rates in *Drosophila pseudoobscura*. *Genetics* **28,** 304–340.

Dobzhansky, T. and Wright, S. (1947). Genetics of natural populations. XV. Rate of diffusion of a mutant gene through a population of *Drosophila pseudoobscura*. *Genetics* **32,** 303–324.

Ehrlich, P. R. and Raven, P. H. (1964). Butterflies and plants: A study in coevolution. *Evolution* **18,** 586–608.

Falcoz, L. (1921). Notes sur divers insectes fungicoles. *Misc. Ent.* **25,** 57–62.

Feeny, P. P. (1968). Effect of oak leaf tannins on larval growth of the winter moth *Operphthera brumata*. *J. Insect Physiol.* **14,** 805–817.
Feeny, P. P. (1970). Seasonal changes in oak leaf tannins and nutrients as a cause of spring feeding of winter moth caterpillars. *Ecology* **51,** 565–581.
Feeny, P. P. (1976). Plant apparency and chemical defence. *Rec. Adv. Phytochem.* **10,** 1–40.
Fisher, R. A. and Ford, E. B. (1947). The spread of a gene in natural conditions in a colony of the moth *Panaxia dominula* (L). *Heredity* **1,** 143–174.
Futuyma, D. J. (1979). *Evolutionary Biology*. Sinauer, Sunderland, Mass.
Futuyma, D. J. (1983). Selective factors in the evolution of host choice by phytophagous insects. In: *Herbivorous Insects: Host-seeking Behavior and Mechanisms* (Ed. by S. Ahmad), pp. 227–244. Academic Press, London.
Futuyma, D. J. and Peterson, S. C. (1985). Genetic variation in the use of resources by insects. *Annu. Rev. Entomol.* **30,** 217–238.
Geyspits, K. F. and Simonenko, N. P. (1970). An experimental analysis of seasonal changes in the photoperiodic reaction of *Drosophila phalerata* Meig. (Diptera, Drosophilidae). *Entomol. Rev. (USSR)* **49,** 46–54 (English translation).
Gilbert, L. E. and Singer, M. C. (1975). Butterfly ecology. *Annu. Rev. Ecol. Syst.* **6,** 365–397.
Gilpin, M. E., Carpenter, M. P. and Pomerantz, M. J. (1986). The assembly of a laboratory community: Multi-species competition in *Drosophila*. In: *Community Ecology* (Ed. by J. Diamond and T. Case), pp. 23–40. Harper and Row, New York.
Godfray, H. C. J. and Ives, A. R. (1988). Stochasticity in invertebrate clutch-size models. *Theor. Pop. Biol.* **33,** 79–101.
Goryshin, N. I., Braun, E. A. and Tyshchenko, G. F. (1973). Device for investigation of daily activity rhythms of insects (in Russian). *Vest. Leningrad. Univ.* **15,** 21–25.
Green, R. F. (1986). Does aggregation prevent competitive exclusion? A response to Atkinson and Shorrocks. *Am. Nat.* **128,** 301–304.
Grimaldi, D. (1985). Niche separation and competitive coexistence in mycophagous *Drosophila* (Diptera: Drosophilidae). *Proc. Entomol. Soc. Wash.* **87,** 498–511.
Grimaldi, D. A. (1986). The *Chymomyza aldrichii* species group (Diptera: Drosophilidae): Relationships, new neotropical species, and the evolution of some sexual traits. *J. N.Y. Entomol. Soc.* **94(3),** 342–371.
Grimaldi, D. A. (1987). Phylogenetics and taxonomy of *Zygothryca* (Diptera: Drosophilidae). *Bull. Am. Museum Nat. Hist.* **186**(2), 107–268.
Grimaldi, D. and Jaenike, J. (1983). The Diptera breeding on skunk cabbage, *Symplocarpus foetidus* (Araceae). *J. N.Y. Entomol. Soc.* **91,** 83–89.
Grimaldi, D. and Jaenike, J. (1984). Competition in natural populations of mycophagous *Drosophila*. *Ecology* **64,** 1113–1120.
Hammond, P. M. and Lawrence, J. F. (1989). Appendix: Mycophagy in insects: Summary. In: *Insect–Fungus Interactions* (Ed. by N. Wilding, N. M. Collins, P. M. Hammond and J. F. Webber), pp. 275–324. Academic Press, London.
Hanski, I. (1989). Fungivory: Fungi, insects and ecology. In: *Insect–Fungus Interactions* (Ed. by N. Wilding, N. M. Collins, P. M. Hammond and J. F. Webber), pp. 25–68. Academic Press, London.
Hey, J. and Houle, D. (1987). Habitat choice in the *Drosophila affinis* subgroup. *Heredity* **58,** 463–471.
Hoffman, A. A. (1985). Effects of experience on oviposition and attraction in *Drosophila*: Comparing apples and oranges. *Am. Nat.* **126,** 41–51.
Hoffman, A. A. and Turelli, M. (1985). Distribution of *Drosophila melanogaster* on alternative resources: Effects of experience and starvation. *Am. Nat.* **126,** 662–679.

House, H. L. (1974). Nutrition. In. *The Physiology of Insecta*, 2nd edn (Ed. by M. Rockstein), Vol. 5, pp. 1–62. Academic Press, London.

Hummel, H. K., van Delden, W. and Drent, R. H. (1979). Estimation of some population parameters of *Drosophila limbata* V. Roser in a greenhouse. *Oecologia* **41,** 135–143.

Huston, M. (1979). A general hypothesis of species diversity. *Am. Nat.* **113,** 81–101.

Ives, A. R. and May, R. M. (1985). Competition within and between species in a patchy environment: Relations between microscopic and macroscopic models. *J. Theor. Biol.* **115,** 65–92.

Jaenike, J. (1978). Resource predictability and niche breadth in the *Drosophila quinaria* species group. *Evolution* **32,** 676–678.

Jaenike, J. (1982). Environmental modification of oviposition behavior in *Drosophila. Am. Nat.* **119,** 784–802.

Jaenike, J. (1985a). Parasite pressure and the evolution of amanitin tolerance in *Drosophila. Evolution* **39,** 1295–1301.

Jaenike, J. (1985b). Genetic and environmental determinants of food preference in *Drosophila tripunctata. Evolution* **39,** 362–369.

Jaenike, J. (1986a). Intraspecific variation for resource use in *Drosophila. Biol. J. Linn. Soc.* **27,** 47–56.

Jaenike, J. (1986b). Genetic complexity of host-selection behavior in *Drosophila. Proc. Natl Acad. Sci.* **83,** 2148–2151.

Jaenike, J. (1988a). Parasitism and mating success in *Drosophila testacea. Am. Nat.* **131,** 774–780.

Jaenike, J. (1988b). Effects of early adult experience on host selection in insects: Some experimental and theoretical results. *J. Insect Behav.* **1,** 3–16.

Jaenike, J. (1989). Genetic population structure of *Drosophila tripunctata*: Patterns of phenotypic and genotypic variation and covariation of traits affecting resource use. *Evolution* **43,** 1467–1482.

Jaenike, J. and Grimaldi, D. (1983). Genetic variation for host preference within and among populations of *Drosophila tripunctata. Evolution* **37,** 1023–1022.

Jaenike, J. and Selander, R. K. (1979). Ecological generalism in *Drosophila falleni*: Genetic evidence. *Evolution* **33,** 741–748.

Jaenike, J., Grimaldi, D. A., Sluder, A. E. and Greenleaf, A. L. (1983). Alpha-amanitin tolerance in mycophagous *Drosophila. Science* **221,** 165–167.

James, A. C., Jakubczak, J., Riley, M. R. and Jaenike, J. (1988). On the causes of monophagy in *Drosophila quinaria. Evolution* **42,** 626–630.

Janssen, A. (1989). Optimal host selection by *Drosophila* parasitoids in the field. *Funct. Ecol.* **3,** 469–479.

Janssen, A., Driessen, G., de Haan, M. and Roodbol, N. (1988). The impact of parasitoids on natural populations of temperate woodland *Drosophila. Neth. J. Zool.* **38,** 61–73.

Johnston, J. S. and Heed, W. B. (1975). Dispersal *of Drosophila*: The effect of baiting on the behavior and distribution of natural populations. *Am. Nat.* **109,** 207–216.

Johnston, J. S. and Heed, W. B. (1976). Dispersal of desert-adapted *Drosophila*: The saguaro-breeding *D. nigrospiracula. Am. Nat.* **110,** 629–651.

Jones, J. S., Bryant, S. H., Lewontin, R. C., Moore, J. A. and Prout, T. (1981). Gene flow and the geographical distribution of a molecular polymorphism in *Drosophila pseudoobscura. Genetics* **98,** 157–178.

Jones, R. E., Hart, J. R. and Bull, A. D. (1982). Egg distribution and survivorship in the pierid butterfly, *Colias alexandra. Oecologia* **66,** 495–498.

Kambysellis, M. B. and Heed, W. B. (1974). Juvenile hormone induces ovarian development in diapausing cave-dwelling *Drosophila* species. *J. Insect Physiol.* **20,** 1779–1786.

Kearney, J. N. (1983). Selection and utilization of natural substrates as breeding sites by woodland *Drosophila* species. *Entomol. exp. appl.* **33,** 63–70.

Kearney, J. N. and Shorrocks, B. (1981). The utilization of naturally occurring yeasts by *Drosophila* species, using chemically defined substrates. *Biol. J. Linn. Soc.* **15,** 39–56.

Kekic, V., Taylor, C. E. and Jelkovic, M. (1980). Habitat choice and resource specialization by *Drosophila subobscura. Genetika* **12,** 219–225.

Kimura, M. T. (1976). *Drosophila* survey of Hokkaido, XXXII. A field survey of fungus preferences of drosophilid flies in Sapporo. *J. Fac. Sci., Hokkaido Univ., Ser. VI, Zoo.* **20**(3), 288–298.

Kimura, M. T. (1980a). Evolution of food preferences in fungus-feeding *Drosophila*: An ecological study. *Evolution* **34**(5), 1009–1018.

Kimura, M. T. (1980b). Bionomics of Drosophilidae (Diptera) in Hokkaido. IV. *Drosophila sexvittata, D. trivattata,* and *D. alboralis. Zool. Mag.* **89,** 206–209.

Kimura, M. T. and Toda, M. J. (1989). Food preferences and nematode parasitism in mycophagous *Drosophila. Ecol. Res.* **4,** 209–218.

Kimura, M. T., Toda, M. J., Beppu, K. and Watabe, H. (1977). Breeding sites of drosophilid flies in and near Sapporro, northern Japan, with supplementary notes on adult feeding habits. *Kontyu* **45,** 571–582.

Kimura, M. T., Beppu, K., Ichijo, N. and Toda, M. J. (1978). Bionomics of Drosophilidae (Diptera) in Hokkaido. II. *Drosophila testacea. Kontyu* **46**(4), 585–595.

Klopfenstein, P. C. (1971). The ecology, behavior and life cycle of the mycetophilous beetle, *Hadraule blaisdelli* (Casey) (Insecta: Coleoptera: Ciidae). Thesis, Bowling Green State University Ohio.

Kukor, J. J. and Martin, M. M. (1987). Nutritional ecology of fungus-feeding arthropods. In: *Nutritional Ecology of Insects, Mites, Spiders, and Related Invertebrates* (Ed. by F. Slansky Jr and J. G. Rodriguez), pp. 791–814. John Wiley, New York.

Lachaise, D. (1975). Les Drosophilidae des savanes preforestieres de Lamto (Cote-d'Ivoire). III. Le peuplement du Palmier Ronier. *Ann. Univ. Abidjan* **8,** 223–280.

Lachaise, D. and Tsacas, L. (1983). Breeding-sites in tropical African drosophilids. In: *The Genetics and Biology of Drosophila* (Ed. by M. Ahsburner, H. L. Carson and J. N. Thompson, Jr), Vol. 3d, pp. 221–332. Academic Press, London.

Lacy, R. C. (1982). Niche breadth and abundance as determinants of genetic variation in populations of mycophagous drosophilid flies (Diptera: Drosophilidae). *Evolution* **36,** 1265–1275.

Lacy, R. C. (1983). Structure of genetic variation within and between populations of mycophagous *Drosophila. Genetics* **104,** 81–94.

Lacy, R. C. (1984a). Ecological and genetic responses to mycophagy in Drosophilidae (Diptera). In: *Fungus–Insect Relationships: Perspectives in Ecology and Evolution* (Ed. by Q. Wheeler and M. Blackwell), pp. 286–304. Columbia University Press, New York.

Lacy, R. C. (1984b). Predictability, toxicity, and trophic niche breadth in fungus-feeding *Drosophila* (Diptera). *Ecol. Entomol.* **9,** 43–54.

Lakovaara, S., Saura, A., Koren-Santibanez, S. and Ehrman, L. (1972). Aspects of diapause and its genetics in northern drosophilids. *Hereditas* **70,** 89–96.

Lawrence, J. F. (1989). Mycophagy in the Coleoptera: Feeding strategies and morphological adaptations. In: *Insect–Fungus Interactions* (Ed. by N. Wilding, N. M. Collins, P. M. Hammond and J. F. Webber), pp. 2–23. Academic Press, London.

Lawton, J. H. and Strong, D. R. (1981). Community patterns and competition in folivorous insects. *Am. Nat.* **188,** 317–338.

Lilly, V. G. (1965). Chemical constituents of the fungal cell. I. Elemental constituents and their roles. In: *The Fungi* (Ed. by G. C. Ainsworth and A. S. Sussman), Vol. I, pp. 163–177. Academic Press, London.

Lumme, J. and Lakovaara, S. (1983). Seasonality and diapause in drosophilids. In: *The Genetics and Biology of Drosophila* (Ed. by M. Ahsburner, H. L. Carson and J. N. Thompson, Jr), Vol. 3d, pp. 171–220. Academic Press, London.

Lumme, J. A., Oikarinen, A., Lakovaara, S. and Alatalo, R. (1974). The environmental regulation of adult diapause in *Drosophila littoralis. J. Insect Phys.* **20,** 2023–2033.

Lumme, J., Muona, O. and Orell, M. (1978). Phenology of boreal drosophilids (Diptera). *Ann. Entomol. Fenn.* **44,** 73–85.

Lumme, J., Lakovaara, S., Muona, O. and Jarvinen, O. (1979). Structure of boreal community of drosophilids (Diptera). *Aquilo Ser. Zool.* **20,** 65–73.

Mangel, M. (1987). Oviposition site selection and clutch size in insects. *J. Math. Biol.* **25,** 1–22.

Martin, M. M. (1979). Biochemical implications of insect mycophagy. *Biol. Rev.* **54,** 1–21.

Matos, M., Avelar, T. and Rocha Pite, M. T. (1987). Longevity of *Drosophila simulans, D. subobscura,* and *D. phalerata* (Diptera, Drosophilidae) throughout the year under semi-natural conditions in Portugal. *Ann. Ent. Soc. Fr.* **23,** 161–168.

Montague, J. R. (1985). Body size, reproductive biology, and dispersal behaviour among artificial baits in *D. falleni. Dros. Info. Serv.* **61,** 123–126.

Montague, J. R. (1989). The ecology of Hawaiian flower-breeding drosophilids. 2. Adult dispersions and reproductive ecology. *Am. Nat.* **133,** 71–82.

Montague, J. R. and Jaenike, J. (1985). Nematode parasitism in natural populations of mycophagous drosophilids. *Ecology* **66,** 624–626.

Moth, J. J. and Barker, J. S. F. (1975). Micronized fluorescent dusts for marking *Drosophila* adults. *J. Nat. Hist.* **9,** 393–396.

Mueller, L. D. (1987). Evolution of accelerated senescence in laboratory populations of *Drosophila. Proc. Natl Acad. Sci.* **84,** 1974–1977.

Muona, O. and Lumme, J. (1981). Geographical variation in the reproductive cycle and photoperiodic diapause of *Drosophila phalerata* and *D. transversa* (Drosophilidae: Diptera). *Evolution* **35,** 158–167.

Oakeshott, J. G., Vacek, D. C. and Anderson, P. H. (1989). Effects of microbial floras on the distributions of some domestic *Drosophila* spp. across fruit resources. *Oecologia* **78,** 537–541.

Ohenoja, E. and Koistinen, R. (1984). Fruiting body production of larger fungi in Finland. 2. Edible fungi in northern Finland 1976–1978. *Ann. Bot. Fenn.* **21,** 357–366.

Ohenoja, E. and Metsanheimo, R. (1982). Phenology and fruiting body production of macrofungi in subarctic Finnish Lapland. In: *Arctic and Alpine Mycology* (Ed. by G. A. Laursen and J. F. Ammirati), pp. 390–409. University of Washington Press, Seattle, Wash.

Okada, T. (1962). Breeding sap preference of the drosophilid flies. *Jap. J. Appl. Entomol. Zool.* **6,** 216–229.

Papaj, D. R. and Rausher, M. D. (1983). Individual variation in host location by phytophagous insects. In: *Herbivorous Insects: Host-seeking Behaviors and Mechanisms* (Ed. by S. Ahmad), pp. 77–123, Academic Press, London.

Parker, G. A. (1978). Evolution of competitive mate searching. *Ann. Rev. Entomol.* **23,** 173–196.

Parker, G. A. and Begon, M. (1986). Optimal egg size and clutch size: Effects of environmental and maternal phenotype. *Am. Nat.* **128,** 579–592.

Parker, G. A. and Courtney, S. P. (1984). Models of clutch size in insect oviposition. *Theor. Popul. Biol.* **26,** 27–48.

Parsons, P. (1977). Lek behaviour in *Drosophila (Hirtodrosophila) polypori* Malloch-an Australian rainforest species. *Evolution* **31,** 223–225.

Parsons, P. (1982). Evolutionary ecology of Australian *Drosophila*, a species analysis. *Evol. Biol.* **14,** 297–350.

Phillips, D. S. and Hubbard, S. F. (1989). Controlled covariance in the Atkinson-Shorrocks model of multispecies competition. *J. Theor. Biol.* **141,** 137–131.

Pianka, E. R. (1970). On r- and K-selection. *Am. Nat.* **104,** 592–597.

Pilson, D. and Rausher, M. D. (1988). Clutch size adjustment by a swallowtail butterfly. *Nature* **333,** 361–363.

Rausher, M. D. (1984). Tradeoffs in performance on different hosts: Evidence from within- and between-site variation in the beetles *Deloyala guttata. Evolution* **38,** 582–595.

Rhoades, R. F. and Cates, R. G. (1976). Toward a general theory of plant antiherbivore chemistry. *Rec. Adv. Phytochem.* **10,** 168–213.

Riel, P. (1921). Les insectes parasites des champignons I. clevages et première liste de Diptères fungicoles. *Ann. Soc. Linn. Lyon,* **67,** 37–44.

Rocha Pite, M. T., Matos, M. and Avelar, T. (1985). Daily fecundity of *Drosophila simulans, D. subobscura*, and *D. phalerata* (Diptera: Drosophilidae) throughout the year under semi-natural conditions. *Biol. Soc. Port. Ent.* **2**(suppl.), 7–26.

Rosewell, J. and Shorrocks, B. (1987). The implication of survival rates in natural populations of *Drosophila*: capture–recapture experiments on domestic species. *Biol. J. Linn. Soc.* **32,** 373–384.

Sabath, M. D., Richmond, R. C. and Torrela, R. M. (1973). Temperature-mediated seasonal color changes in *Drosophila putrida. Am. Midl. Nat.* **90,** 509–512.

Shorrocks, B. (1977). An ecological classification of European *Drosophila* species. *Oecologia* **26,** 335–345.

Shorrocks, B. and Charlesworth, P. (1980). The distribution and abundances of the British fungal-feeding *Drosophila. Ecol. Entomol.* **5,** 61–78.

Shorrocks, B. and Charlesworth, P. (1982). A field study of the association between the stinkhorn *Phallus impudicus* Pers. and the British fungal-breeding *Drosophila. Biol. J. Linn. Soc.* **17,** 301–318.

Shorrocks, B. and Nigro, L. (1981). Microdistribution and habitat selection in *Drosophila subobscura* Collin. *Biol. J. Linn. Soc.* **16(4),** 293–301.

Shorrocks, B. and Rosewell, J. (1986). Guild size in *Drosophila*: A simulation model. *J. Anim. Ecol.* **55,** 527–541.

Shorrocks, B. and Rosewell, J. (1987). Spatial patchiness and community structure: Coexistence and guild size of drosophilids on ephemeral resources. *Br. Ecol. Soc. Symp.* **27,** 29–51.

Shorrocks, B. and Rosewell, J. (1988). Aggregation does prevent competitive exclusion: A response to Green. *Am. Nat.* **131,** 765–771.

Shorrocks, B., Rosewell, J., Edwards, K. and Atkinson, W. D. (1984). Interspecific competition is not a major organizing force in many insect communities. *Nature* **310,** 310–312.

Singer, M. C. (1982). Quantification of host preference by manipulation of oviposition behavior in the butterfly *Euphydryas editha. Oecologia* **52,** 224–229.

Singer, M. C. (1983). Determinants of multiple host use by a phytophagous insect population. *Evolution* **37,** 389–403.
Skinner, S. W. (1985). Clutch size as an optimal foraging problem for insects. *Behav. Ecol. Sociobiol.* **17,** 231–238.
Slatkin, M. (1987). Gene flow and the geographic structure of natural populations. *Science* **236,** 787–792.
Smith, K. G. V. (1956). On the Diptera associated with stinkhorn (*Phallus impudicus*) with notes on other insects and invertebrates found on this fungus. *Proc. R. Ent. Soc.* **A31,** 49–55.
Spieth, H. T. (1987). The *Drosophila* fauna of a native California forest (Diptera: Drosophilidae). *Pan-Pacific Entomol.* **63**(3), 247–255.
Spieth, H. T. and Heed, W. B. (1975). The *Drosophila pinicola* species group. *Pan-Pacific Entomol.* **51**(4), 287–295.
Starmer, W. T. (1981). A comparison of *Drosophila* habitats according to the physiological attributes of the associated yeast communities. *Ecology* **35,** 38–52.
Strong, D. R., Lawton, J. H. and Southwood, T. R. E. (1984). *Insects on Plants.* Harvard University Press, Cambridge, Mass.
Sturtevant, A. H. (1921). The North American species of *Drosophila. Carngegie Inst. Wash. Publ.* **301,** 1–150.
Sturtevant, A. H. (1942). The classification of the genus *Drosophila* with descriptions of nine new species. *Univ. Texas Publ.* **4213,** 5–66.
Tatar, M. (1989). Swallowtail clutch size reconsidered. *Oikos* **55,** 135–136.
Taylor, C. E. and Powell, J. R. (1978). Habitat choice in natural populations of *Drosophila. Oecologia* **37,** 69–75.
Thompson, J. N. (1988). Evolutionary ecology of the relationship between oviposition preference and performance of offspring in phytophagous insects. *Entomol. exp. appl.* **47,** 3–14.
Throckmorton, L. H. (1975). The phylogeny, ecology and geography of Drosophila. In: *Invertebrates of Great Interest: Handbook of Genetics* (Ed. by R. King), Vol. III, pp. 421–469. Plenum Press, New York.
Timofeeff-Ressovsky, N. W. and Timofeeff-Ressovsky, E. A. (1940). Populationsgenetische Versuche an *Drosophila.* I. Zeitliche und raumliche Verteilung der Individuen einiger *Drosophila*—Arten uber das Geland. *Z. induktive Abstammungs-Vererbungslehre* **79,** 28–49.
Toda, M. J. (1977). Vertical microdistribution of Drosophilidae (Diptera) with various forests in Hokkaido. I. Natural broad-leaved forest. *Jap. J. Ecol.* **27,** 207–214.
Toda, M. J. (1985a). Habitat structure of a drosophilid community at Inuvik, NWT, Canada (Diptera: Drosophilidae). *Can. Entomol.* **117,** 135–137.
Toda, M. J. (1985b). Effects of the 1977 eruption of Mt. Usu on drosophilid flies. In August 1978. *Jap. J. Ecol.* **35,** 235–241.
Toda, M. J., Kimura, M. T. and Enomoto, O. (1984). Bionomics of Drosophilidae (Diptera) in Hokkaido VI. Decayed herbage feeders, with special reference to their reproductive strategies. *Jap. J. Ecol.* **34,** 253–270.
Toda, M. J., Kimura, M. T., Takeda, E., Beppu, K. and Iwao, Y. (1986). Bionomics of Drosophilidae (Diptera) in Hokkaido VIII. *Drosophila confusa* and *D. bifasciata*, with special reference to local variation in annual life cycle. *Kontyu* **54**(1), 33–40.
Turelli, M., Coyne, J. A. and Prout, T. (1984). Resource choice in orchard populations of *Drosophila. Biol. J. Linn. Soc.* **22,** 95–106.

Turelli, M., Burkhard, C. Fong, V., Moore, J., Van Horn, S. and Prout, T. (1986). Does dusting distort *Drosophila* dispersal? *D.I.S.* **63,** 131–132.

Turner, W. B. (1971). *Fungal Metabolites.* Academic Press, London.

Tyler, V. E., Benedict, R. G. and Stuntz, D. E. (1965). Chemotaxonomic significance of urea in the higher fungi. *Lloydia* **28,** 342–353.

Vet, L. E. M. (1983). Host-habitat location through olfactory cues by *Leptopilina clavipes* (Hartig) (Hymenoptera: Eucoilidae), a parasitoid of fungivorous *Drosophila*: The influence of conditioning. *Neth. J. Zool.* **33,** 225–248.

Vet, L. E. M. (1985). Olfactory microhabitat location in some Eucoilid and Alysiine species (Hymenoptera), larval parasitoids of Diptera. *Neth. J. Zool.* **35,** 720–730.

Watabe, H. (1979). *Drosophila* survey of Hokkaido. XXXVI. Seasonal changes in the reproductive condition of wild and domestic species of *Drosophila. J. Fac. Sci. Hokkaido Univ. VI Zool.* **21,** 365–372.

Watabe, H., Kimura, M. T., Toda, M. J. and Iwao, Y. (1985). Bionomics of Drosophilidae (Diptera) in Hokkaido VII. *Drosophila nigromaculata* and *D. Brachynephros. Kontyu* **53**(1), 34–41.

Wheeler, M. R. and Kambysellis, M. P. (1967). Notes on the Drosophilidae of Samoa. *Univ. Tex. Publ.* **6615,** 533–565.

Wheeler, M. R. and Takada, H. (1964). Insects of Micronesia: Drosophilidae. *Insects of Micronesia* **14**(6).

Wheeler, Q. and Blackwell, M. (Eds) (1984). *Fungus–Insect Relationships: Perspectives in Ecology and Evolution.* Columbia University Press, New York.

Wiklund, C. (1975). The evolutionary relationship between adult oviposition preferences and larval host plant range in *Papilio machaon* L. *Oecologia* **18,** 185–197.

Wilding, N., Collins, N. M., Hammond, P. M. and Webber, J. F. (Eds) (1989). *Insect–Fungus Interactions.* Academic Press, London.

Wilson, D. S. and Fudge, J. (1984). Burying beetles: Intraspecific interactions and reproductive success in the field. *Ecol. Entomol.* **9,** 195–204.

Worthen, W. B. (1988). Slugs (*Arion* spp). facilitate mycophagous drosophilids in laboratory and field experiments. *Oikos* **53,** 161–166.

Worthen, W. B. (1989a). Predator-mediated coexistence in laboratory communities of mycophagous *Drosophila* (Diptera: Drosophilidae). *Ecol. Entomol.* **14,** 117–126.

Worthen, W. B. (1989b). Effects of resource density on mycophagous fly dispersal and community structure. *Oikos* **54,** 145–153.

Worthen, W. B. and McGuire, T. R. (1988). A criticism of the aggregation model of coexistence of dipteran species on ephemeral resources. *Am. Nat.* **131,** 453–458.

Worthen, W. B. and McGuire, T. R. (in press). Predictability of ephemeral mushrooms and implications for mycophagous fly communities. *Am. Midl. Nat.*

Index

Advances in Ecological Research
Volumes 1–20

Cumulative List of Titles